Springer Series in Statistics

Springer
Berlin
Heidelberg
New York
Hongkong
London
Milan
Paris
Tokyo

Springer Series in Statistics

Andersen/Borgan/Gill/Keiding: Statistical Models Based on Counting Processes.
Atkinson/Riani: Robust Diagnostic Regression Analysis.
Atkinson/Riani/Cerioli: Exploring Multivariate Data with the Forward Search.
Berger: Statistical Decision Theory and Bayesian Analysis, 2nd edition.
Borg/Groenen: Modern Multidimensional Scaling: Theory and Applications.
Brockwell/Davis: Time Series: Theory and Methods, 2nd edition.
Bucklew: Introduction to Rare Event Simulation.
Chan/Tong: Chaos: A Statistical Perspective.
Chen/Shao/Ibrahim: Monte Carlo Methods in Bayesian Computation.
Coles: An Introduction to Statistical Modeling of Extreme Values.
David/Edwards: Annotated Readings in the History of Statistics.
Devroye/Lugosi: Combinatorial Methods in Density Estimation.
Efromovich: Nonparametric Curve Estimation: Methods, Theory, and Applications.
Eggermont/LaRiccia: Maximum Penalized Likelihood Estimation, Volume I: Density Estimation.
Fahrmeir/Tutz: Multivariate Statistical Modelling Based on Generalized Linear Models, 2nd edition.
Fan/Yao: Nonlinear Time Series: Nonparametric and Parametric Methods.
Farebrother: Fitting Linear Relationships: A History of the Calculus of Observations 1750–1900.
Federer: Statistical Design and Analysis for Intercropping Experiments, Volume I: Two Crops.
Federer: Statistical Design and Analysis for Intercropping Experiments, Volume II: Three or More Crops.
Ghosh/Ramamoorthi: Bayesian Nonparametrics.
Glaz/Naus/Wallenstein: Scan Statistics.
Good: Permutation Tests: A Practical Guide to Resampling Methods for Testing Hypotheses, 2nd edition.
Gouriéroux: ARCH Models and Financial Applications.
Gu: Smoothing Spline ANOVA Models.
Györfi/Kohler/Krzyzak/Walk: A Distribution-Free Theory of Nonparametric Regression.
Haberman: Advanced Statistics, Volume I: Description of Populations.
Hall: The Bootstrap and Edgeworth Expansion.
Härdle: Smoothing Techniques: With Implementation in S.
Harrell: Regression Modeling Strategies: With Applications to Linear Models, Logistic Regression, and Survival Analysis.
Hart: Nonparametric Smoothing and Lack-of-Fit Tests.
Hastie/Tibshirani/Friedman: The Elements of Statistical Learning: Data Mining, Inference, and Prediction.
Hedayat/Sloane/Stufken: Orthogonal Arrays: Theory and Applications.
Heyde: Quasi-Likelihood and its Application: A General Approach to Optimal Parameter Estimation.

(continued after index)

Wolfgang Härdle, Marlene Müller,
Stefan Sperlich, Axel Werwatz

Nonparametric and Semiparametric Models

Springer

Wolfgang Härdle
CASE – Center for Applied Statistics and Economics
Wirtschaftswissenschaftliche Fakultät
Humboldt-Universität zu Berlin
10178 Berlin
Germany
haerdle@wiwi.hu-berlin.de

Axel Werwatz
DIW Berlin
Königin-Luise-Straße 5
14195 Berlin
Germany
awerwatz@diw.de

Marlene Müller
Fraunhofer ITWM
Gottlieb-Daimler-Straße
67663 Kaiserslautern
Germany
Marlene.Mueller@itwm.fhg.de

Stefan Sperlich
Departamento de Economía
Universidad Carlos III de Madrid
C./Madrid, 126
28903 Getafe (Madrid)
Spain
stefan@uc3m.es

Cataloging-in-Publication Data applied for

A catalog record for this book is available from the Library of Congress.

Bibliographic information published by Die Deutsche Bibliothek
Die Deutsche Bibliothek lists this publication in the Deutsche Nationalbibliografie; detailed bibliographic data is available in the Internet at <http://dnb.ddb.de>.

Mathematics Subjects Classification (2000): 62G07, 62G08, 62G20, 62G09, 62G10

ISBN 3-540-20722-8 Springer-Verlag Berlin Heidelberg New York

Springer-Verlag is a part of Springer Science+Business Media
springeronline.com

Printed in Germany

Cover design: *design & production*, Heidelberg
Typesetting by the authors
Printed on acid-free paper

40/3142 – 543210

Preface

The concept of smoothing is a central idea in statistics. Its role is to extract structural elements of variable complexity from patterns of random variation. The nonparametric smoothing concept is designed to simultaneously estimate and model the underlying structure. This involves high dimensional objects, like density functions, regression surfaces or conditional quantiles. Such objects are difficult to estimate for data sets with mixed, high dimensional and partially unobservable variables. The semiparametric modeling technique compromises the two aims, flexibility and simplicity of statistical procedures, by introducing partial parametric components. These (low dimensional) components allow one to match structural conditions like for example linearity in some variables and may be used to model the influence of discrete variables. The flexibility of semiparametric modeling has made it a widely accepted statistical technique.

The aim of this monograph is to present the statistical and mathematical principles of smoothing with a focus on applicable techniques. The necessary mathematical treatment is easily understandable and a wide variety of interactive smoothing examples are given. This text is an e-book; it is a downloadable entity (http://www.i-xplore.de) which allows the reader to recalculate all arguments and applications without reference to a specific software platform. This new technique for proliferation of methods and ideas is specifically designed for the beginner in nonparametric and semiparametric statistics. It is based on the XploRe quantlet technology, developed at Humboldt-Universität zu Berlin.

The text has evolved out of the courses "Nonparametric Modeling" and "Semiparametric Modeling", that the authors taught at Humboldt-Universität zu Berlin, ENSAE Paris, Charles University Prague, and Universidad de Cantabria, Santander. The book divides itself naturally into two parts:

- **Part I: Nonparametric Models**
  histogram, kernel density estimation, nonparametric regression
- **Part II: Semiparametric Models**
  generalized regression, single index models, generalized partial linear models, additive and generalized additive models.

The first part (Chapters 2–4) covers the methodological aspects of nonparametric function estimation for cross-sectional data, in particular kernel smoothing methods. Although our primary focus will be on flexible regression models, a closely related topic to consider is nonparametric density estimation. Since many techniques and concepts for the estimation of probability density functions are also relevant for regression function estimation, we first consider histograms (Chapter 2) and kernel density estimates (Chapter 3) in more detail. Finally, in Chapter 4 we introduce several methods of nonparametrically estimating regression functions. The main part of this chapter is devoted to kernel regression, but other approaches such as splines, orthogonal series and nearest neighbor methods are also covered.

The first part is intended for undergraduate students majoring in mathematics, statistics, econometrics or biometrics. It is assumed that the audience has a basic knowledge of mathematics (linear algebra and analysis) and statistics (inference and regression analysis). The material is easy to utilize since the e-book character of the text allows maximum flexibility in learning (and teaching) intensity.

The second part (Chapters 5–9) is devoted to semiparametric regression models, in particular extensions of the parametric generalized linear model. In Chapter 5 we summarize the main ideas of the generalized linear model (GLM). Typical concepts are the logit and probit models. Nonparametric extensions of the GLM consider either the link function (single index models, Chapter 6) or the index argument (generalized partial linear models, additive and generalized additive models, Chapters 7–9). Single index models focus on the nonparametric error distribution in an underlying latent variable model. Partial linear models take the pragmatic point of fixing the error distribution but let the index be of non- or semiparametric structure. Generalized additive models concentrate on a (lower dimensional) additive structure of the index with fixed link function. This model class balances the difficulty of high-dimensional smoothing with the flexibility of nonparametrics.

In addition to the methodological aspects, the second part also covers computational algorithms for the considered models. As in the first part we focus on cross-sectional data. It is intended to be used by Master and PhD students or researchers.

This book would not have been possible without substantial support from many colleagues and students. It has benefited at several stages from useful remarks and suggestions of our students at Humboldt-Universität zu Berlin, ENSAE Paris and Charles University Prague. We are grateful to Lorens Helmchen, Stephanie Freese, Danilo Mercurio, Thomas Kühn, Ying Chen and Michal Benko for their support in text processing and programming, Caroline Condron for language checking and Pavel Čížek, Zdeněk Hlávka and Rainer Schulz for their assistance in teaching. We are indebted to Joel Horowitz (Northwestern University), Enno Mammen (Universität Heidelberg) and Helmut Rieder (Universität Bayreuth) for their valuable comments on earlier versions of the manuscript. Thanks go also to Clemens Heine, Springer Verlag, for being a very supportive and helpful editor.

Berlin/Kaiserslautern/Madrid, February 2004

Wolfgang Härdle
Marlene Müller
Stefan Sperlich
Axel Werwatz

Contents

List of Figures

List of Tables

Notation

Abbreviations

cdf	cumulative distribution function
df	degrees of freedom
iff	if and only if
i.i.d.	independent and identically distributed
w.r.t.	with respect to
pdf	probability density function
ADE	average derivative estimator
AM	additive model
AMISE	asymptotic MISE
AMSE	asymptotic MSE
APLM	additive partial linear model
ASE	averaged squared error
ASH	average shifted histogram
CHARN	conditional heteroscedastic autoregressive nonlinear
CV	cross-validation
DM	Deutsche Mark
GAM	generalized additive model
GAPLM	generalized additive partial linear model
GLM	generalized linear model

GPLM	generalized partial linear model
ISE	integrated squared error
IRLS	iteratively reweighted least squares
LR	likelihood ratio
LS	least squares
MASE	mean averaged squared error
MISE	mean integrated squared error
ML	maximum likelihood
MLE	maximum likelihood estimator
MSE	mean squared error
PLM	partial linear model
PMLE	pseudo maximum likelihood estimator
RSS	residual sum of squares
S.D.	standard deviation
S.E.	standard error
SIM	single index model
SLS	semiparametric least squares
USD	US Dollar
WADE	weighted average derivative estimator
WSLS	weighted semiparametric least squares

Scalars, Vectors and Matrices

X, Y	random variables
x, y	scalars (realizations of X, Y)
$X_1, \ldots, X_n$	random sample of size n
$X_{(1)}, \ldots, X_{(n)}$	ordered random sample of size n
$x_1, \ldots, x_n$	realizations of $X_1, \ldots, X_n$
$\boldsymbol{X}$	vector of variables
$\boldsymbol{x}$	vector (realizations of $\boldsymbol{X}$)
x_0	origin (of histogram)

h	binwidth or bandwidth
$\widetilde{h}$	auxiliary bandwidth in marginal integration
$\mathbf{H}$	bandwidth matrix
$\mathbf{I}$	identity matrix
$\mathbf{X}$	data or design matrix
$\boldsymbol{Y}$	vector of observations $Y_1, \ldots, Y_n$
β	parameter
$\boldsymbol{\beta}$	parameter vector
$\boldsymbol{e}_0$	first unit vector, i.e. $\boldsymbol{e}_0 = (1, 0, \ldots, 0)^\top$
$\boldsymbol{e}_j$	$(j+1)$th unit vector, i.e. $\boldsymbol{e}_j = (0, \ldots, 0, \underset{j}{1}, 0, \ldots, 0)^\top$
$\mathbf{1}_n$	vector of ones of length n
$\boldsymbol{\mu}$	vector of expectations of $Y_1, \ldots, Y_n$ in generalized models
$\boldsymbol{\eta}$	vector of index values $\boldsymbol{X}_1^\top \beta, \ldots, \boldsymbol{X}_n^\top \beta$ in generalized models
LR	likelihood ratio test statistic
$\boldsymbol{U}$	vector of variables (linear part of the model)
$\boldsymbol{T}$	vector of continuous variables (nonparametric part of the model)
$\boldsymbol{X}_{\underline{\alpha}}$	random vector of all but αth component
$\boldsymbol{X}_{\underline{\alpha j}}$	random vector of all but αth and jth component
$\mathbf{S}$, $\mathbf{S}^P$, $\mathbf{S}_\alpha$	smoother matrices
$\boldsymbol{m}$	vector of regression values $m(\boldsymbol{X}_1), \ldots, m(\boldsymbol{X}_n)$
$\boldsymbol{g}_\alpha$	vector of additive component function values $g_\alpha(\boldsymbol{X}_1), \ldots, g_\alpha(\boldsymbol{X}_n)$

Matrix algebra

$\mathrm{tr}(\mathbf{A})$	trace of matrix $\mathbf{A}$
$\mathrm{diag}(\mathbf{A})$	diagonal of matrix $\mathbf{A}$
$\det(\mathbf{A})$	determinant matrix $\mathbf{A}$
$\mathrm{rank}(\mathbf{A})$	rank of matrix $\mathbf{A}$

$\mathbf{A}^{-1}$	inverse of matrix $\mathbf{A}$
$\|\boldsymbol{u}\|$	norm of vector $\boldsymbol{u}$, i.e. $\sqrt{\boldsymbol{u}^\top \boldsymbol{u}}$

Functions

log	logarithm (base e)
φ	pdf of standard normal distribution
Φ	cdf of standard normal distribution
I	indicator function, i.e. $\mathrm{I}(A) = 1$ if A holds, 0 otherwise
K	kernel function (univariate)
K_h	scaled kernel function, i.e. $K_h(u) = K(u/h)/h$
$\mathcal{K}$	kernel function (multivariate)
$\mathcal{K}_{\mathbf{H}}$	scaled kernel function, i.e. $\mathcal{K}_{\mathbf{H}}(\boldsymbol{u}) = \mathcal{K}(\mathbf{H}^{-1}\boldsymbol{u})/\det(\mathbf{H})$
$\mu_2(K)$	second moment of K, i.e. $\int u^2 K(u)\,du$
$\mu_p(K)$	pth moment of K, i.e. $\int u^p K(u)\,du$
$\|K\|_2^2$	squared L_2 norm of K, i.e. $\int \{K(u)\}^2\,du$
f	probability density function (pdf)
f_X	pdf of X
$f(x,y)$	joint density of X and Y
∇_f	gradient vector (partial first derivatives)
$\mathcal{H}_f$	Hessian matrix (partial second derivatives)
$K \star K$	convolution of K, i.e. $K \star K(u) = \int K(u-v)K(v)\,dv$
$w, \widetilde{w}$	weight functions
m	unknown function (to be estimated)
$m^{(\nu)}$	νth derivative (to be estimated)
ℓ, ℓ_i	log-likelihood, individual log-likelihood
G	known link function
g	unknown link function (to be estimated)
a, b, c	exponential family characteristics in generalized models
V	variance function of Y in generalized models
g_α	additive component (to be estimated)

$g_\alpha^{(\nu)}$	νth derivative (to be estimated)
f_α	pdf of X_α

Moments

EX	mean value of X
$\sigma^2 = \mathrm{Var}(X)$	variance of X, i.e. $\mathrm{Var}(X) = E(X - EX)^2$
$E(Y\|X)$	conditional mean Y given X (random variable)
$E(Y\|X = x)$	conditional mean Y given $X = x$ (realization of $E(Y\|X)$)
$E(Y\|x)$	same as $E(Y\|X = x)$
$\sigma^2(x)$	conditional variance of Y given $X = x$ (realization of $\mathrm{Var}(Y\|X)$)
$E_{X_1} g(X_1, X_2)$	mean of $g(X_1, X_2)$ w.r.t. X_1 only
$\mathrm{med}(Y\|X)$	conditional median Y given X (random variable)
μ	same as $E(Y\|X)$ in generalized models
$V(\mu)$	variance function of Y in generalized models
ψ	nuisance (dispersion) parameter in generalized models
MSE_x	MSE at the point x
$\mathcal{P}_\alpha$	conditional expectation function $E(\bullet\|X_\alpha)$

Distributions

$U[0,1]$	uniform distribution on $[0,1]$
$U[a,b]$	uniform distribution on $[a,b]$
$N(0,1)$	standard normal or Gaussian distribution
$N(\mu, \sigma^2)$	normal distribution with mean μ and variance σ^2
$N(\boldsymbol{\mu}, \boldsymbol{\Sigma})$	multi-dimensional normal distribution with mean $\boldsymbol{\mu}$ and covariance matrix $\boldsymbol{\Sigma}$
χ^2_m	χ^2 distribution with m degrees of freedom
t_m	t-distribution with m degrees of freedom

Estimates

$\widehat{\beta}$	estimated coefficient
$\widehat{\boldsymbol{\beta}}$	estimated coefficient vector
$\widehat{f}_h$	estimated density function
$\widehat{f}_{h,-i}$	estimated density function when leaving out observation i
$\widehat{m}_h$	estimated regression function
$\widehat{m}_{p,h}$	estimated regression function using local polynomials of degree p and bandwidth h
$\widehat{m}_{p,\mathbf{H}}$	estimated multivariate regression function using local polynomials of degree p and bandwidth matrix $\mathbf{H}$

Convergence

$o(\bullet)$	$a = o(b)$ iff $a/b \to 0$ as $n \to \infty$ or $h \to 0$
$O(\bullet)$	$a = O(b)$ iff $a/b \to$ constant as $n \to \infty$ or $h \to 0$
$o_p(\bullet)$	$U = o_p(V)$ iff for all $\epsilon > 0$ holds $P(\lvert U/V\rvert > \epsilon) \to 0$
$O_p(\bullet)$	$U = O_p(V)$ iff for all $\epsilon > 0$ exists $c > 0$ such that $P(\lvert U/V\rvert > c) < \epsilon$ as n is sufficiently large or h is sufficiently small
$\xrightarrow{a.s.}$	almost sure convergence
$\xrightarrow{P}$	convergence in probability
$\xrightarrow{L}$	convergence in distribution
$\approx$	asymptotically equal
$\sim$	asymptotically proportional

Other

$\mathbb{N}$	natural numbers
$\mathbb{Z}$	integers
$\mathbb{R}$	real numbers

$\mathbb{R}^d$	d-dimensional real space
$\propto$	proportional
$\equiv$	constantly equal
#	number of elements of a set
B_j	jth bin, i.e. $[x_0 + (j-1)h, x_0 + jh)$
m_j	bin center of B_j, i.e. $m_j = x_0 + (j - \frac{1}{2})h$

1

Introduction

1.1 Density Estimation

Consider a continuous random variable and its *probability density function* (pdf). The pdf tells you "how the random variable is distributed". From the pdf you cannot only calculate the statistical characteristics as mean and variance, but also the probability that this variable will take on values in a certain interval.

The pdf is, thus, very useful as it characterizes completely the "behavior" of a random variable. This fact might provide enough motivation to study nonparametric density estimation. Moreover nonparametric density estimates can serve as a building block in nonparametric regression estimation, as regression functions are fully characterized through the distribution of two (or more) variables.

The following example, which uses data from the Family Expenditure Survey of each year from 1969 to 1983, gives some illustration of the fact that density estimation has a substantial application in its own right.

Example 1.1.
Imagine that we have to answer the following questions: Is there a change in the structure of the income distribution during the period from 1969 to 1983? (You may recall, that many people argued that the neo-liberal policies of former Prime Minister Margaret Thatcher promoted income inequality in the early 1980s.)

To answer this question, we have estimated the distribution of net-income for each year from 1969 to 1983 both parametrically and nonparametrically. In parametric estimation of the distribution of income we have followed standard practice by fitting a log-normal distribution to the data. We employed the method of kernel density estimation (a generalization of the familiar histogram, as we will soon see) to estimate the income distribution nonparametrically. In the upper graph in Figure 1.1 we have plotted the estimated log-normal densities for each of the 15 years: Note that they are all very similar. On the other hand the analogous plot of the kernel density estimates show a movement of the net-income mode (the maximum of the den-

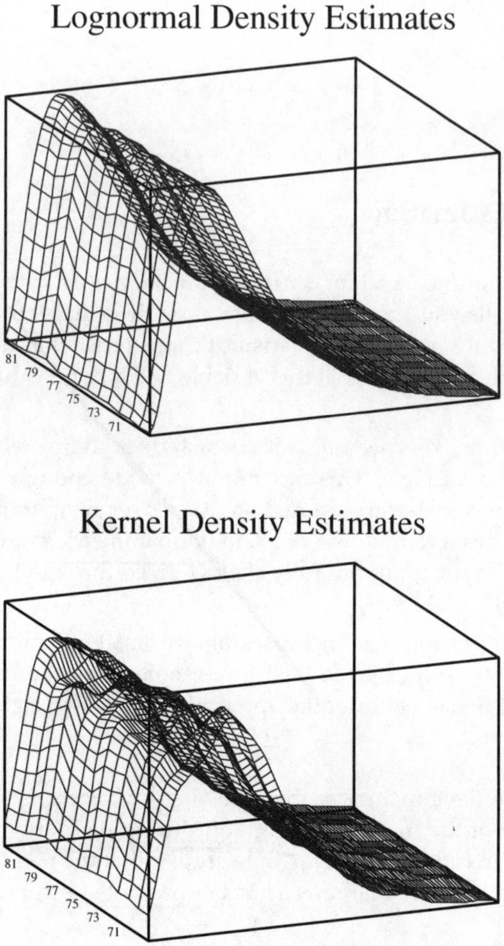

Figure 1.1. Log-normal density estimates (upper graph) versus kernel density estimates (lower graph) of net-income, U.K. Family Expenditure Survey 1969–83 `SPMfesdensities`

sity) to the left (Figure 1.1, lower graph). This indicates that the net-income distribution has in fact changed during this 15 year period. □

1.2 Regression

Let us now consider a typical linear regression problem. We assume that anyone of you has been exposed to the linear regression model where the mean of a dependent variable Y is related to a set of explanatory variables $X_1, X_2, \ldots, X_d$ in the following way:

$$E(Y|\boldsymbol{X}) = X_1\beta_1 + \ldots + X_d\beta_d = \boldsymbol{X}^\top \boldsymbol{\beta}. \tag{1.1}$$

Here $E(Y|\boldsymbol{X})$ denotes the expectation conditional on the vector $\boldsymbol{X} = (X_1, X_2, \ldots, X_d)^\top$ and β_j, $j = 1, 2, \ldots, d$ are unknown coefficients. Defining ε as the deviation of Y from the conditional mean $E(Y|\boldsymbol{X})$:

$$\varepsilon = Y - E(Y|\boldsymbol{X}) \tag{1.2}$$

we can write

$$Y = \boldsymbol{X}^\top \boldsymbol{\beta} + \varepsilon. \tag{1.3}$$

Example 1.2.
To take a specific example, let Y be *log wages* and consider the explanatory variables *schooling* (measured in years), labor market *experience* (measured as AGE − SCHOOL − 6) and *experience squared*. If we assume that, on average, log wages are linearly related to these explanatory variables then the linear regression model applies:

$$E(Y|\text{SCHOOL}, \text{EXP}) = \beta_0 + \beta_1 \cdot \text{SCHOOL} + \beta_2 \cdot \text{EXP} + \beta_3 \cdot \text{EXP}^2. \tag{1.4}$$

Note that we have included an intercept (β_0) in the model. □

The model of equation (1.4) has played an important role in empirical labor economics and is often called *human capital earnings equation* (or *Mincer earnings equation* to honor Jacob Mincer, a pioneer of this line of research). From the perspective of this course, an important characteristic of equation (1.4) is its *parametric* form: the shape of the regression function is governed by the unknown parameters β_j, $j = 1, 2, \ldots, d$. That is, all we have to do in order to determine the linear regression function (1.4) is to estimate the unknown parameters β_j. On the other hand, the parametric regression function of equation (1.4) a priori rules out many conceivable nonlinear relationships between Y and $\boldsymbol{X}$.

Let $m(\text{SCHOOL}, \text{EXP})$ be the true, unknown regression function of log wages on schooling and experience. That is,

$$E(Y|\text{SCHOOL}, \text{EXP}) = m(\text{SCHOOL}, \text{EXP}). \tag{1.5}$$

Suppose that you were assigned the following task: estimate the regression of log wages on schooling and experience as accurately as possible in *one* trial. That is, you are not allowed to change your model if you find that the initial specification does not fit the data well. Of course, you could just go ahead and assume, as we have done above, that the regression you are supposed to estimate has the form specified in (1.4). That is, you assume that

$$m(\text{SCHOOL}, \text{EXP}) = \beta_1 + \beta_2 \cdot \text{SCHOOL} + \beta_3 \cdot \text{EXP} + \beta_4 \cdot \text{EXP}^2,$$

and estimate the unknown parameters by the method of ordinary least squares, for example. But maybe you would not fit this parametric model if we told you that there are ways of estimating the regression function without having to make *any* prior assumptions about its functional form (except that it is a smooth function). Remember that you have just one trial and if the form of $m(\text{SCHOOL}, \text{EXP})$ is very different from (1.4) then estimating the parametric model may give you very inaccurate results.

It turns out that there are indeed ways of estimating $m(\bullet)$ that merely assume that $m(\bullet)$ is a smooth function. These methods are called *nonparametric* regression estimators and part of this course will be devoted to studying nonparametric regression.

Nonparametric regression estimators are very flexible but their statistical precision decreases greatly if we include several explanatory variables in the model. The latter caveat has been appropriately termed *the curse of dimensionality*. Consequently, researchers have tried to develop models and estimators which offer more flexibility than standard parametric regression but overcome the curse of dimensionality by employing some form of *dimension reduction*. Such methods usually combine features of parametric and nonparametric techniques. As a consequence, they are usually referred to as *semiparametric* methods. Further advantages of semiparametric methods are the possible inclusion of categorical variables (which can often only be included in a parametric way), an easy (economic) interpretation of the results, and the possibility of a part specification of a model.

In the following three sections we use the earnings equation and other examples to illustrate the distinctions between parametric, nonparametric and semiparametric regression and we certainly hope that this will whet your appetite for the material covered in this course.

1.2.1 Parametric Regression

Versions of the human capital earnings equation of (1.4) have probably been estimated by more researchers than any other model of empirical economics. For a detailed nontechnical and well-written discussion see Berndt (1991, Chapter 5). Here, we want to point out that:

- Under certain simplifying assumptions, β_2 accurately measures the rate of return to schooling.
- Human capital theory suggests a concave wage-experience profile: rapid human capital accumulation in the early stage of one's labor market career, with rising wages that peak somewhere during midlife and decline thereafter as hours worked and the incentive to invest in human capital decrease. This is the reason for including both EXP and EXP^2 in the model. In order to get a profile as the one envisaged by theory, the estimated value of β_3 should be positive and that of β_4 should be negative.

Table 1.1. Results from OLS estimation for Example 1.2

Dependent Variable: Log Wages			
Variable	Coefficients	S.E.	*t*-values
SCHOOL	0.0898	0.0083	10.788
EXP	0.0349	0.0056	6.185
EXP^2	–0.0005	0.0001	–4.307
constant	0.5202	0.1236	4.209
$R^2 = 0.24$, sample size $n = 534$			

We have estimated the coefficients of (1.4) using ordinary least squares (OLS), using a subsample of the 1985 Current Population Survey (CPS) provided by Berndt (1991). The results are given in Table 1.1.

The estimated rate of return to schooling is roughly 9%. Note that the estimated coefficients of EXP and EXP^2 have the signs predicted by human capital theory. The shape of the wage-schooling (a plot of SCHOOL vs. $0.0898 \cdot \text{SCHOOL}$) and wage-experience (a plot of EXP vs. $0.0349 \cdot \text{EXP} - 0.0005 \cdot \text{EXP}^2$) profiles are given in the left and right graphs of Figure 1.2, respectively.

The estimated wage-schooling relation is linear "by default" since we did not include SCHOOL^2, say, to allow for some kind of curvature within the parametric framework. By looking at Figure 1.2 it is clear that the estimated coefficients of EXP and EXP^2 imply the kind of concave wage-earnings profile predicted by human capital theory.

We have also plotted a graph (Figure 1.3) of the estimated regression surface, i.e. a plot that has the values of the estimated regression function (obtained by evaluating $0.0898 \cdot \text{SCHOOL} + 0.0349 \cdot \text{EXP} - 0.0005 \cdot \text{EXP}^2$ at the observed combinations of schooling and experience) on the vertical axis and schooling and experience on the horizontal axes.

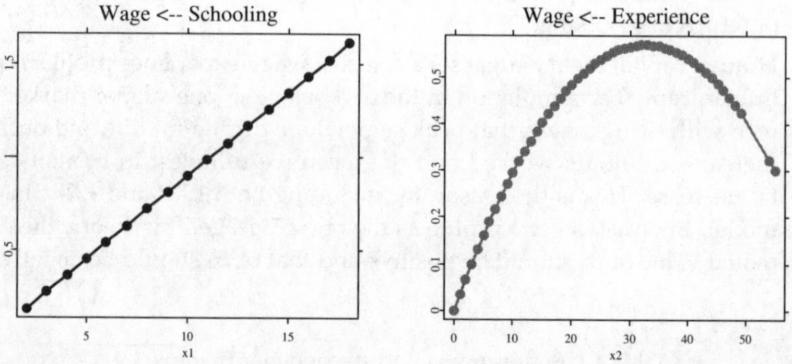

Figure 1.2. Wage-schooling and wage-experience profile SPMcps85lin

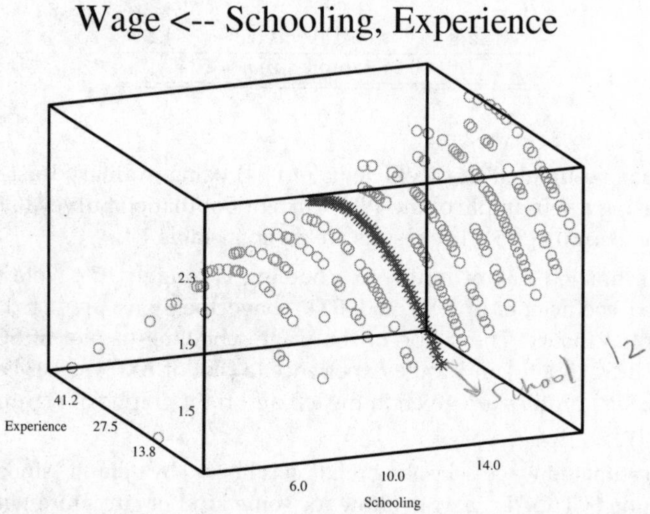

Figure 1.3. Parametrically estimated regression function SPMcps85lin

All of the element curves of the surface appear similar to Figure 1.2 (right) in the direction of experience and like Figure 1.2 (left) in the direction of schooling. To gain a better understanding of the three-dimensional picture we have plotted a single wage-experience profile in three dimensions, fixing schooling at 12 years. Hence, Figure 1.3 highlights the wage-earnings profile for high school graduates.

1.2.2 Nonparametric Regression

Suppose that we want to estimate

$$E(Y|\text{SCHOOL}, \text{EXP}) = m(\text{SCHOOL}, \text{EXP}). \tag{1.6}$$

and we are only willing to assume that $m(\bullet)$ is a smooth function. Nonparametric regression estimators produce an estimate of $m(\bullet)$ at an arbitrary point (SCHOOL $= s$, EXP $= e$) by *locally weighted averaging* over log wages (here s and e denote two arbitrary values that SCHOOL and EXP may take on, such as 12 and 15). Locally weighting means that those values of log wages will be higher weighted for which the corresponding observations of EXP and SCHOOL are close to the point (s, e). Let us illustrate this principle with an example. Let $s = 8$ and $e = 7$ and suppose you can use the four observations given in Table 1.2 to estimate $m(8, 7)$:

Table 1.2. Example observations

Observation	log(WAGES)	SCHOOL	EXP
1	7.31	8	8
2	7.6	16	1
3	7.4	8	6
4	7.8	12	2

In nonparametric regression $m(8, 7)$ is estimated by averaging over the observed values of the dependent variable log wage. But not all values will be given the same weight. In our example, observation 1 will get the most weight since it has values of schooling and experience that are very close to the point where we want to estimate. This makes a lot of sense: if we want to estimate mean log wages for individuals with 8 years of schooling and 7 years of experience then the observed log wage of a person with 8 years of schooling and 8 years of experience seems to be much more informative than the observed log wage of a person with 12 years of schooling and 2 years of experience.

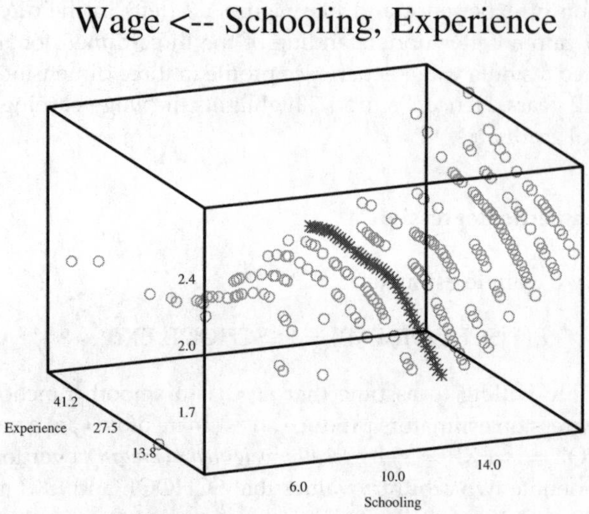

Figure 1.4. Nonparametrically estimated regression function `SPMcps85reg`

Consequently, any reasonable weighting scheme will give more weight to 7.31 than to 7.8 when we average over observed log wages. The exact method of weighting is determined by a weight function that makes precise the idea of weighting nearby observations more heavily. In fact, the weight function might be such that observations that are too far away get zero weight. In our example, observation 2 has values of experience and schooling that are so far away from 8 years of schooling and 7 years of experience that a weight function might assign zero value to the corresponding value of log wages (7.6). It is in this sense that the averaging is local. In Figure 1.4, the surface of nonparametrically estimated values of $m(\bullet)$ are shown. Here, a so-called kernel estimator has been used.

As long as we are dealing with only one regressor, the results of estimating a regression function nonparametrically can easily be displayed in a graph. The following example illustrates this. It relates net-income data, as we considered in Example 1.1, to a second variable that measures household expenditure.

Example 1.3.
Consider for instance the dependence of food expenditure on net-income. Figure 1.5 shows the so-called Engel curve (after the German Economist Engel) of net-income and food share estimated using data from the 1973 Family

Expenditure Survey of roughly 7000 British households. The figure supports the theory of Engel who postulated in 1857:

> ... je ärmer eine Familie ist, einen desto größeren Antheil von der Gesammtausgabe muß zur Beschaffung der Nahrung aufgewendet werden ... (The poorer a family, the bigger the share of total expenditure that has to be used for food.) □

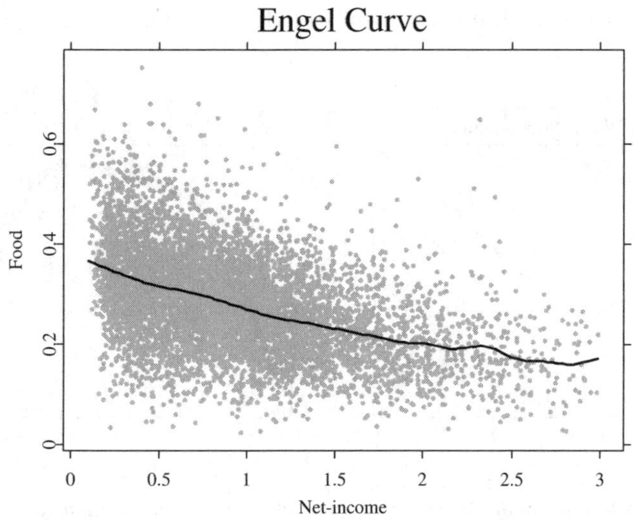

Figure 1.5. Engel curve, U.K. Family Expenditure Survey 1973 `SPMengelcurve2`

1.2.3 Semiparametric Regression

To illustrate semiparametric regression let us return to the human capital earnings function of Example 1.2. Suppose the regression function of log wages on schooling and experience has the following shape:

$$E(Y|\text{SCHOOL}, \text{EXP}) = \alpha + g_1(\text{SCHOOL}) + g_2(\text{EXP}). \tag{1.7}$$

Here $g_1(\bullet)$ and $g_2(\bullet)$ are two unknown, smooth functions and α is an unknown parameter. Note that this model combines the simple additive structure of the parametric regression model (referred to hereafter as the *additive model*) with the flexibility of the nonparametric approach. This is done by not imposing any strong shape restrictions on the functions that determine how schooling and experience influence the mean regression of log wages. The procedure employed to estimate this model will be explained in greater detail later in this course. It should be clear, however, that in order to estimate the unknown functions $g_1(\bullet)$ and $g_2(\bullet)$ nonparametric regression estimators have to be employed. That is, when estimating semiparametric models we usually have to use nonparametric techniques. Hence, we will have to spend a substantial amount of time studying nonparametric estimation if we want to understand how to estimate semiparametric models. For now, we want to focus on the results and compare them with the parametric fit.

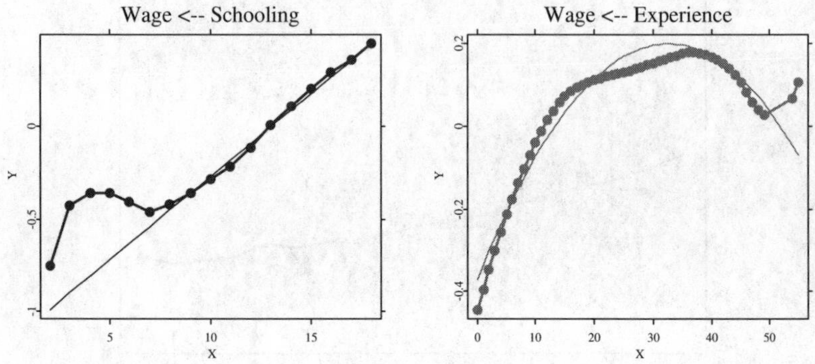

Figure 1.6. Additive model fit vs. parametric fit, wage-schooling (left) and wage-experience (right) `SPMcps85add`

In Figure 1.6 the parametrically estimated wage-schooling and wage-experience profiles are shown as thin lines whereas the estimates of $g_1(\bullet)$ and $g_2(\bullet)$ are displayed as thick lines with bullets. The parametrically estimated wage-school and wage-experience profiles show a good deal of similarity with the estimate of $g_1(\bullet)$ and $g_2(\bullet)$, except for the shape of the curve at extremal values. The good agreement between parametric estimates and additive model fit is also visible from the plot of the estimated regression surface, which is shown in Figure 1.7.

Hence, we may conclude that in this specific example the parametric model is supported by the more flexible nonparametric and semiparametric methods. This potential usefulness of nonparametric and semiparametric techniques for checking the adequacy of parametric models will be illustrated in several other instances in the latter part of this course.

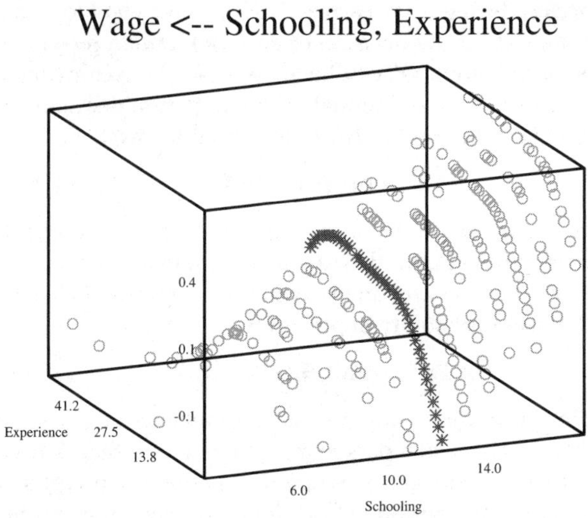

Figure 1.7. Surface plot for the additive model SPMcps85add

Take a closer look at (1.6) and (1.7). Observe that in (1.6) we have to estimate one unknown function of two variables whereas in (1.7) we have to estimate two unknown functions, each a function of one variable. It is in this sense that we have reduced the dimensionality of the estimation problem. Whereas all researchers might agree that additive models like the one in (1.7) are achieving a dimension reduction over completely nonparametric regression, they may not agree to call (1.7) a semiparametric model, as there are no parameters to estimate (except for the intercept parameter α). In the following example we confront a standard parametric model with a more flexible model that, as you will see, truly deserves to be called semiparametric.

Example 1.4.
In the earnings-function example, the dependent variable log wages can principally take on *any* positive value, i.e. the set of values Y is infinite. This may not always be the case. For example, consider the decision of an East-German resident to move to Western Germany and denote the decision variable by Y. In this case, the dependent variable can take on only *two* values,

$$Y = \begin{cases} 1 & \text{if the person can imagine moving to the west,} \\ 0 & \text{otherwise.} \end{cases}$$

We will refer to this as a *binary response* later on. □

In Example 1.2 we tried to estimate the effect of a person's education and work experience on the log wage earned. Now, say we want to find out how these two variables affect the decision of an East German resident to move west, i.e. we want to know $E(Y|\boldsymbol{x})$ where $\boldsymbol{x}$ is a $(d \times 1)$ vector containing all d variables considered to be influential to the migration decision. Since Y is a binary variable (i.e. a Bernoulli distributed variable), we have that

$$E(Y|\boldsymbol{X}) = P(Y = 1|\boldsymbol{X}). \tag{1.8}$$

Thus, the regression of Y on $\boldsymbol{X}$ can be expressed as the probability that a randomly sampled person from the East will migrate to the West, given this person's characteristics collected in the vector $\boldsymbol{X}$. Standard models for $P(Y = 1|\boldsymbol{X})$ assume that this probability depends on $\boldsymbol{X}$ as follows:

$$P(Y = 1|\boldsymbol{X}) = G(\boldsymbol{X}^\top \boldsymbol{\beta}), \tag{1.9}$$

where $\boldsymbol{X}^\top \boldsymbol{\beta}$ is a linear combination of all components of $\boldsymbol{X}$. It aggregates the multiple characteristics of a person into one number (therefore called the *index function* or simply the *index*), where $\boldsymbol{\beta}$ is an unknown vector of coefficients. $G(\bullet)$ denotes any continuous function that maps the real line to the range of $[0, 1]$. $G(\bullet)$ is also called the *link function*, since it links the index $\boldsymbol{X}^\top \boldsymbol{\beta}$ to the conditional expectation $E(Y|\boldsymbol{X})$.

In the context of this lecture, the crucial question is precisely *what* parametric form these two functions take or, more generally, whether they will take any parametric form *at all*. For now we want to compare two models: one that assumes that $G(\bullet)$ is of a known parametric form and one that allows $G(\bullet)$ to be an unknown smooth function.

One of the most widely used fully parametric models applied to the case of binary dependent variables is the *logit model*. The logit model assumes that $G(\boldsymbol{X}^\top \boldsymbol{\beta})$ is the (standard) logistic cumulative distribution function (cdf) for all $\boldsymbol{X}$. Hence, in this case

$$E(Y|\boldsymbol{X}) = P(Y = 1|\boldsymbol{X}) = \frac{1}{\exp(-\boldsymbol{X}^\top \boldsymbol{\beta})}. \tag{1.10}$$

Example 1.5.
In using a logit model, Burda (1993) estimated the effect of various explanatory variables on the migration decision of East German residents. The data for fitting this model were drawn from a panel study of approximately 4,000 East German households in spring 1991. We use a subsample of $n = 402$ observations from the German state "Mecklenburg-Vorpommern" here. Due to space constraints, we merely report the estimated coefficients of three components of the index $\boldsymbol{X}^\top \boldsymbol{\beta}$, as we will refer to these estimates below:

$$\begin{aligned} &\beta_0 + \beta_1 \cdot \text{INC} + \beta_2 \cdot \text{AGE} \\ &= -2.2905 + 0.0004971 \cdot \text{INC} - 0.45499 \cdot \text{AGE} \end{aligned} \tag{1.11}$$

INC and AGE are used to abbreviate the household income and age of the individual. □

Figure 1.8 gives a graphical presentation of the results. Each observation is represented by a "+". As mentioned above, the characteristics of each person are transformed into an index (to be read off the horizontal axis) while the dependent variable takes on one of two values, $Y = 0$ or $Y = 1$ (to be read off the vertical axis). The curve plots estimates of $P(Y = 1|\boldsymbol{X})$, the probability of $Y = 1$ as a function of $\boldsymbol{X}^\top\boldsymbol{\beta}$. Note that the estimates of $P(Y = 1|\boldsymbol{X})$, by assumption, are simply points on the cdf of a standard logistic distribution.

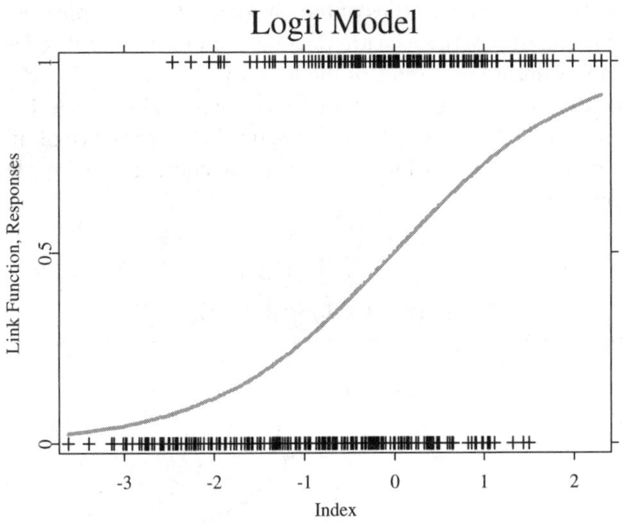

Figure 1.8. Logit fit SPMlogit

We shall continue with Example 1.4 below, but let us pause for a moment to consider the following substantial problem: the logit model, like other parametric models, is based on rather strong functional form (linear index) and distributional assumptions, neither of which are usually justified by economic theory.

The first question to ask before developing alternatives to standard models like the logit model is: what are the consequences of estimating a logit model if one or several of these assumptions are violated? Note that this is a crucial question: if our parametric estimates are largely unaffected by model violations, then there is no need to develop and apply semiparametric models and estimators. Why would anyone put time and effort into a project that promises little return?

One can employ the tools of asymptotic statistical theory to show that violating the assumptions of the logit model leads parameter estimates to being inconsistent. That is, if the sample size goes to infinity, the logit maximum-likelihood estimator (logit-MLE) does not converge to the true parameter value in probability. While it doesn't converge to the true parameter value it does, however, converge to some other value. If this "false" value is close enough to the true parameter value then we may not care very much about this inconsistency.

Consistency is an asymptotic criterion for the performance of an estimator. That is, it looks at the properties of the estimator if the sample size grows without limits. Yet, in practice, we are dealing with finite samples. Unfortunately, the finite-sample properties of the logit maximum-likelihood estimator can not be derived analytically. Hence, we have to rely on simulations to collect evidence of its small-sample performance in the presence of misspecification. We conducted a small simulation in the context of Example 1.4 to which we now return.

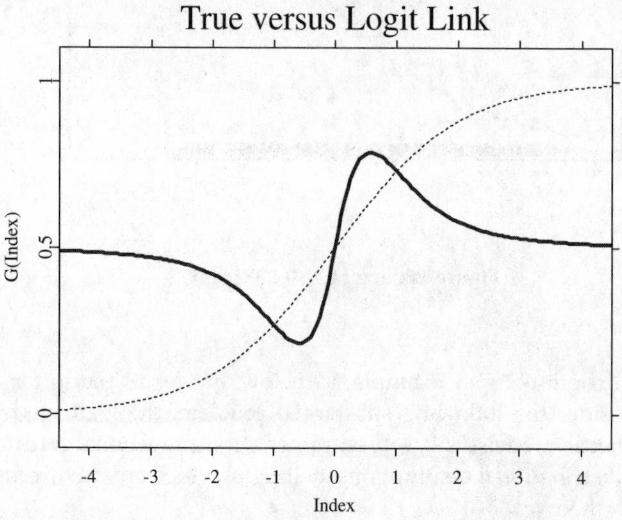

Figure 1.9. Link function of the homoscedastic logit model (thin line) versus the link function of the heteroscedastic model (solid line) `SPMtruelogit`

Example 1.6.
Following Horowitz (1993) we generated data according to a heteroscedastic model with two explanatory variables, INC and AGE. Here we considered heteroscedasticity of the form

$$\mathrm{Var}(\varepsilon|\boldsymbol{X}=x) = \frac{1}{4}\left\{1+(\boldsymbol{x}^\top\boldsymbol{\beta})^2\right\}^2 \cdot \mathrm{Var}(\zeta),$$

where ζ has a (standard) logistic distribution. To give you an impression of how dramatically the *true* heteroscedastic model differs from the *supposed* homoscedastic logit model, we plotted the link functions of the two models as shown in Figure 1.9. □

To add a sense of realism to the simulation, we set the coefficients of these variables equal to the estimates reported in (1.11). Note that the standard logit model introduced above does not allow for heteroscedasticity. Hence, if we apply the standard logit maximum-likelihood estimator to the simulated data, we are estimating under misspecification. We performed 250 replications of this estimation experiment, using the full data set with 402 observations each time. As the estimated coefficients are only identified up to scale, we compared the ratio of the true coefficients, β_{INC}/β_{AGE}, to the ratio of their estimated logit-MLE counterparts, $\widehat{\beta}_{INC}/\widehat{\beta}_{AGE}$. Figure 1.10 shows the sampling distribution of the logit-MLE coefficients, along with the true value (vertical line).

As we have subtracted the true value from each estimated ratio and divided this difference by the true ratio's absolute value, the true ratio is standardized to zero and differences on the horizontal axis can be interpreted as percentage deviations from the truth. In Figure 1.10, the sampling distribution of the estimated ratios is centered around -0.11 which is the percentage deviation from the truth of 11%. Hence, the logit-MLE underestimates the true value.

Now that we have seen how serious the consequences of model misspecification can be, we might want to learn about semiparametric estimators that have desirable properties under more general assumptions than their parametric counterparts. One way to generalize the logit model is the so-called *single index model* (SIM) which keeps the linear form of the index $\boldsymbol{X}^\top\boldsymbol{\beta}$ but allows the function $G(\bullet)$ in (1.9) to be an arbitrary smooth function $g(\bullet)$ (not necessarily a distribution function) that has to be estimated from the data:

$$E(Y|\boldsymbol{X}) = g(\boldsymbol{X}^\top\boldsymbol{\beta}), \tag{1.12}$$

Estimation of the single index model (1.12) proceeds in two steps:

- Firstly, the coefficient vector $\boldsymbol{\beta}$ has to be estimated. Methods to calculate the coefficients for discrete and continuous variables will be covered in depth later.

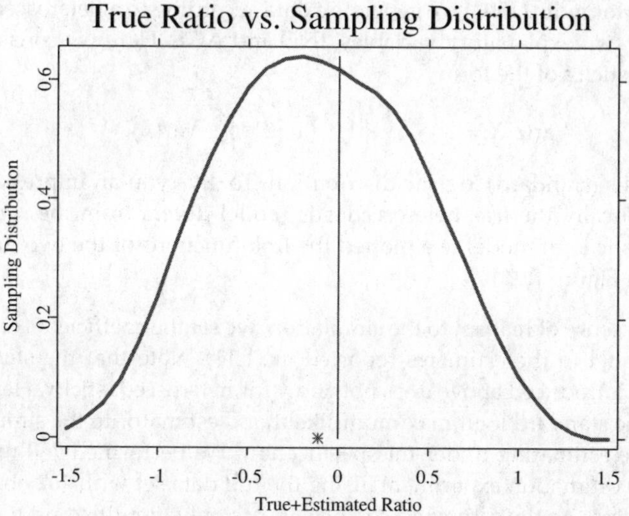

Figure 1.10. Sampling distribution of the ratio of the estimated coefficients (density estimate and mean value indicated as $*$) and the ratio's true value (vertical line) SPMsimulogit

- Secondly, we have to estimate the unknown link function $g(\bullet)$ by nonparametrically regressing the dependent variable Y on the fitted index $X^\top\widehat{\beta}$ where $\widehat{\beta}$ is the coefficient vector we estimated in the first step. To do this, we use again a nonparametric estimator, the kernel estimator we mentioned briefly above.

Example 1.7.
Let us consider what happens if we use $\widehat{\beta}$ from the logit fit and estimate the link function nonparametrically. Figure 1.11 shows this estimated link function. As before, the position of a + sign represents at the same time the values of $X^\top\widehat{\beta}$ and Y of a particular observation, while the curve depicts the estimated link function. □

One additional remark should be made here: As you will soon learn, the shape of the estimated link function (the curve) varies with the so-called bandwidth, a parameter central in nonparametric function estimation. Thus, there is no unique estimate of the link function, and it is a crucial (and difficult) problem of nonparametric regression to find the "best" bandwidth and thus the optimal estimate. Fortunately, there are methods to select an ap-

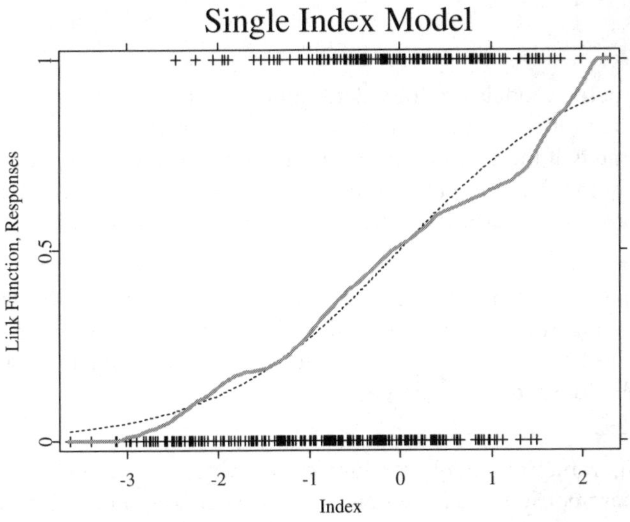

Figure 1.11. Single index versus logit model SPMsim

propriate bandwidth. Here, we have chosen $h = 0.7$ "index units" for the bandwidth. For comparison the shapes of both the single index (solid line) and the logit (dashed line) link functions are shown ins in Figure 1.8. Even though not identical they look rather similar.

Summary

- ⋆ Parametric models are fully determined up to a parameter (vector). The fitted models can easily be interpreted and estimated accurately if the underlying assumptions are correct. If, however, they are violated then parametric estimates may be inconsistent and give a misleading picture of the regression relationship.

- ⋆ Nonparametric models avoid restrictive assumptions of the functional form of the regression function m. However, they may be difficult to interpret and yield inaccurate estimates if the number of regressors is large.

- ⋆ Semiparametric models combine components of parametric and nonparametric models, keeping the easy interpretability of the former and retaining some of the flexibility of the latter.

Part I

Nonparametric Models

2

Histogram

2.1 Motivation and Derivation

Let X be a continuous random variable and f its probability density function (pdf). The pdf tells you "how X is distributed". From the pdf you can calculate the mean and variance of X (if they exist) and the probability that X will take on values in a certain interval. The pdf is, thus, very useful to characterize the distribution of the random variable X.

In practice, the pdf of some observable random variable X is in general unknown. All you have are n observations $X_1, \ldots, X_n$ of X and your task is to use these n values to estimate $f(x)$. We shall assume that the n observations are independent and that they all indeed come from the same distribution, namely $f(x)$. That is, in this and the next chapters we will be concerned with estimating $f(x)$ at a certain value x from *i.i.d.* data (independent and identically distributed).

We will approach this estimation problem without assuming that $f(x)$ has some known functional form except for some unknown parameter(s) that need to be estimated. For instance, we do not assume that $f(x)$ has the well-known form of the normal distribution with unknown parameters μ and σ^2. We will focus on nonparametric ways of estimating $f(x)$ instead. The most commonly used nonparametric density estimator is the *histogram*.

2.1.1 Construction

The construction of a histogram is fairly simple. Suppose you have a random sample $X_1, X_2, \ldots, X_n$ from some unknown continuous distribution.

- Select an origin x_0 and divide the real line into *bins* of *binwidth* h:

$$B_j = [x_0 + (j-1)h, x_0 + jh), \quad j \in \mathbb{Z}.$$

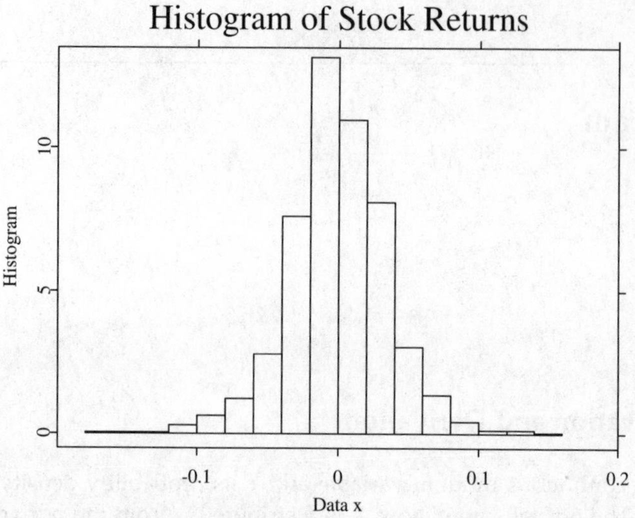

Figure 2.1. Histogram for stock returns data Pagan & Schwert (1990) with binwidth $h = 0.02$ and origin $x_0 = 0$ `SPMhistogram`

- Count how many observations fall into each bin. Denote the number of observations that fall into bin j by n_j.
- For each bin divide the frequency count by the sample size n (to convert them into relative frequencies, the sample analog of probabilities), and by the binwidth h (to make sure that the area under the histogram is equal to one):

$$f_j = \frac{n_j}{nh}.$$

- Plot the histogram by erecting a bar over each bin with height f_j and width h.

More formally, the histogram is given by

$$\widehat{f}_h(x) = \frac{1}{nh} \sum_{i=1}^{n} \sum_{j} \mathrm{I}(X_i \in B_j)\, \mathrm{I}(x \in B_j), \tag{2.1}$$

where

$$\mathrm{I}(X_i \in B_j) = \begin{cases} 1 \text{ if } X_i \in B_j, \\ 0 \text{ otherwise.} \end{cases}$$

Note that formula (2.1) (as well as its corresponding graph, the histogram) gives an estimate of f for all x. Denote by m_j the center of the bin

B_j. It is easy to see from formula (2.1) that the histogram assigns each x in $B_j = [m_j - \frac{h}{2}, m_j + \frac{h}{2})$ the same estimate for f, namely $\widehat{f}_h(m_j)$. This seems to be rather restrictive and inflexible and later on we will see that there is a better alternative.

It can be easily verified that the area of a histogram is indeed equal to one, a property that we certainly require from any reasonable estimator of a pdf. But we can give further motivation for viewing the histogram as an estimator of the pdf of a continuous distribution.

2.1.2 Derivation

Consider Figure 2.2 where the pdf of a random variable X is graphed. The probability that an observation of X will fall into the bin $[m_j - \frac{h}{2}, m_j + \frac{h}{2})$ is given by

$$P\left(X \in \left[m_j - \frac{h}{2}, m_j + \frac{h}{2}\right)\right) = \int_{m_j - \frac{h}{2}}^{m_j + \frac{h}{2}} f(u)\, du \tag{2.2}$$

which is just the shaded area under the density between $m_j - \frac{h}{2}$ and $m_j + \frac{h}{2}$. This area can be approximated by a bar with height $f(m_j)$ and width h (see Figure 2.2). Thus we can write

$$P\left(X \in \left[m_j - \frac{h}{2}, m_j + \frac{h}{2}\right)\right) = \int_{m_j - \frac{h}{2}}^{m_j + \frac{h}{2}} f(u)\, du \approx f(m_j) \cdot h. \tag{2.3}$$

A natural estimate of this probability is the relative frequency of observations in this interval

$$P\left(X \in \left[m_j - \frac{h}{2}, m_j + \frac{h}{2}\right)\right) \approx \frac{1}{n} \# \left\{X_i \in \left[m_j - \frac{h}{2}, m_j + \frac{h}{2}\right)\right\}. \tag{2.4}$$

Combining (2.3) and (2.4) we get

$$\widehat{f}_h(m_j) = \frac{1}{nh} \# \left\{X_i \in \left[m_j - \frac{h}{2}, m_j + \frac{h}{2}\right)\right\}. \tag{2.5}$$

(Here and in the following we use # to denote the cardinality, i.e. the number of elements in a set.)

2.1.3 Varying the Binwidth

The subscript h of $\widehat{f}_h(m_j)$ indicates that the estimate given by the histogram depends on the choice of the binwidth h. Note that it also depends on the

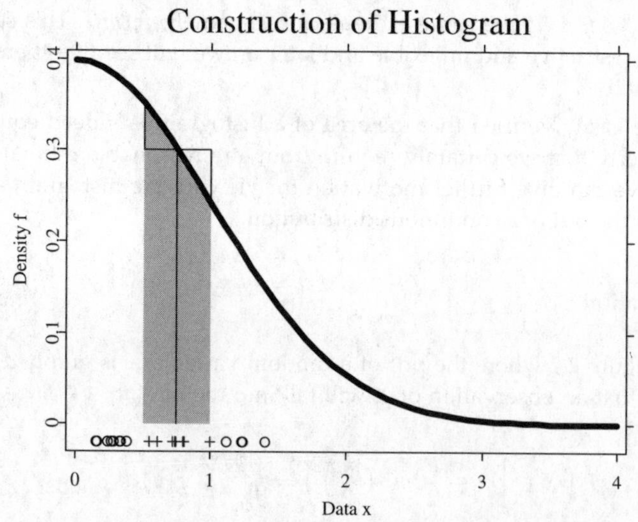

Figure 2.2. Approximation of the area under the pdf over an interval by erecting a rectangle over the interval

choice of the origin even though this dependence is not reflected in the notation. The dependency of the histogram on the origin will be discussed later in Subsection 2.3. To illustrate the effect of the choice of the binwidth on the shape of the histogram consider Figure 2.3 where we have computed and displayed histograms for the stock returns data, corresponding to different binwidths.

Clearly, if we increase h the histogram appears to be smoother but without some reasonable criterion on hand it remains very difficult to say which binwidth provides the "optimal" degree of smoothness.

2.2 Statistical Properties

Let us investigate some of the properties of the histogram as an estimator of the unknown pdf $f(x)$. Suppose that the origin $x_0 = 0$ and that we want to estimate the density at some point $x \in B_j = [(j-1)h, jh)$. The density estimate assigned by the histogram to x is

$$\widehat{f}_h(x) = \frac{1}{nh} \sum_{i=1}^{n} \mathrm{I}(X_i \in B_j). \tag{2.6}$$

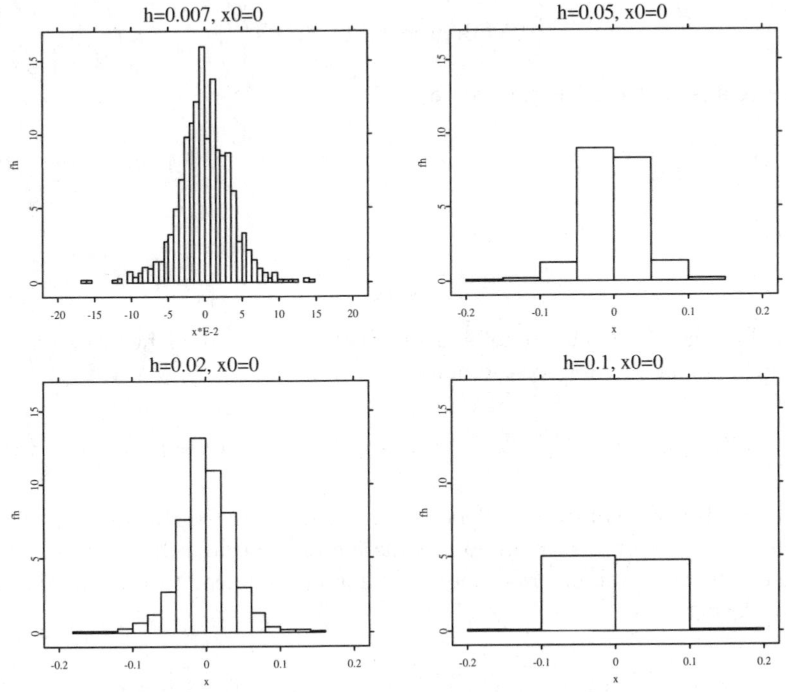

Figure 2.3. Four histograms for the stock returns data with binwidths $h = 0.007$, $h = 0.02$, $h = 0.05$, and $h = 0.1$; origin $x_0 = 0$ SPMhisdiffbin

2.2.1 Bias

Is $E\{\widehat{f}_h(x)\} = f(x)$, i.e. is the histogram an unbiased estimator? Let us calculate $E\{\widehat{f}_h(x)\}$ to find out:

$$E\{\widehat{f}_h(x)\} = \frac{1}{nh} \sum_{i=1}^{n} E\{\mathrm{I}(X_i \in B_j)\}. \tag{2.7}$$

Since the X_i $(i = 1, 2, \ldots, n)$ are i.i.d. random variables it follows that the indicator functions $\mathrm{I}(X_i \in B_j)$ are also i.i.d. random variables, and we can write

$$E\{\widehat{f}_h(x)\} = \frac{1}{nh} n E\{\mathrm{I}(X_i \in B_j)\}. \tag{2.8}$$

It remains to find $E\{\mathrm{I}(X_i \in B_j)\}$. It is straightforward to derive the pdf of the random variable $\mathrm{I}(X_i \in B_j)$:

$$\mathrm{I}(X_i \in B_j) = \begin{cases} 1 \text{ with probability} & \int_{(j-1)h}^{jh} f(u)\,du, \\ 0 \text{ with probability} & 1 - \int_{(j-1)h}^{jh} f(u)\,du. \end{cases} \tag{2.9}$$

Hence it is Bernoulli distributed and

$$E\{\mathrm{I}(X_i \in B_j)\} = P\left\{\mathrm{I}(X_i \in B_j) = 1\right\} = \int_{(j-1)h}^{jh} f(u)\,du\,, \tag{2.10}$$

therefore

$$E\{\widehat{f}_h(x)\} = \frac{1}{h}\int_{(j-1)h}^{jh} f(u)\,du. \tag{2.11}$$

The last term is in general not equal to $f(x)$. Consequently, the histogram is in general not an unbiased estimator of $f(x)$:

$$\mathrm{Bias}\{\widehat{f}_h(x)\} = E\{\widehat{f}_h(x) - f(x)\} = \frac{1}{h}\int_{(j-1)h}^{jh} f(u)\,du - f(x). \tag{2.12}$$

The precise value of the bias depends on the shape of the true density $f(x)$. We can, however, derive an approximative bias formula that allows us to make some general remarks about the situations that lead to a small or large bias. Using

$$\frac{1}{h}\int_{(j-1)h}^{jh} f(u)\,du - f(x) = \frac{1}{h}\left\{\int_{(j-1)h}^{jh} \{f(u) - f(x)\}\,du\right\}. \tag{2.13}$$

and a first-order Taylor approximation of $f(x) - f(u)$ around the center $m_j = (j - \frac{1}{2})h$ of B_j, i.e. $f(u) - f(x) \approx f'(m_j)(u - x)$ yields

$$E\{\widehat{f}_h(x) - f(x)\} \approx f'\left\{m_j\right\}\left\{m_j - x\right\}. \tag{2.14}$$

Note that the absolute value of the approximate bias is increasing in the slope of the true density at the mid point m_j, and that the approximate bias is zero if $x = m_j$.

2.2.2 Variance

Let us calculate the variance of the histogram to see how its volatility depends on the binwidth h:

$$\mathrm{Var}\{\widehat{f}_h(x)\} = \mathrm{Var}\left\{\frac{1}{nh}\sum_{i=1}^{n} \mathrm{I}(X_i \in B_j)\right\}. \tag{2.15}$$

Since the X_i are i.i.d. we can write

$$\text{Var}\{\widehat{f}_h(x)\} = \frac{1}{n^2h^2}\sum_{i=1}^{n}\text{Var}\{\text{I}(X_i \in B_j)\} = \frac{1}{n^2h^2} n\,\text{Var}\{\text{I}(X_i \in B_j)\}. \quad (2.16)$$

From (2.9) we know that $\text{I}(X_i \in B_j)$ is Bernoulli distributed with parameter $p = \int_{B_j} f(u)\,du$. Since the variance of any Bernoulli random variable X is given by $p(1-p)$ we have

$$\text{Var}\{\widehat{f}_h(x)\} = \frac{1}{n^2h^2} n \int_{B_j} f(u)\,du \left(1 - \int_{B_j} f(u)\,du\right).$$

You are asked in Exercise 2.3 to show that the right hand side of (2.17) can be written as the sum of $f(x)/(nh)$ and terms that are of smaller magnitude. That is, you are asked to show that there also exists an approximate formula for the variance of $\text{Var}\{\widehat{f}_h(x)\}$:

$$\text{Var}\{\widehat{f}_h(x)\} \approx \frac{1}{nh} f(x). \quad (2.17)$$

We observe that the variance of the histogram is proportional to $f(x)$ and decreases when nh increases. This implies that increasing h will reduce the variance. We know from (2.14) that increasing h will do the opposite to the bias. So how should we choose h if we want to have a small variance *and* a small bias?

Since variance and bias vary in opposite directions with h, we have to settle for finding the value of h that yields (in some sense) the optimal compromise between variance and bias reduction.

2.2.3 Mean Squared Error

Consider the *mean squared error* (MSE) of the histogram

$$\text{MSE}\{\widehat{f}_h(x)\} = E[\{\widehat{f}_h(x) - f(x)\}^2], \quad (2.18)$$

which can be written as the sum of the variance and the squared bias (this is a general result, not limited to this particular problem)

$$\text{MSE}\{\widehat{f}_h(x)\} = \text{Var}\{\widehat{f}_h(x)\} + [\text{Bias}\{\widehat{f}_h(x)\}]^2. \quad (2.19)$$

Hence, finding the binwidth h that minimizes the MSE might yield a histogram that is neither oversmoothed (i.e. one that overemphasizes variance reduction by employing a relatively large value of h) nor undersmoothed (i.e. one that is overemphasizes bias reduction by using a small binwidth). It can be shown (see Exercise 2.4) that

$$\mathrm{MSE}\{\widehat{f}_h(x)\} = \frac{1}{nh} f(x) + f'\left\{\left(j-\frac{1}{2}\right)h\right\}^2 \left\{\left(j-\frac{1}{2}\right)h - x\right\}^2$$
$$+ o(h) + o\left(\frac{1}{nh}\right), \qquad (2.20)$$

where $o(h)$ and $o\left(\frac{1}{nh}\right)$ denote terms not explicitly written down which are of lower order than h and $\frac{1}{nh}$, respectively.

From (2.20) we can conclude that the histogram converges in mean square to $f(x)$ if we let $h \to 0$, $nh \to \infty$. That is, if we use more and more observations ($n \to \infty$) and smaller and smaller binwidth ($h \to 0$) but do not shrink the binwidth too quickly ($nh \to \infty$) then the MSE of $\widehat{f}_h(x)$ goes to zero. Since convergence in mean square implies convergence in probability, it follows that $\widehat{f}_h(x)$ converges to $f(x)$ in probability or, in other words, that $\widehat{f}_h(x)$ is a consistent estimator of $f(x)$.

Figure 2.4 shows the MSE (at $x = 0.5$) for estimating the density function given in Exercise 2.9 as a function of the binwidth h.

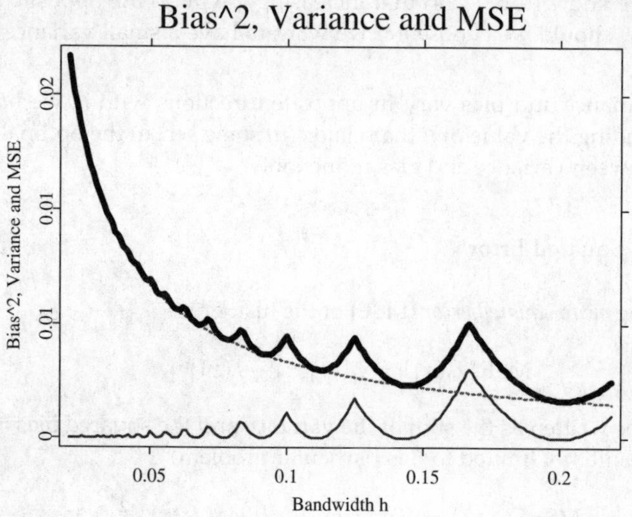

Figure 2.4. Squared bias (thin solid line), variance (dashed line) and MSE (thick line) for the histogram SPMhistmse

2.2.4 Mean Integrated Squared Error

The application of the MSE formula is difficult in practice, since the derived formula for the MSE depends on the unknown density function f both in the variance and the squared bias term. Instead of looking at the accuracy of $\widehat{f}_h(x)$ as an estimator of f at a single point, it might be worthwhile to have a global measure of accuracy. The most widely used global measure of estimation accuracy is the *mean integrated squared error* (MISE):

$$\text{MISE}(\widehat{f}_h) = E\left[\int_{-\infty}^{\infty} \{\widehat{f}_h(x) - f(x)\}^2 \, dx\right] \tag{2.21}$$

$$= \int_{-\infty}^{\infty} E\left[\{\widehat{f}_h(x) - f(x)\}^2\right] dx$$

$$= \int_{-\infty}^{\infty} \text{MSE}\{\widehat{f}_h(x)\} \, dx \tag{2.22}$$

Using (2.20) we can write for any x

$$\begin{aligned}
\text{MISE}(\widehat{f}_h) &\approx \int \frac{1}{nh} f(x)\,dx \\
&\quad + \int \sum_j \text{I}(x \in B_j) \left\{\left(j - \frac{1}{2}\right)h - x\right\}^2 \left[f'\left\{\left(j - \frac{1}{2}\right)h\right\}\right]^2 dx \\
&= \frac{1}{nh} + \sum_j \int_{B_j} \left\{x - \left(j - \frac{1}{2}\right)h\right\}^2 \left\{f'\left[\left(j - \frac{1}{2}\right)h\right]\right\}^2 dx \\
&\approx \frac{1}{nh} + \sum_j \left\{f'\left[\left(j - \frac{1}{2}\right)h\right]\right\}^2 \cdot \int_{B_j} \left\{x - \left(j - \frac{1}{2}\right)h\right\}^2 dx \\
&\approx \frac{1}{nh} + \frac{h^2}{12} \int \{f'(x)\}^2 \, dx = \frac{1}{nh} + \frac{h^2}{12} \|f'\|_2^2, \quad \text{for } h \to 0.
\end{aligned}$$

Here, $\|f'\|_2^2$ denotes the so-called squared L_2-norm of the function f'. Now, the asymptotic MISE, denoted AMISE, is given by

$$\text{AMISE}(\widehat{f}_h) = \frac{1}{nh} + \frac{h^2}{12}\|f'\|_2^2. \tag{2.23}$$

2.2.5 Optimal Binwidth

We are now in a position to employ a precise criterion for selecting an optimal binwidth h: select the binwidth h that minimizes AMISE! Differentiating AMISE with respect to h gives

$$\frac{\partial\{\text{AMISE}(\widehat{f}_h)\}}{\partial h} = \frac{-1}{nh^2} + \frac{1}{6} h \|f'\|_2^2 = 0\,,$$

hence

$$h_0 = \left(\frac{6}{n\|f'\|_2^2} \right)^{1/3} \sim n^{-1/3}, \tag{2.24}$$

where h_0 denotes the optimal binwidth. Looking at (2.24) it becomes clear that we run into a problem if we want to calculate h_0 since $\|f'\|_2^2$ is unknown.

A way out of this dilemma will be described later, when we deal with kernel density estimation. For the moment, assume that we know the true density $f(x)$. More specifically assume a standard normal distribution, i.e.

$$f(x) = \varphi(x) = \frac{1}{\sqrt{2\pi}} \exp\left(-\frac{x^2}{2}\right).$$

In order to calculate h_0 we have to find $\|f'\|_2^2$. Using the fact that for the standard normal distribution $f'(x) = (-x)f(x)$, it can be shown that

$$\|f'\|_2^2 = \frac{1}{\sqrt{2\pi}} \sqrt{\frac{1}{2}} \int x^2 \frac{1}{\sqrt{2\pi}} \frac{1}{\sqrt{\frac{1}{2}}} \exp\left(-x^2\right) dx.$$

Since the term inside the integral is just the formula for computing the variance of a random variable that follows a normal distribution with mean $\mu = 0$ and variance $\sigma^2 = \frac{1}{2}$, we can write

$$\|f'\|_2^2 = \frac{1}{\sqrt{2\pi}} \sqrt{\frac{1}{2}} \cdot \frac{1}{2} = \frac{1}{4\sqrt{\pi}}. \tag{2.25}$$

Using this result we can calculate h_0 for this application:

$$h_0 = \left(\frac{6}{n\|f'\|_2^2} \right)^{1/3} = \left(\frac{6}{n\frac{1}{4\sqrt{\pi}}} \right)^{1/3} = \left(\frac{24\sqrt{\pi}}{n} \right)^{1/3} \approx 3.5\, n^{-1/3}. \tag{2.26}$$

Unfortunately, in practice we do not know f (if we did there would be no point in estimating it). However, (2.26) can serve as a rule-of-thumb binwidth (Scott, 1992, Subsection 3.2.3).

2.3 Dependence of the Histogram on the Origin

In Section 2.1.3 we have already pointed out that the binwidth is not the only parameter that governs shape and appearance of the histogram. Look at Figure 2.5 where four histograms for the stock returns data have been plotted. We have used the same binwidth $h = 0.04$ for each histogram but varied the origin x_0 of the bin grid.

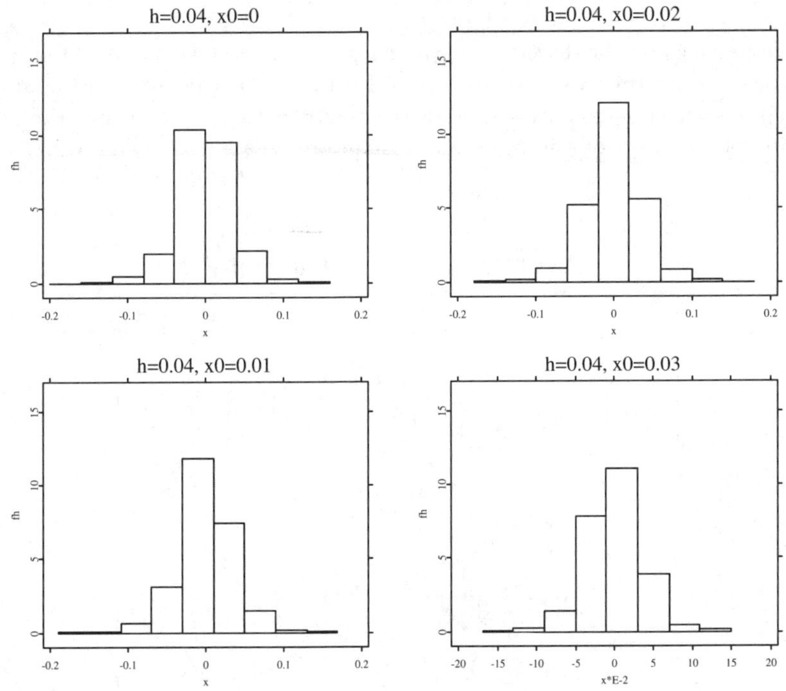

Figure 2.5. Four histograms for the stock returns data corresponding to different origins; binwidth $h = 0.04$ SPMhisdiffori

Even though we use the same data and the same binwidth, the histograms give quite different accounts of some of the key features of the data: whereas all histograms indicate that the true pdf is unimodal, only the upper right histogram suggests a symmetrical pdf. Also, note that the estimates of $f(0)$ differ considerably.

This property of histograms strongly conflicts with the goal of nonparametric statistics to "let the data speak for themselves". Obviously, the same data speak quite differently out of the different histograms. How can we get rid of the dependency of the histogram on the choice of the origin of the bin grid? A natural remedy might be to compute histograms using the same binwidth but different origins and to average over the different histograms. We will consider this technique in the next section.

2.4 Averaged Shifted Histogram

Before we get to the details it might be a good idea to take a look at the end product of the procedure. If you look at Figure 2.6 you can see a "histogram" that has been obtained by averaging over eight histograms corresponding to different origins (four of these eight histograms are plotted in Figure 2.7).

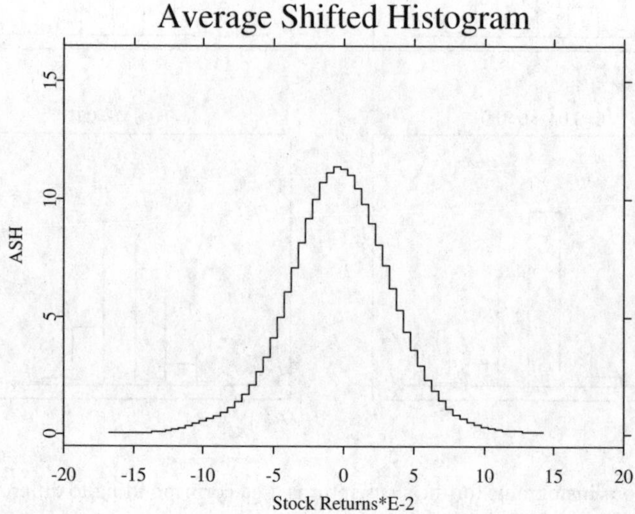

Figure 2.6. Averaged shifted histogram for stock returns data; average of 8 histograms with origins 0, 0.005, 0.01, 0.015, 0.02, 0.025, 0.03, 0.035, 0.04 and binwidth $h = 0.04$ `SPMashstock`

The resulting *averaged shifted histogram* (ASH) is freed from the dependency of the origin and seems to correspond to a smaller binwidth than the histograms from which it was constructed. Even though the ASH can in some sense (which will be made more precise below) be viewed as having a smaller binwidth, you should be aware that it is *not* simply an ordinary histogram with a smaller binwidth (as you can easily see by looking at Figure 2.7 where we graphed an ordinary histogram with a comparable binwidth and origin $x_0 = 0$).

Let us move on to the details. Consider a bin grid corresponding to a histogram with origin $x_0 = 0$ and bins $B_j = [(j-1)h, jh)$, $j \in \mathbb{Z}$, i.e.

$$\ldots B_1 = [0, h), \qquad B_2 = [h, 2h), \qquad B_3 = [2h, 3h), \ldots$$

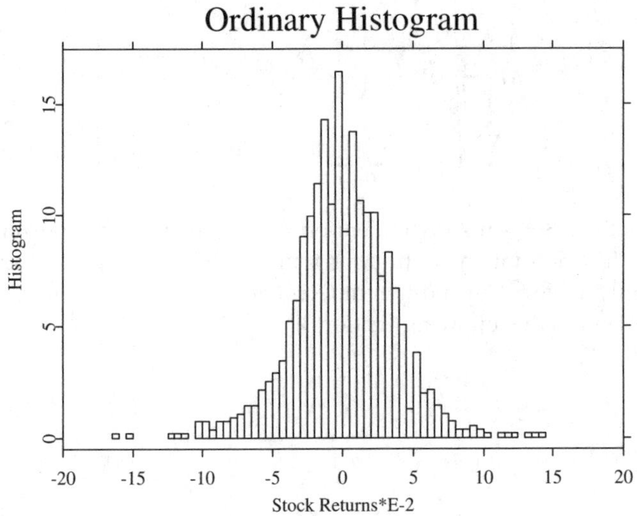

Figure 2.7. Ordinary histogram for stock returns; binwidth $h = 0.005$
SPMhiststock

Let us generate $M-1$ new bin grids by shifting each B_j by the amount lh/M to the right

$$B_{jl} = [(j-1+l/M)h, (j+l/M)h), \quad l \in \{1, \ldots, M-1\}. \tag{2.27}$$

Example 2.1.
As an example take $M = 10$:

$$\ldots B_{11} = [0.1h, 1.1h), \qquad B_{21} = [1.1h, 2.1h), \qquad B_{31} = [2.1h, 3.1h), \ldots$$

$$\ldots B_{12} = [0.2h, 1.2h), \qquad B_{22} = [1.2h, 2.2h), \qquad B_{32} = [2.2h, 3.2h), \ldots$$

$$\vdots$$

$$\ldots B_{19} = [0.9h, 1.9h), \qquad B_{29} = [1.9h, 2.9h), \qquad B_{39} = [2.9h, 3.9h), \ldots$$

Of course if we take $l = 0$ then we get the original bin grid, i.e. $B_j = B_{j0}$. □

Now suppose we calculate a histogram for each of the M bin grids. Then we get M different estimates for f at each x

$$\widehat{f}_{h,l}(x) = \frac{1}{nh} \sum_{i=1}^{n} \left\{ \sum_j \mathrm{I}(X_i \in B_{jl})\, \mathrm{I}(x \in B_{jl}) \right\}. \tag{2.28}$$

The ASH is obtained by averaging over these estimates

$$\widehat{f}_h(x) = \frac{1}{M}\sum_{l=0}^{M-1}\frac{1}{nh}\sum_{i=1}^{n}\left\{\sum_j \mathrm{I}(X_i \in B_{jl})\,\mathrm{I}(x \in B_{jl})\right\} \tag{2.29}$$

$$= \frac{1}{n}\sum_{i=1}^{n}\left\{\frac{1}{Mh}\sum_{l=0}^{M-1}\sum_j \mathrm{I}(X_i \in B_{jl})\,\mathrm{I}(x \in B_{jl})\right\}. \tag{2.30}$$

As $M \to \infty$, the ASH is not dependent on the origin anymore and converts from a step function into a continuous function. This asymptotic behavior can be directly achieved by a different technique: kernel density estimation, studied in detail in the following Section 3.

Bibliographic Notes

Additional material on the histogram can be found in Scott (1992) who in specifically covers rules for the optimal number of bins, goodness-of-fit criteria and multidimensional histograms.

A related density estimator is the frequency polygon which is constructed by interpolating the histogram values $\widehat{f}(m_j)$. This yields a piecewise linear but now continuous estimate of the density function. For details and asymptotic properties see Scott (1992, Chapter 4).

The idea of averaged shifted histograms can be used to motivate the kernel density estimators introduced in the following Chapter 3. For this application we refer to Härdle (1991) and Härdle & Scott (1992).

Exercises

Exercise 2.1. Show that equation (2.13) holds.

Exercise 2.2. Derive equation (2.14).

Exercise 2.3. Show that

$$\text{Var}\{\widehat{f}_h(x)\} = \frac{1}{nh^2} \int_{B_j} f(u)\, du \left(1 - \int_{B_j} f(u)\, du\right) \approx \frac{1}{nh} f(x).$$

Exercise 2.4. Derive equation (2.20).

Exercise 2.5. Prove that for every density function f, which is a step function, i.e.

$$f(x) = \sum_{j=1}^{m} a_j \, \text{I}(x \in A_j) \quad , \quad A_j = [(j-1)h, jh),$$

the histogram $\widehat{f}_h$ defined on the bins $B_j = A_j$ is the maximum likelihood estimate.

Exercise 2.6. Simulate a sample of standard normal distributed random variables and compute an optimal histogram corresponding to the optimal binwidth h_0 in this case.

Exercise 2.7. Consider $f(x) = 2x \cdot \text{I}(x \in [0,1])$ and histograms using binwidths $h = \frac{1}{m}$ for $m = 1, 2, \ldots$ starting at $x_0 = 0$. Calculate

$$\text{MISE}(\widehat{f}_h) = \int_0^1 \text{MSE}\left\{\widehat{f}_h(x)\right\} dx$$

and the optimal binwidth h_0. (Hint: The solution is $\text{MISE}(\widehat{f}_h) = (nh)^{-1} + \frac{1}{3}h^2 - \frac{4}{3}n^{-1} + \frac{1}{3}n^{-1}h^2$.)

Exercise 2.8. Recall that for $\widehat{f}_h(x)$ to be a consistent estimator of $f(x)$ it has to be true that for any $\epsilon > 0$ holds $P(|\widehat{f}_h(x) - f(x)| > \epsilon) \to 0$, i.e. it has to be true that $\widehat{f}_h(x)$ converges in probability. Why is it sufficient to show that $\text{MSE}\{\widehat{f}_h(x)\}$ converges to 0?

Exercise 2.9. Compute $\|f'\|^2$ for

$$f(x) = \frac{2}{3}\left[\left(\frac{x}{2} + 1\right) \text{I}\{x \in [-2, 0)\} + (1 - x)\, \text{I}\{x \in [0, 1)\}\right]$$

and derive the MISE optimal binwidth.

Exercise 2.10. Explain in detail why for the standard normal pdf $f(x) = \varphi(x)$ we obtain

$$\|f'\|_2^2 = \frac{1}{\sqrt{2\pi}}\sqrt{\frac{1}{2}} \cdot \frac{1}{2} = \frac{1}{4\sqrt{\pi}}\,.$$

Exercise 2.11. The optimal binwidth h_0 that minimizes AMISE for $N(0,1)$ is $h_0 = \left(24\sqrt{\pi}\right)^{1/3} n^{-1/3}$. How does this rule of thumb change for $N(0,\sigma^2)$ and $N(\mu,\sigma^2)$?

Exercise 2.12. How would the formula for the histogram change if we based it on intervals of the form $[m_j - h, m_j + h)$ instead of $[m_j - \frac{h}{2}, m_j + \frac{h}{2})$?

Exercise 2.13. Show that the histogram $\widehat{f}_h(x)$ is a maximum likelihood estimator of $f(x)$ for an arbitrary discrete distribution, supported by $\{0,1,\dots\}$, if one considers $h = 1$ and $B_j = [j, j+1), j = 0, 1, \dots$.

Exercise 2.14. Consider an exponential distribution with parameter λ.

a) Compute the bias, the variance, and the AMISE of $\widehat{f}_h$.

b) Compute the optimal binwidth h_0 that minimizes AMISE.

Summary

---

* A histogram with binwidth h and origin x_0 is defined by

$$\widehat{f}_h(x) = \frac{1}{nh} \sum_{i=1}^{n} \sum_{j} \mathrm{I}(X_i \in B_j)\, \mathrm{I}(x \in B_j)$$

where $B_j = [x_0 + (j-1)h, x_0 + jh)$ and $j \in \mathbb{Z}$.

---

* The bias of a histogram is

$$E\{\widehat{f}_h(x) - f(x)\} \approx f'\left\{\left(j - \frac{1}{2}\right) h\right\}\left\{(j - \frac{1}{2})h - x\right\}.$$

---

* The variance of a histogram is $\mathrm{Var}\{\widehat{f}_h(x)\} \approx \frac{1}{nh} f(x)$.

---

* The asymptotic MISE is given by $\mathrm{AMISE}(\widehat{f}_h) = \frac{1}{nh} + \frac{h^2}{12}\|f'\|_2^2$.

---

* The optimal binwidth h_0 that minimizes AMISE is

$$h_0 = \left(\frac{6}{n\|f'\|_2^2}\right)^{1/3} \sim n^{-1/3}.$$

---

* The optimal binwidth h_0 that minimizes AMISE for $N(0,1)$ is

$$h_0 \approx 3.5\, n^{-1/3}.$$

---

* The averaged shifted histogram (ASH) is given by

$$\widehat{f}_h(x) = \frac{1}{n} \sum_{i=1}^{n} \left\{ \frac{1}{Mh} \sum_{l=0}^{M-1} \sum_{j} \mathrm{I}(X_i \in B_{jl})\, \mathrm{I}(x \in B_{jl}) \right\}.$$

The ASH is less dependent on the origin as the ordinary histogram.

3

Nonparametric Density Estimation

3.1 Motivation and Derivation

3.1.1 Introduction

Contrary to the treatment of the histogram in statistics textbooks we have shown that the histogram is more than just a convenient tool for giving a graphical representation of an empirical frequency distribution. It is a serious and widely used method for estimating an unknown pdf. Yet, the histogram has some shortcomings and hopefully this chapter will persuade you that the method of *kernel density estimation* is in many respects preferable to the histogram.

Recall that the shape of the histogram is governed by two parameters: the binwidth h, and the origin of the bin grid, x_0. We have seen that the averaged shifted histogram is a slick way to free the histogram from the dependence on the choice of an origin. You may recall that we have not had similar success in providing a convincing and applicable rule for choosing the binwidth h. There is no choice-of-an-origin problem in kernel density estimation but you will soon discover that we will run into the binwidth-selection problem again. Hopefully, the second time around we will be able to give a better answer to this challenge.

Even if the ASH seemed to solve the choice-of-an-origin problem the histogram retains some undesirable properties:

- The histogram assigns each x in $[m_j - \frac{h}{2}, m_j + \frac{h}{2})$ the same estimate for f, namely $\widehat{f}_h(m_j)$. This seems to be overly restrictive.
- The histogram is not a continuous function, but has jumps at the boundaries of the bins. It is not differentiable at the jumps and has zero derivative elsewhere. This leads to the ragged appearance of the histogram

which is especially undesirable if we want to estimate a smooth, continuous pdf.

3.1.2 Derivation

Recall that our derivation of the histogram was based on the intuitively plausible idea that

$$\frac{1}{n \cdot \text{interval length}} \#\{\text{observations that fall into a small interval } \underline{\text{containing}}\ x\}$$

is a reasonable way of estimating $f(x)$. The kernel density estimator can be viewed as being based on an idea that is even more plausible and has the added advantage of freeing the estimator from the problem of choosing the origin of a bin grid. The idea goes as follows: a reasonable way to estimate $f(x)$ is to calculate

$$\frac{1}{n \cdot \text{interval length}} \#\{\text{observations that fall into a small interval } \underline{\text{around}}\ x\} \tag{3.1}$$

Note the subtle but important difference to the construction of the histogram: this time the calculation is based on an interval placed around x, not an interval containing x which is placed around some bin center m_j, determined by the choice of the origin x_0. In another deviation from the construction of the histogram we will take the interval length to be $2h$. That is, we consider intervals of the form $[x-h, x+h)$. (Recall that in Chapter 2 we had intervals of length h only.) Hence, we can write

$$\widehat{f}_h(x) = \frac{1}{2hn} \# \{X_i \in [x-h, x+h)\} \tag{3.2}$$

This formula can be rewritten if we use a weighting function, called the uniform kernel function

$$K(u) = \frac{1}{2} \mathrm{I}(|u| \le 1), \tag{3.3}$$

and let $u = (x - X_i)/h$. That is, the uniform kernel function assigns weight $1/2$ to each observation X_i whose distance from x (the point at which we want to estimate the pdf) is not bigger than h. Points farther away from x get zero weight because the indicator function $\mathrm{I}(|u| \le 1)$ is by definition equal to 0 for all values of the scaled distance $u = (x - X_i)/h$ that are bigger than 1.

Then we can write (3.2) as

$$\widehat{f}_h(x) = \frac{1}{nh} \sum_{i=1}^{n} K\left(\frac{x - X_i}{h}\right) \tag{3.4}$$

$$= \frac{1}{nh} \sum_{i=1}^{n} \frac{1}{2} \mathrm{I}\left(\left|\frac{x - X_i}{h}\right| \le 1\right) \tag{3.5}$$

It is especially apparent from (3.5) that all we have done so far is to formalize (3.1).

Note from (3.5) that for each observation that falls into the interval $[x - h, x + h)$ the indicator function takes on the value 1, and we get a contribution to our frequency count. But each contribution is weighted equally (namely by a factor of one), no matter how close the observation X_i is to x (provided that it is within h of x). Maybe we should give more weight to contributions from observations very close to x than to those coming from observations that are more distant.

For instance, consider the formula

$$\widehat{f}_h(x) = \frac{1}{2nh}\sum_{i=1}^{n}\frac{3}{2}\left\{1-\left(\frac{x-X_i}{h}\right)^2\right\}\mathrm{I}\left(\left|\frac{x-X_i}{h}\right|\leq 1\right) \tag{3.6}$$

$$= \frac{1}{nh}\sum_{i=1}^{n} K\left(\frac{x-X_i}{h}\right), \tag{3.7}$$

where $K(\bullet)$ is a shorthand for a different weighting function, the Epanechnikov kernel

$$K(u) = \frac{3}{4}(1-u^2)\,\mathrm{I}(|u|\leq 1).$$

Table 3.1. Kernel functions

Kernel	$K(u)$
Uniform	$\frac{1}{2}\,\mathrm{I}(\lvert u\rvert \leq 1)$
Triangle	$(1-\lvert u\rvert)\,\mathrm{I}(\lvert u\rvert \leq 1)$
Epanechnikov	$\frac{3}{4}(1-u^2)\,\mathrm{I}(\lvert u\rvert \leq 1)$
Quartic (Biweight)	$\frac{15}{16}(1-u^2)^2\,\mathrm{I}(\lvert u\rvert \leq 1)$
Triweight	$\frac{35}{32}(1-u^2)^3\,\mathrm{I}(\lvert u\rvert \leq 1)$
Gaussian	$\frac{1}{\sqrt{2\pi}}\exp(-\frac{1}{2}u^2)$
Cosine	$\frac{\pi}{4}\cos(\frac{\pi}{2}u)\,\mathrm{I}(\lvert u\rvert \leq 1)$

If you look at (3.6) it will be clear that one could think of the procedure as a slick way of counting the number of observations that fall into the interval around x, where contributions from X_i that are close to x are weighted more than those that are further away. The latter property is shared by the

Epanechnikov kernel with many other kernels, some of which we introduce in Table 3.1. Figure 3.1 displays some of the kernel functions.

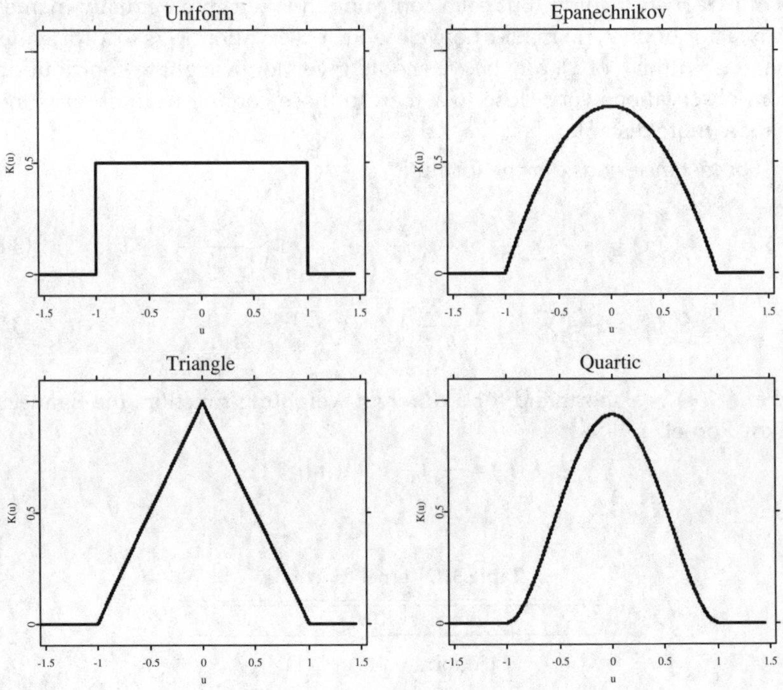

Figure 3.1. Some kernel functions: Uniform (top left), Triangle (bottom left), Epanechnikov (top right), Quartic (bottom right) `SPMkernel`

Now we can give the following general form of the kernel density estimator of a probability density f, based on a random sample $X_1, X_2, \ldots, X_n$ from f:

$$\widehat{f}_h(x) = \frac{1}{n} \sum_{i=1}^{n} K_h(x - X_i), \tag{3.8}$$

where

$$K_h(\bullet) = \frac{1}{h} K(\bullet / h). \tag{3.9}$$

$K(\bullet)$ is some kernel function like those given in Table 3.1 and h denotes the bandwidth. Note that the term *kernel function* refers to the weighting function K, whereas the term *kernel density estimator* refers to formula (3.8).

3.1.3 Varying the Bandwidth

Similar to the histogram, h controls the smoothness of the estimate and the choice of h is a crucial problem. Figure 3.2 shows density estimates for the stock returns data using the Quartic kernel and different bandwidths.

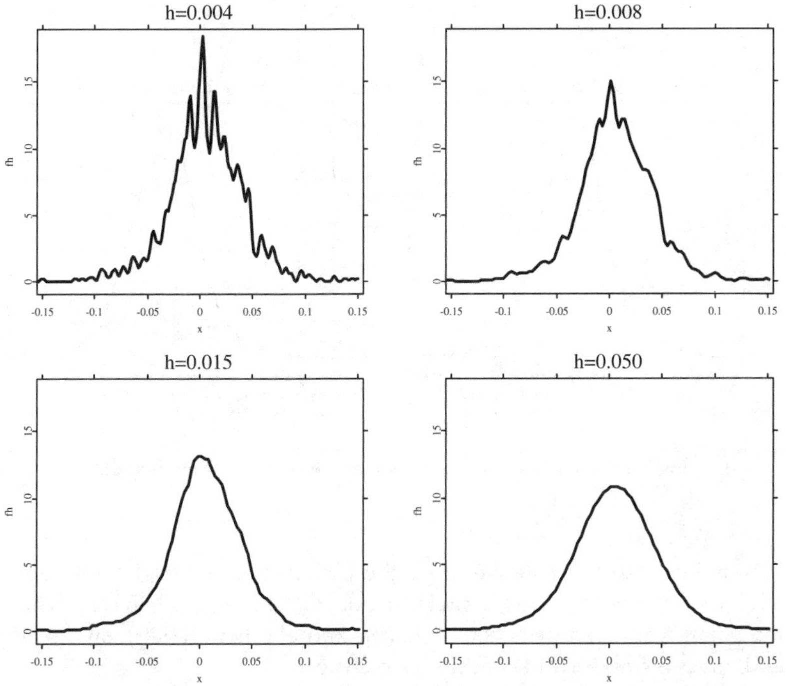

Figure 3.2. Four kernel density estimates for the stock returns data with bandwidths $h = 0.004$, $h = 0.008$, $h = 0.015$, and $h = 0.05$ SPMdensity

Again, it is hard to determine which value of h provides the optimal degree of smoothness without some formal criterion at hand. This problem will be handled further below, especially in the Section 3.3.

3.1.4 Varying the Kernel Function

Kernel functions are usually probability density functions, i.e. they integrate to one and $K(u) \geq 0$ for all u in the domain of K. An immediate consequence of $\int K(u)du = 1$ is $\int \widehat{f}_h(x)dx = 1$, i.e. the kernel density estimator is a pdf, too. Moreover, $\widehat{f}_h$ will inherit all the continuity and differentiability properties of K. For instance, if K is ν times continuously differentiable then the same will hold true for $\widehat{f}_h$. On a more intuitive level this "inheritance property" of $\widehat{f}_h$ is reflected in the smoothness of its graph. Consider Figure 3.3 where, for the same data set (stock returns) and a given value of $h = 0.018$, kernel density estimates have been graphed using different kernel functions.

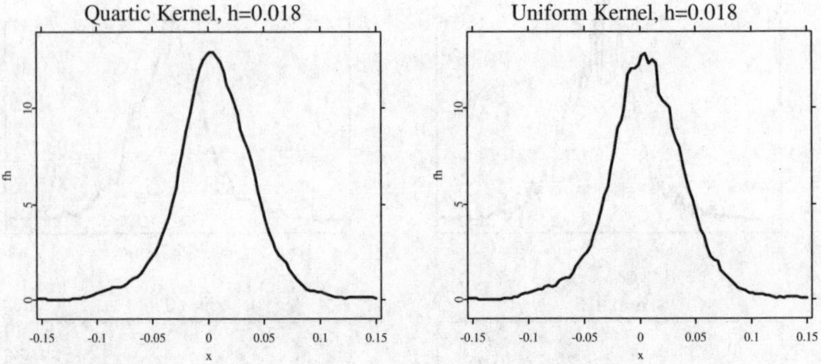

Figure 3.3. Different kernels for estimation `SPMdenquauni`

Note how the estimate based on the Uniform kernel (right) reflects the box shape of the underlying kernel function with its ragged behavior. The estimate that employed the smooth Quartic kernel function (left), on the other hand, gives a smooth and continuous picture.

Differences are not confined to estimates that are based on kernel functions that are continuous or non-continuous. Even among estimates based on continuous kernel functions there are considerable differences in smoothness (for the same value of h) as you can confirm by looking at Figure 3.4. Here, for $h = 0.18$ density estimates are graphed for income data from the Family Expenditure Survey, using the Epanechnikov kernel (left) and the Triweight kernel (right), respectively. There is quite a difference in the smoothness of the graphs of the two estimates.

You might wonder how we will ever solve this dilemma: on one hand we will be trying to find an optimal bandwidth h but obviously a given value of h does not guarantee the same degree of smoothness if used with different kernel functions. We will come back to this problem in Section 3.4.2.

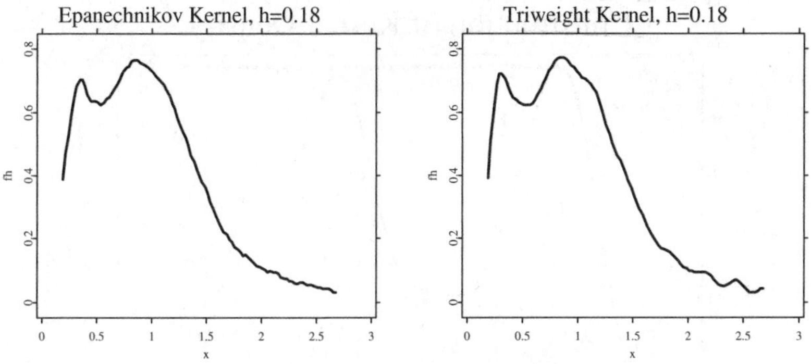

Figure 3.4. Different continuous kernels for estimation `SPMdenepatri`

3.1.5 Kernel Density Estimation as a Sum of Bumps

Before we turn to the statistical properties of kernel density estimators let us present another view on kernel density estimation that provides both further motivation as well as insight into how the procedure works. Look at Figure 3.5 where the kernel density estimate for an artificial data set is shown along with individual rescaled kernel functions.

What do we mean by a rescaled kernel function? The rescaled kernel function is simply

$$\frac{1}{nh} K\left(\frac{x - X_i}{h}\right) = \frac{1}{n} K_h\left(x - X_i\right).$$

Note that while the area under the density estimate is equal to one, the area under each rescaled kernel function is equal to (using integration by substitution)

$$\int \frac{1}{nh} K\left(\frac{x - X_i}{h}\right) dx = \frac{1}{nh} \int K(u) h \, du = \frac{1}{nh} h \int K(u) \, du = \frac{1}{n}.$$

Let us rewrite (3.8):

$$\widehat{f}_h(x) = \frac{1}{n} \sum_{i=1}^{n} K_h\left(x - X_i\right) = \sum_{i=1}^{n} \frac{1}{nh} K\left(\frac{x - X_i}{h}\right). \tag{3.10}$$

Obviously $\widehat{f}_h(x)$ can be written as the sum over the rescaled kernel functions. Figure 3.5 gives a nice graphical representation of this summation over the rescaled kernels. Graphically, the rescaled kernels are the little "bumps",

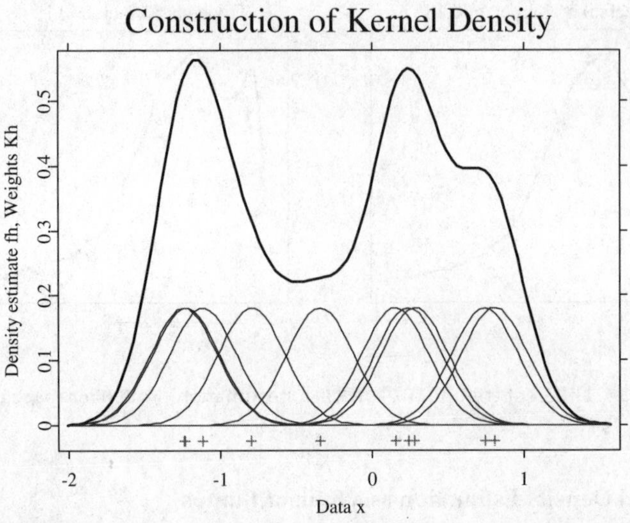

Figure 3.5. Kernel density estimate as a sum of bumps SPMkdeconstruct

where each bump is centered at one of the observations. At a given x we find $\widehat{f}_h(x)$ by vertically summing over the bumps. Obviously, different values of h change the appearance of the bumps and, as a consequence, the appearance of their sum $\widehat{f}_h$.

3.2 Statistical Properties

Let us now turn to the statistical properties of kernel density estimators. We are interested in the mean squared error since it combines squared bias and variance.

3.2.1 Bias

For the bias we have

$$\begin{aligned} \text{Bias}\{\widehat{f}_h(x)\} &= E\{\widehat{f}_h(x)\} - f(x) \qquad (3.11)\\ &= \frac{1}{n}\sum_{i=1}^{n} E\{K_h(x - X_i)\} - f(x) \\ &= E\{K_h(x - X)\} - f(x) \end{aligned}$$

such that

$$\text{Bias}\{\widehat{f}_h(x)\} = \int \frac{1}{h} K\left(\frac{x-u}{h}\right) f(u)\, du \; - \; f(x) \,. \tag{3.12}$$

Using the variable $s = \frac{u-x}{h}$, the symmetry of the kernel, i.e. $K(-s) = K(s)$, and a second-order Taylor expansion of $f(u)$ around x it can be shown (see Exercise 3.11) that

$$\text{Bias}\{\widehat{f}_h(x)\} = \frac{h^2}{2} f''(x)\, \mu_2(K) + o(h^2), \quad \text{as } h \to 0. \tag{3.13}$$

Here we denote $\mu_2(K) = \int s^2 K(s)\, ds$.

Observe from (3.13) that the bias is proportional to h^2. Hence, we have to choose a small h to reduce the bias. Moreover, $\text{Bias}\{\widehat{f}_h(x)\}$ depends on $f''(x)$, the curvature of the density at x. The effects of this dependence are illustrated in Figure 3.6 where the dashed lines mark $E\{\widehat{f}_h(\bullet)\}$ and the solid line the true density $f(\bullet)$. The bias is thus given by the vertical difference between the dashed and the solid line.

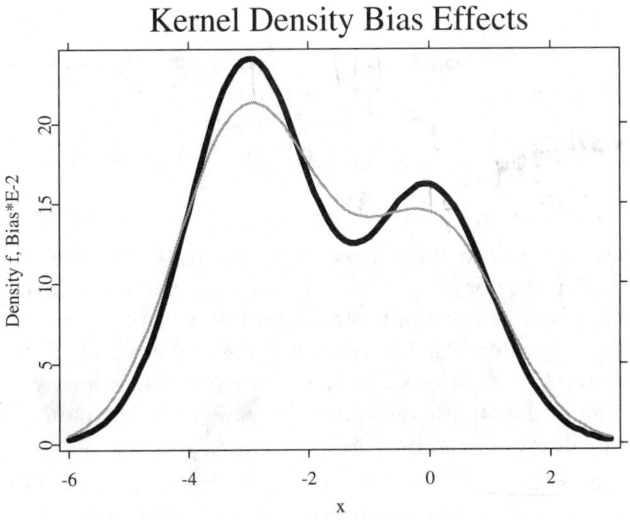

Figure 3.6. Bias effects SPMkdebias

Note that in "valleys" of f the bias is positive since $f'' > 0$ around a local minimum of f. Consequently, the dashed line is always above the solid line. Near peaks of f the opposite is true. The magnitude of the bias depends on the curvature of f, reflected in the absolute value of f''. Obviously, large values of $|f''|$ imply large values of $\text{Bias}\{\widehat{f}_h(x)\}$.

3.2.2 Variance

For the variance we calculate

$$\begin{aligned}\text{Var}\{\widehat{f}_h(x)\} &= \text{Var}\left\{\frac{1}{n}\sum_{i=1}^{n} K_h(x-X_i)\right\} \qquad (3.14)\\ &= \frac{1}{n^2}\sum_{i=1}^{n}\text{Var}\left\{K_h(x-X_i)\right\}\\ &= \frac{1}{n}\text{Var}\left\{K_h(x-X)\right\}\\ &= \frac{1}{n}\left(E\{K_h^2(x-X)\} - [E\{K_h(x-X)\}]^2\right)\end{aligned}$$

Using

$$\begin{aligned}\frac{1}{n}E\{K_h^2(x-X)\} &= \frac{1}{n}\frac{1}{h^2}\int K^2\left(\frac{x-u}{h}\right) f(u)\,du,\\ E\{K_h(x-X)\} &= f(x) + o(h)\end{aligned}$$

and similar variable substitution and Taylor expansion arguments as in the derivation of the bias, it can be shown (see Exercise 3.13) that

$$\text{Var}\{\widehat{f}_h(x)\} = \frac{1}{nh}\|K\|_2^2 f(x) + o\left(\frac{1}{nh}\right), \quad \text{as } nh \to \infty. \qquad (3.15)$$

Here, $\|K\|_2^2$ is shorthand for $\int K^2(s)\,ds$, the squared L_2 norm of K.

Notice that the variance of the kernel density estimator is nearly proportional to nh^{-1}. Hence, in order to make the variance small we have to choose a fairly large h. Large values of h mean bigger intervals $[x-h, x+h)$, more observations in each interval and hence more observations that get non-zero weight in the sum $\sum_{i=1}^{n} K_h(x-X_i)$. But, as you may recall from the analysis of the properties of the sample mean in basic statistics, using more observations in a sum will produce sums with less variability.

Similarly, for a given value of h (be it large or small), increasing the sample size n will decrease $\frac{1}{nh}$ and therefore reduce the variance. But this makes sense because having a greater total number of observations means that, on average, there will be more observations in each interval $[x-h, x+h)$.

Also observe that the variance is increasing in $\|K\|_2^2$. This term will be rather small for flat kernels such as the Uniform kernel. Intuitively speaking, we might say that smooth and flat kernels will produce less volatile estimates in repeated sampling since in each sample all realizations are given roughly equal weight.

3.2.3 Mean Squared Error

We have already seen for the histograms that choosing the bandwidth h is a crucial problem in nonparametric (density) estimation. The kernel density estimator is no exception. If we look at formulae (3.13) and (3.15) we can see that we face the familiar trade-off between variance and bias. We would surely like to keep both variance and bias small but increasing h will lower the variance while it will raise the bias (decreasing h will do the opposite). Minimizing the MSE, i.e. the sum between variance and squared bias (cf. (2.19)), represents a compromise between over- and undersmoothing. Figure 3.7 puts variance, bias, and MSE onto one graph.

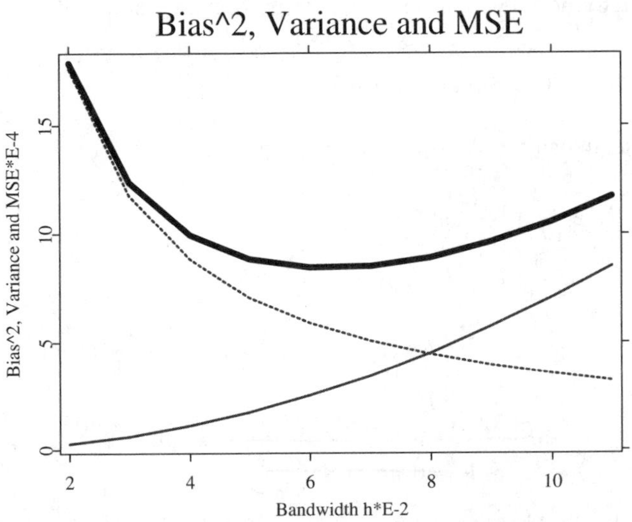

Figure 3.7. Squared bias part (thin solid), variance part (thin dashed) and MSE (thick solid) for kernel density estimate SPMkdemse

Moreover, looking at the MSE provides a way of assessing whether the kernel density estimator is consistent. Recall that convergence in mean square implies convergence in probability which is consistency. (3.13) and (3.15) yield

$$\mathrm{MSE}\{\widehat{f}_h(x)\} = \frac{h^4}{4} f''(x)^2 \mu_2(K)^2 + \frac{1}{nh}\|K\|_2^2 f(x) + o(h^4) + o\left(\frac{1}{nh}\right). \quad (3.16)$$

By looking at (3.16) we can see that the MSE of the kernel density estimator goes to zero as $h \to 0$ and $nh \to \infty$ goes to infinity. Hence, the kernel density estimator is indeed consistent. Unfortunately, by looking at (3.16) we can also observe that the MSE depends on f and f'', both functions being unknown in practice. If you derive the value of h that is minimizing the MSE (call it $h_{opt}(x)$) you will discover that both $f(x)$ and $f''(x)$ do not drop out in the process of deriving $h_{opt}(x)$. Consequently, $h_{opt}(x)$ is not applicable in practice unless we find a way of obtaining suitable substitutes for $f(x)$ and $f''(x)$. Note further that $h_{opt}(x)$ depends on x and is thus a local bandwidth.

In the case of the histogram we were able to reduce the dimensionality of the problem (in some sense) by using the MISE instead of the MSE, the former having the added advantage of being a global rather than a local measure of estimation accuracy. Hence, in the following subsections we will turn our attention to the MISE and derive the MISE-optimal bandwidth.

3.2.4 Mean Integrated Squared Error

For the kernel density estimator the MISE is given by

$$\begin{aligned} \text{MISE}(\widehat{f}_h) &= \int \text{MSE}\{\widehat{f}_h(x)\}\,dx \qquad (3.17)\\ &= \frac{1}{nh}\|K\|_2^2 \int f(x)\,dx \\ &\quad + \frac{h^4}{4}\{\mu_2(K)\}^2 \int \{f''(x)\}^2 dx + o\left(\frac{1}{nh}\right) + o(h^4) \\ &= \frac{1}{nh}\|K\|_2^2 + \frac{h^4}{4}\{\mu_2(K)\}^2 \, \|f''\|_2^2 + o\left(\frac{1}{nh}\right) + o(h^4), \\ &\quad \text{as } h \to 0, nh \to \infty. \end{aligned}$$

Ignoring higher order terms an approximate formula for the MISE, called AMISE, can be given as

$$\text{AMISE}(\widehat{f}_h) = \frac{1}{nh}\|K\|_2^2 + \frac{h^4}{4}\{\mu_2(K)\}^2 \|f''\|_2^2. \qquad (3.18)$$

Differentiating the AMISE with respect to h and solving the first-order condition for h yields the AMISE optimal bandwidth

$$h_{opt} = \left(\frac{\|K\|_2^2}{\|f''\|_2^2 \, \{\mu_2(K)\}^2 n} \right)^{1/5} \sim n^{-1/5}. \qquad (3.19)$$

Apparently, the problem of having to deal with unknown quantities has not been solved completely as h_{opt} still depends on $\|f''\|_2^2$. At least we can use

h_{opt} to get a further theoretical result regarding the statistical properties of the kernel density estimator. Inserting h_{opt} into (3.18) gives

$$\text{AMISE}(\widehat{f}_{h_{opt}}) = \frac{5}{4}\left(\|K\|_2^2\right)^{4/5}\left(\mu_2(K)\|f''\|_2\right)^{2/5} n^{-4/5} \sim n^{-4/5} \tag{3.20}$$

where we indicated that the terms preceeding $n^{-4/5}$ are constant with respect to n. Obviously, if we let the sample size get larger and larger, AMISE is converging at the rate $n^{-4/5}$. If you take the AMISE optimal bandwidth of the histogram (2.24) and plug it into (2.23) you will find out that for the histogram AMISE is converging at the slower rate of $n^{-2/3}$, giving yet another reason for why the kernel density estimator is superior to the histogram.

3.3 Smoothing Parameter Selection

We have still not found a way to select the bandwidth that is both applicable in practice as well as theoretically desirable. In the following two subsections we will introduce two of the most frequently used methods of bandwidth selection, the plug-in method and the method of cross-validation. The treatments of both methods will describe one representative of each method. In the case of the plug-in method we will focus on the "quick & dirty" plug-in method introduced by Silverman. With regard to cross-validation we will focus on least squares cross-validation. For more complete treatments of plug-in and cross-validation methods of bandwidth selection, see e.g. Härdle (1991) and Park & Turlach (1992).

3.3.1 Silverman's Rule of Thumb

Generally speaking, plug-in methods derive their name from their underlying principle: if you have an expression involving an unknown parameter, replace the unknown parameter with an estimate. Take (3.19) as an example. The expression on the right hand side involves the unknown quantity $\|f''\|_2^2$. Suppose we knew or assumed that the unknown density f belongs to the family of normal distributions with mean μ and variance σ^2 then we have

$$\|f''\|_2^2 = \sigma^{-5} \int \{\varphi''(x)\}^2 \, dx \tag{3.21}$$

$$= \sigma^{-5} \frac{3}{8\sqrt{\pi}} \approx 0.212\,\sigma^{-5}, \tag{3.22}$$

where $\varphi(\bullet)$ denotes the pdf of the standard normal distribution. It remains to replace the unknown standard deviation σ by an estimator $\widehat{\sigma}$, such as

$$\widehat{\sigma} = \sqrt{\frac{1}{n-1}\sum_{i=1}^{n}(x_i - \bar{x})^2}.$$

To apply (3.19) in practice we have to choose a kernel function. Taking the Gaussian kernel (which is identical to the standard normal pdf and will therefore be denoted by φ, too) we get the following "rule-of-thumb" bandwidth $\widehat{h}_{rot}$

$$\widehat{h}_{rot} = \left(\frac{\|\varphi\|_2^2}{\|\widehat{f''}\|_2^2 \, \mu_2^2(\varphi)\, n} \right)^{1/5} \tag{3.23}$$

$$= \left(\frac{4\widehat{\sigma}^5}{3n} \right)^{1/5} \approx 1.06\, \widehat{\sigma} n^{-1/5}, \tag{3.24}$$

with $\|\widehat{f''}\|_2^2 = \widehat{\sigma}^{-5} \frac{3}{8\sqrt{\pi}}$.

You may object by referring to what we said at the beginning of Chapter 2. Isn't assuming normality of $f(x)$ just the opposite of the philosophy of nonparametric density estimation? Yes, indeed. If we knew that X had a normal distribution then we could estimate its density much easier and more efficiently if we simply estimate μ with the sample mean and σ^2 with the sample variance, and plug these estimates into the formula of the normal density.

What we achieved by working under the normality assumption is an explicit, applicable formula for bandwidth selection. In practice, we do not know whether X is normally distributed. If it is, then $\widehat{h}_{opt}$ in (3.24) gives the optimal bandwidth. If not, then $\widehat{h}_{opt}$ in (3.24) will give a bandwidth not too far from the optimum if the distribution of X is not too different from the normal distribution (the "reference distribution"). That's why we refer to (3.24) as a *rule-of-thumb bandwidth* that will give reasonable results for all distributions that are unimodal, fairly symmetric and do not have tails that are too fat.

A practical problem with the rule-of-thumb bandwidth is its sensitivity to outliers. A single outlier may cause a too large estimate of σ and hence implies a too large bandwidth. A more robust estimator is obtained from the interquartile range

$$R = X_{[0.75n]} - X_{[0.25n]}, \tag{3.25}$$

i.e. we simply calculate the sample interquartile range from the 75%-quantile $X_{[0.75n]}$ (upper quartile) and the 25%-quantile $X_{[0.25n]}$ (lower quartile). Still assuming that the true pdf is normal we know that $X \sim N(\mu, \sigma^2)$ and $Z = \frac{X-\mu}{\sigma} \sim N(0,1)$. Hence, asymptotically

$$\begin{aligned} R &= X_{[0.75n]} - X_{[0.25n]} \\ &= (\mu + \sigma Z_{[0.75n]}) - (\mu + \sigma Z_{[0.25n]}) \\ &= \sigma(Z_{[0.75n]} - Z_{[0.25n]}) \\ &\approx \sigma(0.67 - (-0.67)) = 1.34\,\sigma, \end{aligned} \tag{3.26}$$

and thus

$$\widehat{\sigma} = \frac{R}{1.34}. \tag{3.27}$$

This relation can be plugged into (3.24) to give

$$\widehat{h}_{rot} = 1.06 \,\frac{R}{1.34}\, n^{-1/5} \approx 0.79\, \widehat{R}\, n^{-1/5}. \tag{3.28}$$

We can combine (3.24) and (3.28) into a "better rule of thumb"

$$\widehat{h}_{rot} = 1.06 \min\left\{\widehat{\sigma}, \frac{R}{1.34}\right\} n^{-1/5}. \tag{3.29}$$

Again, both (3.24) and (3.29) will work quite well if the true density resembles the normal distribution but if the true density deviates substantially from the shape of the normal distribution (by being multimodal for instance) we might be considerably misled by estimates using the rule-of-thumb bandwidths.

3.3.2 Cross-Validation

As mentioned earlier, we will focus on least squares cross-validation. To get started, consider an alternative distance measure between $\widehat{f}$ and f, the *integrated squared error* (ISE):

$$\text{ISE}(h) = \text{ISE}\{\widehat{f}_h\} = \int \{\widehat{f}_h - f\}(x)^2\, dx = \int \{\widehat{f}_h(x) - f(x)\}^2\, dx. \tag{3.30}$$

Comparing (3.30) with the definition of the MISE you will notice that, as the name suggests, the MISE is indeed the expected value of the ISE. Our aim is to choose a value for h that will make the ISE as small as possible. Let us rewrite (3.30)

$$\text{ISE}(h) = \int \widehat{f}_h^2(x)\, dx - 2\int \{\widehat{f}_h f\}(x)\, dx + \int f^2(x)\, dx. \tag{3.31}$$

Apparently, $\int f^2(x)dx$ does not depend on h and can be ignored as far as minimization over h is concerned. Moreover, $\int \widehat{f}_h^2(x)dx$ can be calculated from the data. This leaves us with one term that depends on h and involves the unknown quantity f.

If we look at this term more closely, we observe that $\int\{\widehat{f}_h f\}(x)\,dx$ is the expected value of $\widehat{f}_h(X)$, where the expectation is calculated w.r.t. an independent random variable X. We can estimate this expected value by

$$\widehat{E\{\widehat{f}_h(X)\}} = \frac{1}{n}\sum_{i=1}^{n}\widehat{f}_{h,-i}(X_i), \tag{3.32}$$

where

$$\widehat{f}_{h,-i}(x) = \frac{1}{n-1}\sum_{j=1,i\neq j}^{n} K_h(x - X_j). \tag{3.33}$$

Here $\widehat{f}_{h,-i}(x)$ is the leave-one-out estimator. As the name of this estimator suggests the ith observation is not used in the calculation of $\widehat{f}_{h,-i}(X_i)$. This way we ensure that the observations used for calculating $\widehat{f}_{h,-i}(\bullet)$ are independent of X_i, the observation at which we estimate $\widehat{f}_{h,-i}(x)$ in (3.32). (See also Exercise 3.15).

Let us repeat the formula of the integrated squared error (ISE), the criterion function we seek to minimize with respect to h:

$$\mathrm{ISE}(h) = \int \widehat{f}_h^2(x)\,dx - 2E\{\widehat{f}_h(X)\} + \int f^2(x)\,dx. \tag{3.34}$$

As pointed out above, we do not have to worry about the third term of the sum since it does not depend on h. Hence, we might as well bring it to the left side of the equation and consider the criterion

$$\mathrm{ISE}(h) - \int f^2(x)\,dx = \int \widehat{f}_h^2(x)\,dx - 2E\{\widehat{f}_h(X)\}. \tag{3.35}$$

Now we can reap the fruits of the work done above and plug in (3.32) and (3.33) for estimating $E\{\widehat{f}_h(X)\}$. This gives the so-called *cross-validation criterion*

$$CV(h) = \int \widehat{f}_h^2(x)\,dx - \frac{2}{n(n-1)}\sum_{i=1}^{n}\sum_{j=1,i\neq j}^{n} K_h(X_i - X_j). \tag{3.36}$$

We have almost everything in place for an applicable formula that allows us to calculate an optimal bandwidth from a set of observations. It remains to replace $\int \widehat{f}_h^2(x)\,dx$ by a term that employs sums rather than an integral. It can be shown (Härdle, 1991, p. 230ff) that

$$\int \widehat{f}_h^2(x)\,dx = \frac{1}{n^2 h}\sum_{i=1}^{n}\sum_{j=1}^{n} K\star K\left(\frac{X_j - X_i}{h}\right), \tag{3.37}$$

where $K\star K(u)$ is the convolution of K, i.e. $K\star K(u) = \int K(u-v)K(v)\,dv$. Inserting (3.37) into (3.36) gives the following criterion to minimize w.r.t. h

$$CV(h) = \frac{1}{n^2 h} \sum_i \sum_j K \star K \left(\frac{X_j - X_i}{h} \right) - \frac{2}{n(n-1)} \sum_i \sum_{j \neq i} K_h(X_i - X_j). \quad (3.38)$$

Thus, we have found a way to choose a bandwidth based on a reasonable criterion without having to make any assumptions about the family to which the unknown density belongs.

A nice feature of the cross-validation method is that the selected bandwidth automatically adapts to the smoothness of $f(\bullet)$. This is in contrast to plug-in methods like Silverman's rule-of-thumb or the refined methods presented in Subsection 3.3.3. Moreover, the cross-validation principle can analogously be applied to other density estimators (different from the kernel method). We will also see these advantages later in the context of regression function estimation.

Finally, it can be shown that the bandwidth selected by minimizing CV fulfills an optimality property. Denote the bandwidth selected by the cross-validation criterion by $\widehat{h}_{cv}$ and assume that the density f is a bounded function. Stone (1984) proved that this bandwidth is asymptotically optimal in the following sense

$$\frac{\text{ISE}(\widehat{h}_{cv})}{\min_h \text{ISE}(h)} \xrightarrow{a.s.} 1,$$

where $a.s.$ indicates convergence with probability 1 (almost sure convergence). In other words, this means that the ISE for $\widehat{h}_{cv}$ asymptotically coincides with the bandwidth which minimizes ISE$(\bullet)$, i.e. the ISE optimal bandwidth.

3.3.3 Refined Plug-in Methods

With Silverman's rule-of-thumb we introduced in Subsection 3.3.1 the simplest possible plug-in bandwidth. Recall that essentially we assumed a normal density for a simple calculation of $\|f''\|^2$. This procedure yields a relatively good estimate of the optimal bandwidth if the true density function f is nearly normal. However, if this is not the case (as for multimodal densities) Silverman's rule-of-thumb will fail dramatically. A natural refinement consists of using nonparametric estimates for $\|f''\|^2$ as well. A further refinement is the use of a better approximation to MISE. The following approaches apply these ideas.

In contrast to the cross-validation method plug-in bandwidth selectors try to find a bandwidth that minimizes MISE. This means we are looking at another optimality criteria than these from the previous section.

A common method of assessing the quality of a selected bandwidth $\widehat{h}$ is to compare it with h_{opt}, the MISE optimal bandwidth, in relative value. We say that the convergence of $\widehat{h}$ to h_{opt} is of order $O_p(n^{-\alpha})$ if

$$n^{\alpha} \frac{\widehat{h} - h_{opt}}{h_{opt}} \xrightarrow{P} T,$$

where T is some random variable (independent of n). If $\alpha = \frac{1}{2}$ then this is usually called $\sqrt{n}$-convergence and this rate of convergence is also the best achievable as Hall & Marron (1991) have shown.

Park & Marron (1990) proposed to estimate $\|f''\|_2^2$ in MISE by using a nonparametric estimate of f and taking the second derivative from this estimate. Suppose we use a bandwidth g here, then the second derivative of $\widehat{f}_g$ can be computed as

$$\widehat{f}_g''(x) = \frac{1}{ng^3} \sum_{i=1}^{n} K'' \left(\frac{x - X_i}{g} \right).$$

Of course, this will yield a bandwidth choice problem as well, and we only transfered our problem to bandwidth selection for the second derivative. However, we can now use a rule-of-thumb bandwidth in this first stage. A further problem occurs due to the bias of $\|\widehat{f}_g''(x)\|_2^2$ which can be overcome by using a bias corrected estimate

$$\widehat{\|f''\|_2^2} = \|\widehat{f}_g''\|_2^2 - \frac{1}{ng^5} \|K''\|_2^2. \tag{3.39}$$

Using this to replace $\|f''\|_2^2$ and optimizing w.r.t. h in AMISE yields the bandwidth selector

$$\widehat{h}_{pm} = \left(\frac{\|K\|_2^2}{\widehat{\|f''\|_2^2}\, \mu_2^2(K)\, n} \right)^{1/5}. \tag{3.40}$$

Park & Marron (1990) showed that $\widehat{h}_{pm}$ has a relative rate of convergence to h_{opt} of order $O_p(n^{-4/13})$ which means a rate of convergence $\alpha = \frac{4}{13} < \frac{1}{2}$ to the optimal bandwidth h_{opt}. The performance of $\widehat{h}_{pm}$ in simulation studies is usually quite good. A disadvantage is that for small bandwidths, the estimator $\widehat{\|f''\|_2^2}$ may give negative results.

3.3.4 An Optimal Bandwidth Selector?!

In Subsection 3.3.2 we introduced

$$\frac{\text{ISE}(\widehat{h})}{\min_h \text{ISE}(h)} \xrightarrow{a.s.} 1. \tag{3.41}$$

as a criterion of asymptotic optimality for a bandwidth selector $\widehat{h}$. This property was fulfilled by the least squares cross-validation criterion CV which tries to minimize ISE.

Most of the other existing bandwidth choice methods attempt to minimize MISE. A condition analogous to (3.41) for MISE is usually much more complicated to prove. Hence, most of the literature is concerned with investigating the relative rate of convergence of a selected bandwidth $\widehat{h}$ to h_{opt}. Fan & Marron (1992) derived a Fisher-type lower bound for the relative errors of a bandwidth selector. It is given by

$$\sigma_f^2 = \frac{4}{25}\left(\frac{\int f^{(4)}(x)^2 f(x)\,dx}{\|f''\|_2^2} - 1\right).$$

Considering the relative order of convergence to h_{opt} as a criterion, the *best* selector should fulfill

$$\sqrt{n}\left(\frac{\widehat{h} - h_{opt}}{h_{opt}}\right) \xrightarrow{L} N(0, \sigma_f^2).$$

The biased cross validation method of Hall, Sheather, Jones & Marron (1991) has this property, however, this selector is only superior for very large samples. Another $\sqrt{n}$-convergent method is the smoothed cross-validation method but this selector pays with a larger asymptotic variance.

In summary: *one* best method does *not* exist! Moreover, even asymptotically optimal criteria may show bad behavior in simulations. See the bibliographic notes for references on such simulation studies. As a consequence, we recommend determining bandwidths by different selection methods and comparing the resulting density estimates.

3.4 Choosing the Kernel

3.4.1 Canonical Kernels and Bandwidths

To discuss the choice of the kernel we will consider equivalent kernels, i.e. kernel functions that lead to exactly the same kernel density estimator. Consider a kernel function $K(\bullet)$ and the following modification:

$$K_\delta(\bullet) = \delta^{-1} K(\bullet/\delta).$$

Now compare the kernel density estimate $\widehat{f}_h$ using kernel K and bandwidth h with a kernel density estimate $\widetilde{f}_{\widetilde{h}}$ using K_δ and bandwidth $\widetilde{h}$. It is easy to derive that

$$\widehat{f}_h(x) = \frac{1}{nh}\sum_{i=1}^{n} K\left(\frac{x - X_i}{h}\right) = \frac{1}{n\widetilde{h}\delta}\sum_{i=1}^{n} K\left(\frac{x - X_i}{\widetilde{h}\delta}\right) = \widetilde{f}_{\widetilde{h}}(x)$$

if the relation

$$\widetilde{h}\delta = h$$

holds. This means, all rescaled versions K_δ of a kernel function K are equivalent if the bandwidth is adjusted accordingly.

Different values of δ correspond to different members of an equivalence class of kernels. We will now show how Marron & Nolan (1988) use the equivalence class idea to uncouple the problems of choosing h and K. Recall the AMISE criterion, i.e.

$$\text{AMISE} = \frac{1}{nh}\|K\|_2^2 + \frac{h^4}{4}\|f''\|_2^2\mu_2^2(K). \tag{3.42}$$

We rewrite this formula for some equivalence class of kernel functions K_δ:

$$\text{AMISE}(K_\delta) = \frac{1}{nh}\|K_\delta\|_2^2 + \frac{h^4}{4}\|f''\|_2^2\mu_2^2(K_\delta). \tag{3.43}$$

In each of the two components of this sum there is a term involving K_δ, namely $\|K_\delta\|_2^2$ in the left component, and $\mu_2^2(K_\delta)$ in the right component. The idea for separating the problems of choosing h and K is to find δ such that

$$\|K_\delta\|_2^2 = \mu_2^2(K_\delta).$$

This is fulfilled (see Exercise 3.7) if

$$\delta_0 = \left(\frac{\|K\|_2^2}{\mu_2^2(K)}\right)^{1/5}. \tag{3.44}$$

The value δ_0 is called the *canonical bandwidth* corresponding to the kernel function K.

What happens to AMISE if we use the very member that corresponds to δ_0, namely the kernel K_{δ_0}? By construction, for δ_0 we have

$$\frac{1}{\delta_0}\|K\|_2^2 = \delta_0^4\mu_2^2(K) = T(K),$$

or equivalently, cf. (3.44),

$$T(K) = \frac{1}{\delta_0}\|K\|_2^2 = \left\{\|K\|_2^8\,\mu_2^2(K)\right\}^{1/5}. \tag{3.45}$$

Hence, AMISE becomes

$$\text{AMISE}[K_{\delta_0}] = \left\{ \frac{1}{nh} + \frac{1}{4} h^4 \|f''\|_2^2 \right\} T(K). \tag{3.46}$$

Obviously, there is only one term left that involves K, and this term is merely a multiplicative constant. This has an interesting implication: Even though $T(K)$ is not the same for different kernels, it does not matter for the asymptotic behavior of AMISE (since it is just a multiplicative constant). Hence, AMISE will be asymptotically equal for different equivalence classes if we use K_{δ_0} to represent each class. To put it differently, using K_{δ_0} ensures that the degree of smoothness is asymptotically equal for different equivalence classes when we use the same bandwidth.

Because of these unique properties Marron and Nolan call K_{δ_0} the *canonical kernel* of an equivalence class. Table 3.2 gives the canonical bandwidths δ_0 for selected (equivalence classes of) kernels.

Table 3.2. δ_0 for different kernels

Kernel		δ_0
Uniform	$\left(\frac{9}{2}\right)^{1/5}$	$\approx$ 1.3510
Epanechnikov	$15^{1/5}$	$\approx$ 1.7188
Quartic	$35^{1/5}$	$\approx$ 2.0362
Triweight	$\left(\frac{9450}{143}\right)^{1/5}$	$\approx$ 2.3122
Gaussian	$\left(\frac{1}{4\pi}\right)^{1/10}$	$\approx$ 0.7764

3.4.2 Adjusting Bandwidths across Kernels

In Subsection 3.1.4 we saw that the smoothness of two kernel density estimates with the same bandwidth but different kernel functions may be quite different. To get estimates based on two different kernel functions that have about the same degree of smoothness, we have to adjust one of the bandwidths by multiplying with an adjustment factor.

These adjustment factors can be easily computed from the canonical bandwidths. Suppose now that we have estimated an unknown density f using some kernel K^A and bandwidth h_A (A might stand for Epanechnikov, for instance). We consider estimating f with a different kernel, K^B (B might stand for Gaussian, say). Now we ask ourselves: what bandwidth h_B should we use in the estimation with kernel K^B when we want to get approximately the same degree of smoothness as we had in the case of K^A and h_A? The answer is given by the following formula:

$$h_B = h_A \frac{\delta_0^B}{\delta_0^A}. \tag{3.47}$$

That is, we have to multiply h_A by the ratio of the canonical bandwidths δ_0^B and δ_0^A from Table 3.2.

Example 3.1.
As an example, suppose we want to compare an estimate based on the Epanechnikov kernel and bandwidth h_E with an estimate based on the Gaussian kernel. What bandwidth h_G should we use in the estimation with the Gaussian kernel? Using the values for δ_0^E and δ_0^G given in Table 3.2:

$$h_G = \frac{\delta_0^G}{\delta_0^E} h_E = \frac{0.7764}{1.7188} h_E \approx 0.452\, h_E.$$

An analogous calculation for Gaussian and Quartic kernel yields

$$h_Q = \frac{\delta_0^Q}{\delta_0^G} h_G = \frac{2.0362}{0.7764} h_G \approx 2.623\, h_G,$$

which can be used to derive the rule-of-thumb bandwidth for the Quartic kernel. □

The scaling factors δ_0 are also useful for finding an optimal kernel function (see Exercise 3.6). We turn your attention to this problem in the next section.

3.4.3 Optimizing the Kernel

Recall that if we use canonical kernels the AMISE depends on K only through a multiplicative constant $T(K)$ and we have effectively separated the choice of h from the choice of K.

A question of immediate interest is to find the kernel that minimizes $T(K)$ (this, of course, will be the kernel that minimizes AMISE with respect to K). Epanechnikov (1969, the person, not the kernel) has shown that under all nonnegative kernels with compact support, the kernel of the form

$$K(u) = \frac{3}{4}\left(\frac{1}{15^{1/5}}\right)\left\{1 - \left(\frac{u}{15^{1/5}}\right)^2\right\} \mathrm{I}\left(|u| \leq 15^{1/5}\right) \tag{3.48}$$

minimizes the function $T(K)$. You might recognize (3.48) as the canonical Epanechnikov kernel.

Does this mean that one should always use the Epanechnikov kernel? Before we can answer this question we should compare the values of $T(K)$ of

Table 3.3. Efficiency of kernels

Kernel	$T(K)$	$T(K)/T(K_{Epa})$
Uniform	0.3701	1.0602
Triangle	0.3531	1.0114
Epanechnikov	0.3491	1.0000
Quartic	0.3507	1.0049
Triweight	0.3699	1.0595
Gaussian	0.3633	1.0408
Cosine	0.3494	1.0004

other kernels with the value of $T(K)$ for the Epanechnikov kernel. Table 3.3 shows that using, say, the Quartic kernel will lead to an increase in $T(K)$ of less than half a percent.

After all, we can conclude that for practical purposes the choice of the kernel function is almost irrelevant for the efficiency of the estimate.

3.5 Confidence Intervals and Confidence Bands

In order to derive confidence intervals or confidence bands for $\widehat{f}_h(x)$ we have to know its sampling distribution. The distribution for finite sample sizes is not known but the following result concerning the asymptotic distribution of $\widehat{f}_h(x)$ can be derived. Suppose that f'' exists and $h = cn^{-1/5}$. Then

$$n^{2/5}\left\{\widehat{f}_h(x) - f(x)\right\} \xrightarrow{L} N\left(\underbrace{\frac{c^2}{2}f''(x)\mu_2(K)}_{b_x}, \underbrace{\frac{1}{c}f(x)\|K\|_2^2}_{v_x^2}\right), \tag{3.49}$$

where $N(b_x, v_x^2)$ denotes the normal distribution with mean b_x and variance v_x^2. Denoting the $(1-\frac{\alpha}{2})$ quantile of the standard normal distribution with $z_{1-\frac{\alpha}{2}}$ and probabilities with P, we get

$$\begin{aligned}
1-\alpha &\approx P\left(b_x - z_{1-\frac{\alpha}{2}}v_x \le n^{2/5}\{\widehat{f}_h(x) - f(x)\} \le b_x + z_{1-\frac{\alpha}{2}}v_x\right)\\
&= P\Big(\widehat{f}_h(x) - n^{-2/5}\{b_x + z_{1-\frac{\alpha}{2}}v_x\}\\
&\qquad \le f(x) \le \widehat{f}_h(x) - n^{-2/5}\{b_x - z_{1-\frac{\alpha}{2}}v_x\}\Big).
\end{aligned}$$

Employing the relation $h = cn^{-1/5}$ we get an asymptotic confidence interval for $f(x)$:

$$\left[\widehat{f}_h(x) - \frac{h^2}{2} f''(x)\mu_2(K) - z_{1-\frac{\alpha}{2}} \sqrt{\frac{f(x)\|K\|_2^2}{nh}}, \right. \tag{3.50}$$

$$\left. \widehat{f}_h(x) - \frac{h^2}{2} f''(x)\mu_2(K) + z_{1-\frac{\alpha}{2}} \sqrt{\frac{f(x)\|K\|_2^2}{nh}} \right].$$

Unfortunately, the interval boundaries still depend on $f(x)$ and $f''(x)$. If h is small relative to $n^{-1/5}$ we can neglect the second term of each boundary. Replacing $f(x)$ with $\widehat{f}_h(x)$ gives an approximate confidence interval that is applicable in practice

$$\left[\widehat{f}_h(x) - z_{1-\frac{\alpha}{2}} \sqrt{\frac{\widehat{f}_h(x)\|K\|_2^2}{nh}}, \widehat{f}_h(x) + z_{1-\frac{\alpha}{2}} \sqrt{\frac{\widehat{f}_h(x)\|K\|_2^2}{nh}} \right]. \tag{3.51}$$

Note that this is a confidence interval for $f(x)$ and not the entire density f. Confidence bands for f have only been derived under some rather restrictive assumptions. Suppose that f is a density on $[0,1]$ and given that certain regularity conditions are satisfied then for $h = n^{-\delta}$, $\delta \in (\frac{1}{5}, \frac{1}{2})$, and for all $x \in [0,1]$ the following formula has been derived by Bickel & Rosenblatt (1973)

$$\lim_{n\to\infty} P\left(\widehat{f}_h(x) - \left\{ \frac{\widehat{f}_h(x)\|K\|_2^2}{nh} \right\}^{1/2} \left\{ \frac{z}{(2\delta \log n)^{1/2}} + d_n \right\}^{1/2} \le f(x) \right.$$

$$\left. \le \widehat{f}_h(x) + \left\{ \frac{\widehat{f}_h(x)\|K\|_2^2}{nh} \right\}^{1/2} \left\{ \frac{z}{(2\delta \log n)^{1/2}} + d_n \right\}^{1/2} \right)$$

$$= \exp\{-2\exp(-z)\}, \tag{3.52}$$

with

$$d_n = (2\delta \log n)^{1/2} + (2\delta \log n)^{-1/2} \log\left(\frac{1}{2\pi} \frac{\|K'\|_2}{\|K\|_2} \right).$$

A confidence band for a given significance level α can be found by searching the value of z that satisfies

$$\exp\{-2\exp(-z)\} = 1 - \alpha.$$

For instance, if we take $\alpha = 0.05$ then $z \approx 3.663$.

When using nonparametric density estimators in practice, another important question arises: Can we find a *parametric* estimate that describes the data in a sufficiently satisfactory manner? That is, given a specific sample, could we justify the use of a parametric density function? Note that using parametric estimates is computationally far less intensive. Moreover, many important properties, e.g. the moments and derivatives, of parametric density functions are usually well known so they can easily be manipulated for analytical purposes. To verify whether a parametric density function describes the data accurately enough in a statistical sense, we can make use of confidence bands.

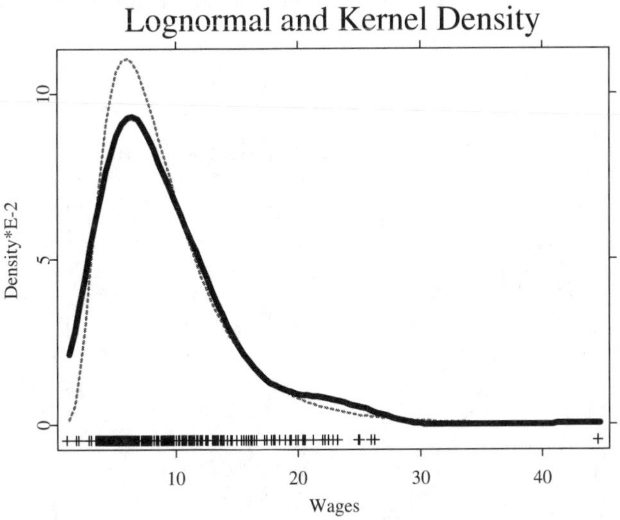

Figure 3.8. Parametric (lognormal, thin line) versus nonparametric density estimate for average hourly earnings (Quartic kernel, $h = 5$, thick solid line) `SPMcps85dist`

Example 3.2.
For an example, consider Figure 3.8. Here, we have used a sample of 534 randomly selected U.S. workers (Berndt, 1991) taken from the May 1985 Current Population Survey (CPS). The value of each worker's average hourly earnings is marked by a + on the abscissa. The nonparametrically estimated density of these values is shown by the solid line. Using the quartic kernel, the bandwidth was set to $h = 5$.

Now, as with many size distributions, the nonparametric density estimate closely resembles the lognormal distribution. Thus, we calculated the estimated parameters of the lognormal distribution,

$$\widehat{\mu} = \frac{1}{n}\sum_{i=1}^{n}\log(Y_i), \quad \widehat{\sigma} = \sqrt{\frac{1}{n}\sum_{i=1}^{n}(\log(Y_i) - \widehat{\mu})^2}$$

to fit a parametric estimate to the data. The parametric estimate is represented by the dotted line in Figure 3.8. □

Note that the nonparametrically estimated density is considerably flatter than its parametric counterpart. Now, let us see whether the sample provides some justification for using the lognormal density. To this end, we computed the 95% confidence bands around our nonparametric estimate, as shown in Figure 3.9.

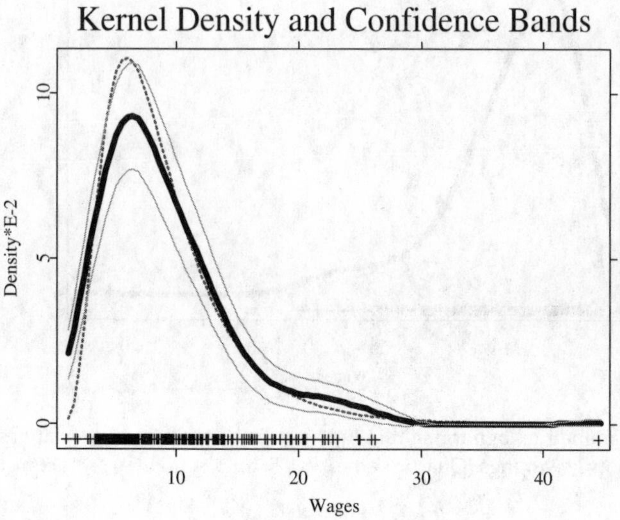

Figure 3.9. Confidence intervals versus parametric (lognormal, thin line) and kernel (thick solid line) density estimates for average hourly earnings SPMcps85dist

We found that in the neighborhood of the mode, the parametric density exceeds the upper limit of the confidence band. Hence, in a strict sense the data reject the lognormal distribution as the "true" distribution of average hourly earnings (at the 5% level). Yet, it is quite obvious from the picture that the lognormal density captures the shape of the nonparametrically estimated density quite well. Hence, if all we are interested in are qualitative features of the underlying distribution like skewness or unimodality the lognormal distribution seems to work sufficiently well.

Let us remark that checking whether the parametric density estimate does not exceed the confidence bands is a very conservative test for its correct specification. In contrast to fully parametric approaches, it is often possible to find a nonparametric test yielding better rates of convergence than the nonparametric density estimate. Moreover, there exist more powerful model checks than looking at confidence bands. Therefore, as nonparametric testing is an exhaustive topic on its own, we will not discuss nonparametric tests in detail in this book. Instead, we restrict ourselves to the presentation of the main ideas and for further details refer to the bibliographic notes.

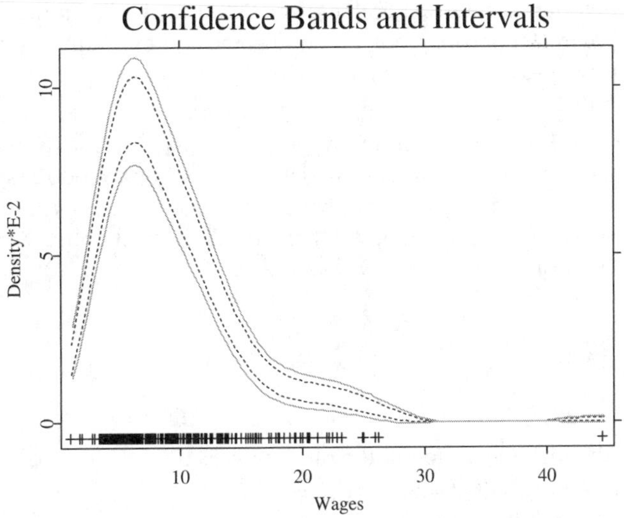

Figure 3.10. Confidence bands (solid lines) versus confidence intervals (dashed lines) for average hourly earnings `SPMcps85dist`

In the problem considered in the preceeding paragraph, we were concerned with how well the lognormal distribution works as an estimate of the true density for the entire range of possible values of average hourly earnings. This global check of the adequacy of the lognormal was provided by comparing it with confidence *bands* around the nonparametrically estimated density. Suppose that instead we are merely interested in how well the lognormal fits at a single given value of average hourly earnings. Then, the proper comparison confronts the lognormal density at this point with a confidence *interval* around the nonparametric estimate. As confidence intervals, by construction, only refer to a single point, they are narrower (at this point) than a confidence band which is supposed to hold simultaneously at many points. Figure 3.10 shows both the asymptotic confidence bands (solid line) and the asymptotic confidence intervals (broken line) of our kernel estimate. Both are computed for the same 95% significance level. Note that, as expected, the confidence intervals are narrower than the confidence bands.

3.6 Multivariate Kernel Density Estimation

Often one is not only interested in estimating one-dimensional densities, but also multivariate densities. Recall e.g. the U.K. Family Expenditure data where we have in fact observations for net-income and expenditures on different goods, such as housing, fuel, food, clothing, durables, transport, alcohol & tobacco etc.

Consider a d-dimensional random vector $\boldsymbol{X} = (X_1, \ldots, X_d)^\top$ where X_1, $\ldots$, X_d are one-dimensional random variables. Drawing a random sample of size n in this setting means that we have n observations for each of the d random variables, $X_1, \ldots, X_d$. Suppose that we collect the ith observation of each of the d random variables in the vector $\boldsymbol{X}_i$:

$$\boldsymbol{X}_i = \begin{pmatrix} X_{i1} \\ \vdots \\ X_{id} \end{pmatrix}, \quad i = 1, \ldots, n,$$

where X_{ij} is the ith observation of the random variable X_j. Our goal now is to estimate the probability density of $\boldsymbol{X} = (X_1, \ldots, X_d)^\top$, which is just the joint pdf f of the random variables $X_1, \ldots, X_d$

$$f(\boldsymbol{x}) = f(x_1, \ldots, x_d). \tag{3.53}$$

From our previous experience with the one-dimensional case we might consider adapting the kernel density estimator to the d-dimensional case, and write

$$\begin{aligned} \widehat{f}_h(\boldsymbol{x}) &= \frac{1}{n} \sum_{i=1}^{n} \frac{1}{h^d} \mathcal{K}\left(\frac{\boldsymbol{x} - \boldsymbol{X}_i}{h}\right) \qquad (3.54) \\ &= \frac{1}{n} \sum_{i=1}^{n} \frac{1}{h^d} \mathcal{K}\left(\frac{x_1 - X_{i1}}{h}, \ldots, \frac{x_d X_{id}}{h}\right), \end{aligned}$$

$\mathcal{K}$ denoting a multivariate kernel function operating on d arguments. Note, that (3.54) assumes that the bandwidth h is the same for each component. If we relax this assumption then we have a vector of bandwidths $\boldsymbol{h} = (h_1, \ldots, h_d)^\top$ and the multivariate kernel density estimator becomes

$$\widehat{f}_h(\boldsymbol{x}) = \frac{1}{n}\sum_{i=1}^{n}\frac{1}{h_1\dots h_d}\mathcal{K}\left(\frac{x_1 - X_{i1}}{h_1},\dots,\frac{x_d - X_{id}}{h_d}\right). \tag{3.55}$$

What form should the multidimensional kernel $\mathcal{K}(\boldsymbol{u}) = \mathcal{K}(u_1,\dots,u_d)$ take on? The easiest solution is to use a *multiplicative* kernel

$$\mathcal{K}(\boldsymbol{u}) = K(u_1)\cdot\ldots\cdot K(u_d), \tag{3.56}$$

where K denotes a univariate kernel function. In this case (3.55) becomes

$$\widehat{f}_h(\boldsymbol{x}) = \frac{1}{n}\sum_{i=1}^{n}\left\{\prod_{j=1}^{d} h_j^{-1} K\left(\frac{x_j - X_{ij}}{h_j}\right)\right\}. \tag{3.57}$$

Example 3.3.
To get a better understanding of what is going on here let us consider the two-dimensional case where $\boldsymbol{X} = (X_1, X_2)^\top$. In this case (3.55) becomes

$$\begin{aligned}\widehat{f}_h(\boldsymbol{x}) &= \frac{1}{n}\sum_{i=1}^{n}\frac{1}{h_1}\frac{1}{h_2}\mathcal{K}\left(\frac{x_1 - X_{i1}}{h_1},\frac{x_2 - X_{i2}}{h_2}\right)\\ &= \frac{1}{n}\sum_{i=1}^{n}\frac{1}{h_1}\frac{1}{h_2}K\left(\frac{x_1 - X_{i1}}{h_1}\right)K\left(\frac{x_2 - X_{i2}}{h_2}\right).\end{aligned} \tag{3.58}$$

Each of the n observations is of the form (X_{i1}, X_{i2}), where the first component gives the value that the random variable X_1 takes on at the ith observation and the second component does the same for X_2. For illustrational purposes, let us take K to be the Epanechnikov kernel. Then we get

$$\begin{aligned}\widehat{f}_h(\boldsymbol{x}) &= \widehat{f}_{h_1,h_2}(x_1, x_2)\\ &= \frac{1}{n}\sum_{i=1}^{n}\frac{1}{h_1}\frac{1}{h_2}K\left(\frac{x_1 - X_{i1}}{h_1}\right)K\left(\frac{x_2 - X_{i2}}{h_2}\right)\\ &= \frac{1}{n}\sum_{i=1}^{n}\frac{1}{h_1}\frac{1}{h_2}\cdot\frac{3}{4}\left\{1-\left(\frac{x_1 - X_{i1}}{h_1}\right)^2\right\}\mathrm{I}\left(\left|\frac{x_1 - X_{i1}}{h_1}\right| \le 1\right)\\ &\quad\cdot\frac{3}{4}\left\{1-\left(\frac{x_1 - X_{i2}}{h_2}\right)^2\right\}\mathrm{I}\left(\left|\frac{x_2 - X_{i2}}{h_2}\right| \le 1\right).\end{aligned}$$

Note that we get a contribution to the sum for observation i only if X_{i1} falls into the interval $[x_1 - h_1, x_1 + h_1)$ and if X_{i2} falls into the interval $[x_2 - h_2, x_2 + h_2)$. If even one of the two components fails to fall into the respective interval then one of the indicator functions takes the value 0 and consequently the observation does not enter the frequency count. □

Note that for kernels with support $[-1, 1]$ (as the Epanechnikov kernel) observations in a cube around $\boldsymbol{x}$ are used to estimate the density at the point

x. An alternative is to use a true multivariate function $\mathcal{K}(\boldsymbol{u})$, as e.g. the multivariate Epanechnikov

$$\mathcal{K}(\boldsymbol{u}) \propto (1 - \boldsymbol{u}^\top \boldsymbol{u})\, \mathrm{I}(\boldsymbol{u}^\top \boldsymbol{u} \leq 1),$$

where $\propto$ denotes proportionality. These multivariate kernels can be obtained from univariate kernel functions by taking

$$\mathcal{K}(\boldsymbol{u}) \propto K(\|\boldsymbol{u}\|), \tag{3.59}$$

where $\|\boldsymbol{u}\| = \sqrt{\boldsymbol{u}^\top \boldsymbol{u}}$ denotes the Euclidean norm of the vector $\boldsymbol{u}$. Kernels of the form (3.59) use observations from a circle around $\boldsymbol{x}$ to estimate the pdf at $\boldsymbol{x}$. This type of kernel is usually called *spherical* or *radial-symmetric* since $\mathcal{K}(\boldsymbol{u})$ has the same value for all $\boldsymbol{u}$ on a sphere around zero.

Example 3.4.
Figure 3.11 shows the contour lines from a bivariate product and a bivariate radial-symmetric Epanechnikov kernel, on the left and right hand side respectively. □

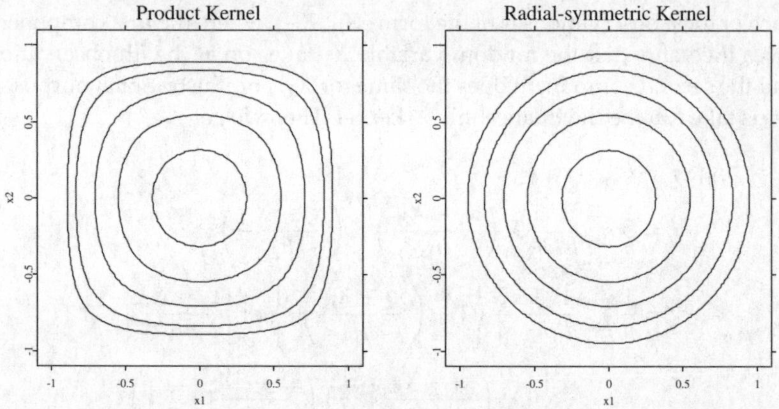

Figure 3.11. Contours from bivariate product (left) and bivariate radial-symmetric (right) Epanechnikov kernel with equal bandwidths SPMkernelcontours

Note that the kernel weights in Figure 3.11 correspond to equal bandwidth in each direction, i.e. $\boldsymbol{h} = (h_1, h_2)^\top = (1,1)^\top$. When we use different bandwidths, the observations around $\boldsymbol{x}$ in the density estimate $\widehat{f}_{\boldsymbol{h}}(\boldsymbol{x})$ will be used with different weights in both dimensions.

Example 3.5.
The contour plots of product and radial-symmetric Epanechnikov weights with different bandwidths, i.e. $\mathcal{K}_{\boldsymbol{h}}(\boldsymbol{u}) = \mathcal{K}(u_1/h_1, u_2/h_2)/(h_1 h_2)$, are shown in Figure 3.12. Here we used $h_1 = 1$ and $h_2 = 0.5$ which naturally includes fewer observations in the second dimension. □

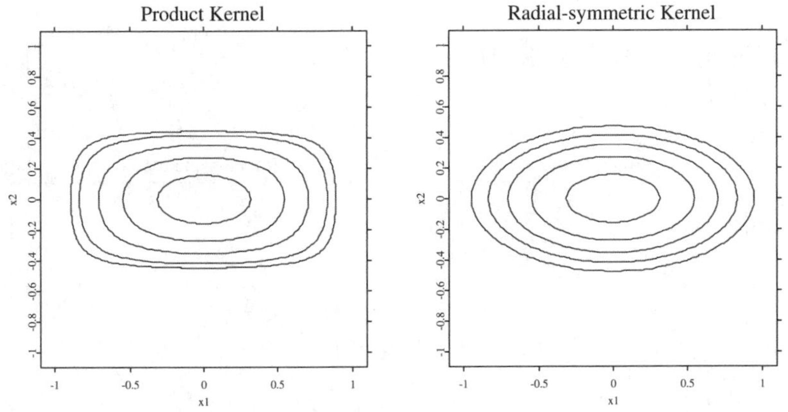

Figure 3.12. Contours from bivariate product (left) and bivariate radial-symmetric (right) Epanechnikov kernel with different bandwidths `SPMkernelcontours`

A very general approach is to use a *bandwidth matrix* **H** (nonsingular). The general form for the multivariate density estimator is then

$$\widehat{f}_{\mathbf{H}}(\boldsymbol{x}) = \frac{1}{n}\sum_{i=1}^{n} \frac{1}{\det(\mathbf{H})}\mathcal{K}\left\{\mathbf{H}^{-1}(\boldsymbol{x} - \mathbf{X}_i)\right\} = \frac{1}{n}\sum_{i=1}^{n} \mathcal{K}_{\mathbf{H}}(\boldsymbol{x} - \mathbf{X}_i), \tag{3.60}$$

see Silverman (1986) and Scott (1992). Here we used the short notation

$$\mathcal{K}_{\mathbf{H}}(\bullet) = \frac{1}{\det(\mathbf{H})}\mathcal{K}(\mathbf{H}^{-1}\bullet)$$

analogously to K_h in the one-dimensional case. A bandwidth matrix includes all simpler cases as special cases. An equal bandwidth h in all dimensions as in (3.54) corresponds to $\mathbf{H} = h\mathbf{I}_d$ where $\mathbf{I}_d$ denotes the $d \times d$ identity matrix. Different bandwidths as in (3.55) are equivalent to $\mathbf{H} = \mathrm{diag}(h_1, \ldots, h_d)$, the diagonal matrix with elements $h_1, \ldots, h_d$.

What effect has the inclusion of off-diagonal elements? We will see in Subsection 3.6.2 that a good rule of thumb is to use a bandwidth matrix proportional to $\widehat{\boldsymbol{\Sigma}}^{-1/2}$ where $\widehat{\boldsymbol{\Sigma}}$ is the covariance matrix of the data. Hence, using such a bandwidth corresponds to a transformation of the data, so that they have an identity covariance matrix. As a consequence we can use bandwidth matrices to adjust for correlation between the components of $\boldsymbol{X}$. We have plotted the contour curves of product and radial-symmetric Epanechnikov weights with bandwidth matrix

$$\mathbf{H} = \begin{pmatrix} 1 & 0.5 \\ 0.5 & 1 \end{pmatrix}^{1/2},$$

i.e. $\mathcal{K}_{\mathbf{H}}(\boldsymbol{u}) = \mathcal{K}(\mathbf{H}^{-1}\boldsymbol{u}) / \det(\mathbf{H})$, in Figure 3.13.

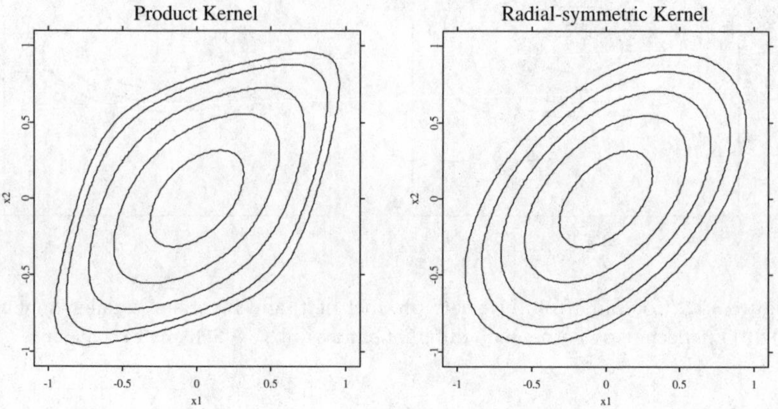

Figure 3.13. Contours from bivariate product (left) and bivariate radial-symmetric (right) Epanechnikov kernel with bandwidth matrix `SPMkernelcontours`

In the following subsection we will consider statistical properties of bias, variance, the issue of bandwidth selection and applications for this estimator.

3.6.1 Bias, Variance and Asymptotics

First let us mention that as a consequence of the standard assumption

$$\int \mathcal{K}(\boldsymbol{u})\, d\boldsymbol{u} = 1 \tag{3.61}$$

the estimate $\widehat{f}_{\mathbf{H}}$ is a density function, i.e. $\int \widehat{f}_{\mathbf{H}}(\boldsymbol{x})\, d\boldsymbol{x} = 1$. Also, the estimate is consistent in any point $\boldsymbol{x}$:

$$\widehat{f}_{\mathbf{H}}(x) = \frac{1}{n}\sum_{i=1}^{n} \mathcal{K}_{\mathbf{H}}(\boldsymbol{X}_i - \boldsymbol{x}) \stackrel{P}{\longrightarrow} f(\boldsymbol{x}), \tag{3.62}$$

see e.g. Ruppert & Wand (1994). The derivation of MSE and MISE is principally analogous to the one-dimensional case. We will only sketch the asymptotic expansions and hence just move on to the derivation of AMISE.

A detailed derivation of the components of AMISE can be found in Scott (1992) or Wand & Jones (1995) and the references therein. As in the univariate case we use a second order Taylor expansion. Here and in the following formulae we denote with ∇_f the gradient and with $\mathcal{H}_f$ the Hessian matrix of second partial derivatives of a function (here f). Then the Taylor expansion of $f(\bullet)$ around $\boldsymbol{x}$ is

$$f(\boldsymbol{x}+\boldsymbol{t}) = f(\boldsymbol{x}) + \boldsymbol{t}^\top \nabla_f(\boldsymbol{x}) + \frac{1}{2}\boldsymbol{t}^\top \mathcal{H}_f(\boldsymbol{x})\boldsymbol{t} + o(\boldsymbol{t}^\top \boldsymbol{t}),$$

see Wand & Jones (1995, p. 94). This leads to the expectation

$$\begin{aligned} E\widehat{f}_{\mathbf{H}}(\boldsymbol{x}) &= \int \mathcal{K}_{\mathbf{H}}(\boldsymbol{u}-\boldsymbol{x})\, f(\boldsymbol{u})\, d\boldsymbol{u} \\ &= \int \mathcal{K}(\boldsymbol{s})\, f(\boldsymbol{x}+\mathbf{H}\boldsymbol{s})\, d\boldsymbol{s} \\ &\approx \int \mathcal{K}(\boldsymbol{s})\, \{f(\boldsymbol{x}) + \boldsymbol{s}^\top \mathbf{H}^\top \nabla_f(\boldsymbol{x}) \\ &\qquad + \frac{1}{2}\boldsymbol{s}^\top \mathbf{H}^\top \mathcal{H}_f(\boldsymbol{x})\mathbf{H}\boldsymbol{s}\}\, d\boldsymbol{s}. \end{aligned} \tag{3.63}$$

If we assume additionally to (3.61)

$$\int \boldsymbol{u}\mathcal{K}(\boldsymbol{u})\, d\boldsymbol{u} = 0_d, \tag{3.64}$$

$$\int \boldsymbol{u}\boldsymbol{u}^\top \mathcal{K}(\boldsymbol{u})\, d\boldsymbol{u} = \mu_2(\mathcal{K})\mathbf{I}_d, \tag{3.65}$$

then (3.63) yields $E\widehat{f}_{\mathbf{H}}(\boldsymbol{x}) - f(\boldsymbol{x}) \approx \frac{1}{2}\mu_2(\mathcal{K})\,\mathrm{tr}\{\mathbf{H}^\top \mathcal{H}_f(\boldsymbol{x})\mathbf{H}\}$, hence

$$\mathrm{Bias}\{\widehat{f}_{\mathbf{H}}(\boldsymbol{x})\} \approx \frac{1}{4}\mu_2^2(\mathcal{K}) \int \left[\mathrm{tr}\{\mathbf{H}^\top \mathcal{H}_f(\boldsymbol{x})\mathbf{H}\}\right]^2 d\boldsymbol{x}. \tag{3.66}$$

For the variance we find

$$\begin{aligned} \mathrm{Var}\left\{\widehat{f}_{\mathbf{H}}(\boldsymbol{x})\right\} &= \frac{1}{n}\int \{\mathcal{K}_{\mathbf{H}}(\boldsymbol{u}-\boldsymbol{x})\}^2\, d\boldsymbol{u} - \frac{1}{n}\left\{E\widehat{f}_{\mathbf{H}}(\boldsymbol{x})\right\}^2 \\ &\approx \int \frac{1}{n\,\det(\mathbf{H})}\mathcal{K}^2(\boldsymbol{s})\, f(\boldsymbol{x}+\mathbf{H}\boldsymbol{s})\, d\boldsymbol{s} \\ &\approx \int \frac{1}{n\,\det(\mathbf{H})}\mathcal{K}^2(\boldsymbol{s})\left\{f(\boldsymbol{x}) + \boldsymbol{s}^\top \mathbf{H}^\top \nabla_f(\boldsymbol{x})\right\} d\boldsymbol{s} \\ &\approx \frac{1}{n\,\det(\mathbf{H})}\|\mathcal{K}\|_2^2\, f(\boldsymbol{x}), \end{aligned} \tag{3.67}$$

with $\|\mathcal{K}\|_2^2$ denoting the d-dimensional squared L_2-norm of $\mathcal{K}$. Therefore we have the following AMISE formula for the multivariate kernel density estimator

$$\text{AMISE}(\mathbf{H}) = \frac{1}{4}\mu_2^2(\mathcal{K}) \int \left[\text{tr}\{\mathbf{H}^\top \mathcal{H}_f(\boldsymbol{x})\mathbf{H}\}\right]^2 \, d\boldsymbol{x} + \frac{1}{n \det(\mathbf{H})}\|\mathcal{K}\|_2^2. \quad (3.68)$$

Let us now turn to the problem of how to choose the AMISE optimal bandwidth. Again this is the bandwidth which balances bias-variance tradeoff in AMISE. Denote h a scalar, such that $\mathbf{H} = h\mathbf{H}_0$ and $\det(\mathbf{H}_0) = 1$. Then AMISE can be written as

$$\text{AMISE}(\mathbf{H}) = \frac{1}{4}\, h^4\, \mu_2^2(\mathcal{K}) \int \left[\text{tr}\{\mathbf{H}_0^\top \mathcal{H}_f(\boldsymbol{x})\mathbf{H}_0\}\right]^2 \, d\boldsymbol{x} + \frac{1}{nh^d}\|\mathcal{K}\|_2^2.$$

If we only allow changes in h the optimal orders for the smoothing parameter h and AMISE are

$$h_{opt} \sim n^{-1/(4+d)}, \quad \text{AMISE}(h_{opt}\mathbf{H}_0) \sim n^{-4/(4+d)}.$$

Hence, the multivariate density estimator has a slower rate of convergence compared to the univariate one, in particular when d is large.

If we consider $\mathbf{H} = h\mathbf{I}_d$ (the same bandwidth in all d dimensions) and we fix the sample size n, then the AMISE optimal bandwidth has to be considerably larger than in the one-dimensional case to make sure that the estimate is reasonably smooth. Some ideas of comparable sample sizes to reach the same quality of the density estimates over different dimensions can be found in Silverman (1986, p. 94) and Scott & Wand (1991). Moreover, the computational effort of this technique increases with the number of dimensions d. Therefore, multidimensional density estimation is usually not applied if $d \geq 5$.

3.6.2 Bandwidth Selection

The problem of an automatic, data-driven choice of the bandwidth $\mathbf{H}$ has actually more importance for the multivariate than for the univariate case. In one or two dimensions it is easy to choose an appropriate bandwidth interactively just by looking at the plot of density estimates for different bandwidths. But how can this be done in three, four or more dimensions? Here arises the problem of graphical representation which we address in the next subsection. As in the one-dimensional case,

- plug-in bandwidths, in particular rule-of-thumb bandwidths, and
- cross-validation bandwidths

are used. We will introduce generalizations for Silverman's rule-of-thumb and least squares cross-validation to show the analogy with the one-dimensional bandwidth selectors.

Rule-of-thumb Bandwidth

Rule-of-thumb bandwidth selection gives a formula arising from the optimal bandwidth for a reference distribution. Obviously, the pdf of a multivariate normal distribution $N_d(\boldsymbol{\mu}, \boldsymbol{\Sigma})$ is a good candidate for a reference distribution in the multivariate case. Suppose that the kernel $\mathcal{K}$ is multivariate Gaussian, i.e. the pdf of $N_d(0, \mathbf{I})$. Note that $\mu_2(\mathcal{K}) = 1$ and $\|\mathcal{K}\|_2^2 = 2^{-d}\pi^{-d/2}$ in this case. Hence, from (3.68) and the fact that

$$\int [\text{tr}\{\mathbf{H}^\top \mathcal{H}_f(x)\mathbf{H}\}]^2 \, dx = \frac{1}{2^{d+2}\pi^{d/2}\det(\boldsymbol{\Sigma})^{1/2}} \left[2\,\text{tr}(\mathbf{H}^\top \boldsymbol{\Sigma}^{-1}\mathbf{H})^2 + \{\text{tr}(\mathbf{H}^\top \boldsymbol{\Sigma}^{-1}\mathbf{H})\}^2\right],$$

cf. Wand & Jones (1995, p. 98), we can easily derive rule-of-thumb formulae for different assumptions on $\mathbf{H}$ and $\boldsymbol{\Sigma}$.

In the simplest case, i.e. that we consider $\mathbf{H}$ and $\boldsymbol{\Sigma}$ to be diagonal matrices $\mathbf{H} = \text{diag}(h_1, \ldots, h_d)$ and $\boldsymbol{\Sigma} = \text{diag}(\sigma_1^2, \ldots, \sigma_d^2)$, this leads to

$$\widetilde{h}_j = \left(\frac{4}{d+2}\right)^{1/(d+4)} n^{-1/(d+4)}\sigma_j. \tag{3.69}$$

Note that this formula coincides with Silverman's rule of thumb in the case $d = 1$, see (3.24) and Silverman (1986, p. 45). Replacing the σ_js with estimates and noting that the first factor is always between 0.924 and 1.059, we arrive at Scott's rule:

$$\widehat{h}_j = n^{-1/(d+4)}\widehat{\sigma}_j, \tag{3.70}$$

see Scott (1992, p. 152).

It is not possible to derive the rule-of-thumb for general $\mathbf{H}$ and $\boldsymbol{\Sigma}$. However, (3.69) shows that it might be a good idea to choose the bandwidth matrix $\mathbf{H}$ proportional to $\boldsymbol{\Sigma}^{1/2}$. In this case we get as a generalization of Scott's rule:

$$\widehat{\mathbf{H}} = n^{-1/(d+4)}\widehat{\boldsymbol{\Sigma}}^{1/2}. \tag{3.71}$$

We remark that this rule is equivalent to applying a Mahalanobis transformation to the data (to transform the estimated covariance matrix to identity), then computing the kernel estimate with Scott's rule (3.70) and finally retransforming the estimated pdf back to the original scale.

Principally all plug-in methods for the one-dimensional kernel density estimation can be extended to the multivariate case. However, in practice this is cumbersome, since the derivation of asymptotics involves multivariate derivatives and higher order Taylor expansions.

Cross-validation

As we mentioned before, the cross-validation method is fairly independent of the special structure of the parameter or function estimate. Considering the bandwidth choice problem, cross-validation techniques allow us to adapt to a wider class of density functions f than the rule-of-thumb approach. (Remember that the rule-of-thumb bandwidth is optimal for the reference pdf, hence it will fail for multimodal densities for instance.)

Recall, that in contrast to the rule-of-thumb approach, least squares cross-validation for density estimation does not estimate the $MISE$ optimal but the ISE optimal bandwidth. Here we approximate the integrated squared error

$$\begin{aligned}\text{ISE}(\mathbf{H}) &= \int \{\widehat{f}_{\mathbf{H}}(\boldsymbol{x}) - f(\boldsymbol{x})\}^2 \, d\boldsymbol{x} \\ &= \int \widehat{f}^2_{\mathbf{H}}(\boldsymbol{x}) \, d\boldsymbol{x} - 2 \int \widehat{f}_{\mathbf{H}}(\boldsymbol{x}) f(\boldsymbol{x}) \, d\boldsymbol{x} + \int f^2(\boldsymbol{x}) \, d\boldsymbol{x}. \end{aligned} \tag{3.72}$$

Apparently, this is the same formula as in the one-dimensional case and with the same arguments the last term of (3.72) can be ignored. The first term can again be easily calculated from the data. Hence, only the second term of (3.72) is unknown and must be estimated. However, observe that $\int \widehat{f}_{\mathbf{H}}(\boldsymbol{x}) f(\boldsymbol{x}) \, d\boldsymbol{x} = E\widehat{f}_{\mathbf{H}}(\boldsymbol{X})$, where the only new aspect now is that $\boldsymbol{X}$ is d-dimensional. The resulting expectation, though, is a scalar. As in (3.32) we estimate this term by a leave-one-out estimator

$$\widehat{E\widehat{f}_{\mathbf{H}}(\boldsymbol{X})} = \frac{1}{n} \sum_{i=1}^{n} \widehat{f}_{\mathbf{H},-i}(\boldsymbol{X}_i)$$

where

$$\widehat{f}_{\mathbf{H},-i}(\boldsymbol{x}) = \frac{1}{n-1} \sum_{i \neq j, j=1}^{n} \mathcal{K}_{\mathbf{H}}(\boldsymbol{X}_j - \boldsymbol{x})$$

is simply the multivariate version of (3.33). Also, the multivariate generalization of (3.37) is straightforward, which yields the multivariate cross-validation criterion as a perfect generalization of CV in the one-dimensional case:

$$CV(\mathbf{H}) = \frac{1}{n^2 \det(\mathbf{H})} \sum_{i=1}^{n} \sum_{j=1}^{n} \mathcal{K} \star \mathcal{K} \left\{ \mathbf{H}^{-1}(\mathbf{X}_j - \mathbf{X}_i) \right\} - \frac{2}{n(n-1)} \sum_{i=1}^{n} \sum_{\substack{j=1 \\ j \neq i}}^{n} \mathcal{K}_{\mathbf{H}}(\mathbf{X}_j - \mathbf{X}_i).$$

The difficulty comes in the fact that the bandwidth is now a $d \times d$ matrix $\mathbf{H}$. In the most general case this means to minimize over $d(d+1)/2$ parameters. Still, if we assume to $\mathbf{H}$ a diagonal matrix, this remains a d-dimensional optimization problem. This holds for other cross-validation approaches, too.

3.6.3 Computation and Graphical Representation

Consider now the problem of graphically displaying our multivariate density estimates. Assume first $d = 2$. Here we are still able to show the density estimate in a three-dimensional plot. This is particularly useful if the estimated function can be rotated interactively on the computer screen. For a two-dimensional presentation a contour plot gives often more insight into the structure of the data.

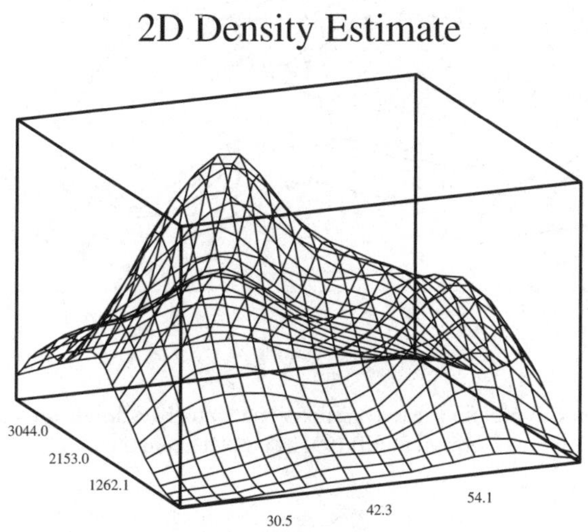

Figure 3.14. Two-dimensional density estimate for age and household income from East German SOEP 1991 SPMdensity2D

Example 3.6.
Figures 3.14 and 3.15 display such a two-dimensional density estimate

$$\widehat{f}_h(x) = \widehat{f}_h(x_1, x_2)$$

for two explanatory variables on East-West German migration intention in Spring 1991, see Example 1.4. We use the subscript h to indicate that we used a diagonal bandwidth matrix $\mathbf{H} = \operatorname{diag}(h)$. Aside from some categorical variables on an educational level, professional status, existence of relations to Western Germany and regional dummies, our data set contains observations on age, household income and environmental satisfaction. In Figure 3.14 we plotted the joint density estimate for age and household income. Additionally Figure 3.15 gives a contour plot of this density estimate. It is easily observed that the age distribution is considerably left skewed. □

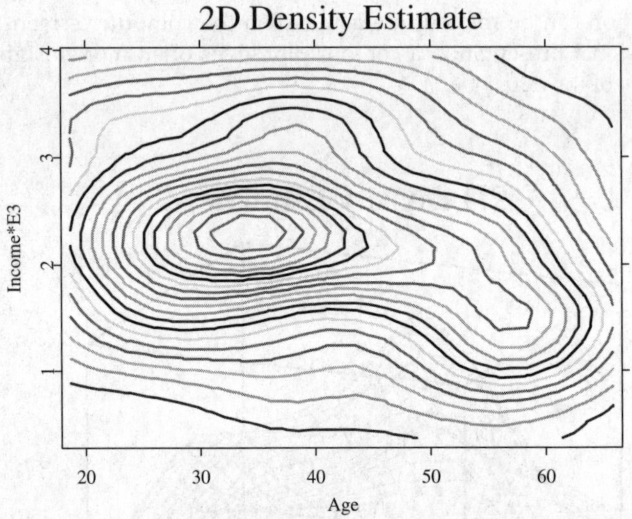

Figure 3.15. Contour plot for the two-dimensional density estimate for age and household income from East German SOEP 1991 `SPMcontour2D`

Here and in the following plots the bandwidth was chosen according to the general rule of thumb (3.71), which tends to oversmooth bimodal structures of the data. The kernel function is always the product Quartic kernel.

Consider now how to display three- or even higher dimensional density estimates. One possible approach is to hold one variable fixed and to plot the density function only in dependence of the other variables. For three-dimensional data this gives three plots: x_1, x_2 vs. $\widehat{f}_h(x_1, x_2, x_3)$, x_1, x_3 vs. $\widehat{f}_h(x_1, x_2, x_3)$ and x_2, x_3 vs. $\widehat{f}_h(x_1, x_2, x_3)$.

Example 3.7.
We display this technique in Figure 3.16 for data from a credit scoring sample, using duration of the credit, household income and age as variables (Fahrmeir & Tutz, 1994). The title of each panel indicates which variable is held fixed at which level. □

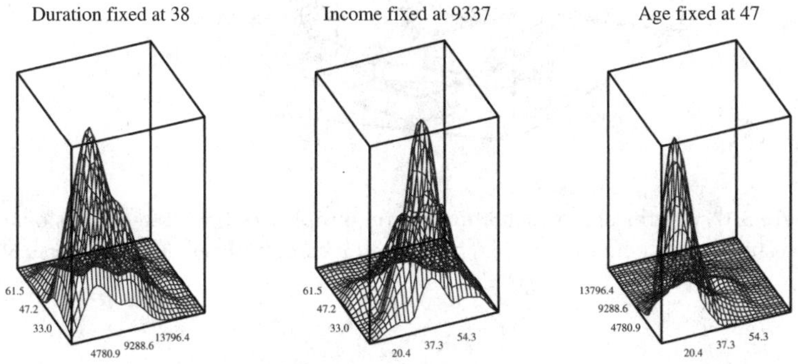

Figure 3.16. Two-dimensional intersections for the three-dimensional density estimate for credit duration, household income and age `SPMslices3D`

Example 3.8.
Alternatively, we can plot contours of the density estimate, now in three dimensions, which means three-dimensional surfaces. Figure 3.17 shows this for the credit scoring data. In the original version of this plot, red, green and blue surfaces show the values of the density estimate at the levels (in percent) indicated on the right. Colors and the possibility of rotating the contours on the computer screen ease the exploration of the data structures significantly. □

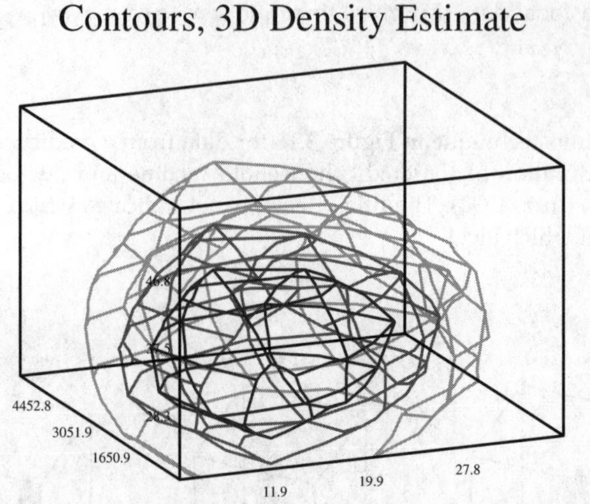

Figure 3.17. Graphical representation by contour plot for the three-dimensional density estimation for credit duration, household income and age SPMcontour3D

Bibliographic Notes

For alternative texts on kernel density estimation we refer to the monographs by Silverman (1986), Härdle (1990), Scott (1992) and Wand & Jones (1995).

A particular field of interest and ongoing research is the matter of bandwidth selection. In addition to what we have presented, a variety of other cross-validation approaches and refined plug-in bandwidth selectors have been proposed.

In particular, the following methods are based on the cross-validation idea: Pseudo-likelihood cross-validation (Habbema, Hermans & van den Broek, 1974; Duin, 1976), biased cross-validation (Scott & Terrell, 1987; Cao, Cuevas & González Manteiga, 1994) and smoothed cross-validation (Hall, Marron & Park, 1992). The latter two approaches also attempt to find MISE optimal bandwidths. Hence their performance is also assessed by relative convergence to the MISE optimal bandwidth h_{opt}. A detailed treatment of many cross-validation procedures can be found in Park & Turlach (1992).

Regarding other refined plug-in bandwidth selectors, the methods of Sheather and Jones (Sheather & Jones, 1991) and Hall, Sheather, Jones, and Marron (Hall, Sheather, Jones & Marron, 1991) should be mentioned, as they have have good asymptotic properties ($\sqrt{n}$-convergence). A number of authors provide extensive simulation studies for smoothing parameter selection, we want to mention in particular Marron (1989), Jones, Marron & Sheather (1996), Park & Turlach (1992), and Cao, Cuevas & González Manteiga (1994).

A alternative approach is introduced by Chaudhuri & Marron (1999)'s SiZer (significance zero crossings of derivatives) which tries to directly find features of a curve, such as bumps and valleys. At the same time it is a tool for visualizing the estimated curve at a range of different bandwidth values. SiZer provides thus a way around the issue of smoothing parameter selection.

Exercises

Exercise 3.1. Calculate the exact values of

$$\int K^2(u)\,du \quad \text{and} \quad \int u^2 K(u)\,du$$

for the Gaussian, Epanechnikov and Quartic kernels.

Exercise 3.2. Show that for the $N(0,\sigma^2)$ normal density holds

$$\|f''\|_2^2 \approx 0.212 \cdot \sigma^{-5}.$$

Exercise 3.3. Show that the density estimate $\widehat{f}_h(x) = \frac{1}{n}\sum_{i=1}^n K_h(x - X_i)$ is itself a pdf if the kernel K is one (i.e. if $\int K(u)\,du = 1$).

Exercise 3.4. The statistical bureau of Borduria questions $n = 100$ households about their income. The young econometrician Tintin proposes to use these data $x_1, \ldots x_n$ for a nonparametric estimate of the income density $f(x)$. Tintin suggests computing a confidence interval for his kernel density estimate

$$\widehat{f}_h(x) = \frac{1}{n}\sum_{i=1}^{n} K_h(x - X_i)$$

with kernel function K. Explain how this can be done. Simulate this on the basis of a lognormal sample with parameters $\mu = 2$ and $\sigma^2 = 0.5$.

Exercise 3.5. Explain the constant 1.34 in the "better rule of thumb" (3.29).

Exercise 3.6. Show that for a rescaled kernel $K_\delta(\bullet) = \delta^{-1}K(\bullet/\delta)$

$$\int K_\delta(u)\,du = 1.$$

Exercise 3.7. Derive formula (3.44) for the canonical bandwidth.

Exercise 3.8. Derive the formula for $CV(h)$ in the case $K = \varphi$ (Gaussian Kernel).

Exercise 3.9. Show that

$$\mathrm{Var}\left(\widehat{f}_h''(x)\right) \sim n^{-1}h^{-5} \quad \text{as } nh \to \infty.$$

Exercise 3.10. Simulate a mixture of normal densities with pdf

$$0.3\,\varphi(x) + \frac{0.7}{0.3}\,\varphi\left(\frac{x-1}{0.3}\right)$$

and plot the density and its estimate with a cross-validated bandwidth.

Exercise 3.11. Show how to get from equation (3.12) to (3.13).

Exercise 3.12. Compute δ_0 for the Quartic kernel.

Exercise 3.13. One possible way to construct a multivariate kernel $\mathcal{K}(\mathbf{u})$ is to use a one-dimensional kernel $K(u)$. This relationship is given by the formula (3.59), i.e.

$$\mathcal{K}(\boldsymbol{u}) = cK(\|\boldsymbol{u}\|),$$

where $\|\boldsymbol{u}\| = \sqrt{\boldsymbol{u}^\top \boldsymbol{u}}$. Find an appropriate constant c for a two-dimensional a) Gaussian, b) Epanechnikov, and c) Triangle kernel.

Exercise 3.14. Show that $\widehat{f}_h(x) \xrightarrow{a.s.} f(x)$. Assume that f possesses a second derivative and $\|K\| < \infty$.

Exercise 3.15. Explain why averaging over the leave-one-out estimator (3.32) is the appropriate way to estimate the expected value of $\widehat{f}_h(X)$ w.r.t. an independent random variable X.

Exercise 3.16. Show that

$$\int y \sum_{i=1}^{n} K_h(x - X_i) K_h(y - Y_i)\, dy = \sum_{i=1}^{n} K_h(x - X_i) Y_i.$$

Summary

- ⋆ Kernel density estimation is a generalization of the histogram. The kernel density estimate at point x

$$\widehat{f}_h(x) = \frac{1}{nh} \sum_{i=1}^{n} K\left(\frac{x - X_i}{h}\right)$$

corresponds to the histogram bar height for the bin $[x - h/2, x + h/2)$ if we use the uniform kernel.

- ⋆ The bias and variance of the kernel density estimator are

$$\text{Bias}\{\widehat{f}_h(x)\} \approx \frac{h^2}{2} f''(x)\, \mu_2(K), \quad \text{Var}\{\widehat{f}_h(x)\} \approx \frac{1}{nh} \|K\|_2^2 f(x).$$

- ⋆ The AMISE of the kernel density estimator is

$$\text{AMISE}(\widehat{f}_h) = \frac{1}{nh} \|K\|_2^2 + \frac{h^4}{4} \{\mu_2(K)\}^2 \|f''\|_2^2.$$

- ⋆ By using the normal distribution as a reference distribution for calculating $\|f''\|_2^2$ we get Silverman's rule-of-thumb bandwidth

$$\widehat{h} = 1.06 \cdot \widehat{\sigma} \cdot n^{-1/5},$$

which assumes the kernel to be Gaussian. Other plug-in bandwidths can be found by using more sophisticated replacements for $\|f''\|_2^2$.

- ⋆ When using ISE as a goodness-of-fit criterion for $\widehat{f}_h$ we can derive the least squares cross-validation criterion for bandwidth selection:

$$CV(h) = \frac{1}{n^2 h} \sum_{i,j} K \star K\left(\frac{X_j - X_i}{h}\right) - \frac{2}{n(n-1)} \sum_{i=1}^{n} \sum_{i \neq j} K_h(X_i - X_j).$$

★ The concept of canonical kernels allows us to separate the kernel choice from the bandwidth choice. We find a canonical bandwidth δ_0 for each kernel function K which gives us the equivalent degree of smoothing. This equivalence allows one to adjust bandwidths from different kernel functions to obtain approximately the same value of AMISE. For bandwidth h_A and kernel K^A the bandwidth

$$h_B = h_A \frac{\delta_0^B}{\delta_0^A}$$

is the equivalent bandwidth for kernel K^B. So for instance, Silverman's rule-of-thumb bandwidth has to be adjusted by a factor of 2.623 for using it with the Quartic kernel.

★ The asymptotic normality of the kernel density estimator

$$n^{2/5}(\widehat{f}_{h_n}(x) - f(x)) \xrightarrow{L} N\left(\frac{c^2}{2} f''(x)\mu_2(K), \frac{1}{c} f(x)\|K\|_2^2\right),$$

allows us to compute confidence intervals for f. Confidence bands can be computed as well, although under more restrictive assumptions on f.

★ The kernel density estimator for univariate data can be easily generalized to the multivariate case

$$\widehat{f}_{\mathbf{H}}(\boldsymbol{x}) = \frac{1}{n} \sum_{i=1}^{n} \frac{1}{\det(\mathbf{H})} \mathcal{K}\left\{\mathbf{H}^{-1}(\mathbf{X}_i - \boldsymbol{x})\right\},$$

where the bandwidth matrix $\mathbf{H}$ now replaces the bandwidth parameter. The multivariate kernel is typically chosen to be a product or radial-symmetric kernel function. Asymptotic properties and bandwidth selection are analogous, but more cumbersome. Canonical bandwidths can be used as well to adjust between different kernel functions.
A special problem is the graphical display of multivariate density estimates. Lower dimensional intersections, projections or contour plot may display only part of the features of a density function.

4

Nonparametric Regression

4.1 Univariate Kernel Regression

An important question in many fields of science is the relationship between two variables, say X and Y. Regression analysis is concerned with the question of how Y (the *dependent* variable) can be explained by X (the *independent* or *explanatory* or *regressor* variable). This means a relation of the form

$$Y = m(X),$$

where $m(\bullet)$ is a function in the mathematical sense. In many cases theory does not put any restrictions on the form of $m(\bullet)$, i.e. theory does not say whether $m(\bullet)$ is linear, quadratic, increasing in X, etc.. Hence, it is up to empirical analysis to use data to find out more about $m(\bullet)$.

4.1.1 Introduction

Let us consider an example from Economics. Suppose Y is expenditure on potatoes and X is net-income. If we draw a graph with quantity of potatoes on the vertical axis and income on the horizontal axis then we have drawn an *Engel curve*. Apparently, Engel curves relate optimal quantities of a commodity to income, holding prices constant. If we derive the Engel curve analytically, then it takes the form $Y = m(X)$, where Y denotes the quantity of potatoes bought at income level X. Depending on individual preferences several possibilities arise:

- The Engel curve slopes upward, i.e. $m(\bullet)$ is an increasing function of X. As income increases the consumer is buying more of the good. In this case the good is said to be *normal*. A special case of an upward sloping Engel curve is a straight line through the origin, i.e. $m(\bullet)$ is a linear function.

- The Engel curve slopes downward or eventually slopes downward (after sloping upward first) as income grows. If the Engel curve slopes downward the good is said to be *inferior*.

Are potatoes inferior goods? There is just one way to find out: collect appropriate data and estimate an Engel curve for potatoes. We can interpret the statement "potatoes are inferior" in the sense that, on average, consumers will buy fewer potatoes if their income grows while prices are held constant. The principle that theoretic laws usually do not hold in every individual case but merely on average can be formalized as

$$y_i = m(x_i) + \varepsilon_i, \quad i = 1, \ldots, n, \tag{4.1}$$

$$E(Y|X = x) = m(x). \tag{4.2}$$

Equation (4.1) says that the relationship $Y = m(X)$ doesn't need to hold exactly for the ith observation (household) but is "disturbed" by the random variable ε. Yet, (4.2) says that the relationship holds on average, i.e. the expectation of Y on the condition that $X = x$ is given by $m(x)$. The goal of the empirical analysis is to use a finite set of observations (x_i, y_i), $i = 1, \ldots, n$ to estimate $m(\bullet)$.

Example 4.1.
In Figure 4.1, we have $n = 7125$ observations of net-income and expenditure on food expenditures (not only potatoes), taken from the Family Expenditure Survey of British households in 1973. Graphically, we try to fit an (Engel) curve to the scatterplot of food versus net-income. Clearly, the graph of the estimate of $m(\bullet)$ will not run through every point in the scatterplot, i.e. we will not be able to use this graph to perfectly predict food consumption of every household, given that we know the household's income. But this does not constitute a serious problem (or any problem at all) if you recall that our theoretical statement refers to average behavior. □

Let us point out that, in a parametric approach, it is often assumed that $m(x) = \alpha + \beta \cdot x$, and the problem of estimating $m(\bullet)$ is reduced to the problem of estimating α and β. But note that this approach is not useful in our example. After all, the alleged shape of the Engel curve for potatoes, upward sloping for smaller income levels but eventually downward sloping as income is increased, is ruled out by the specification $m(x) = \alpha + \beta \cdot x$. The nonparametric approach does not put such prior restrictions on $m(\bullet)$. However, as we will see below, there is a price to pay for this flexibility.

Conditional Expectation

In this section we will recall two concepts that you should already be familiar with, *conditional expectation* and *conditional expectation function*. However,

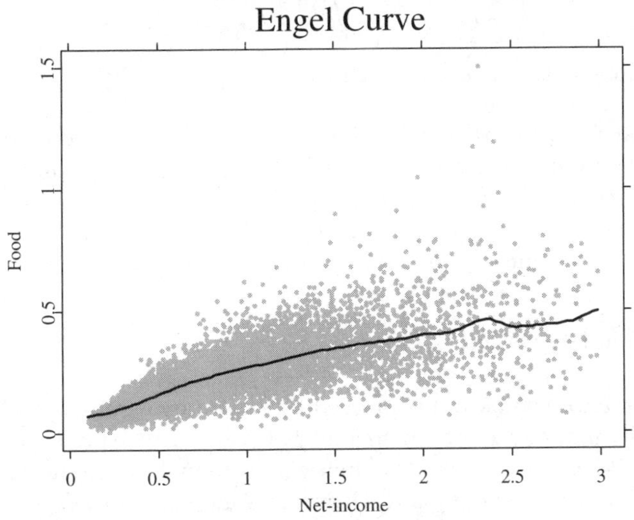

Figure 4.1. Nadaraya-Watson kernel regression, $h = 0.2$, U.K. Family Expenditure Survey 1973 SPMengelcurve1

these concepts are central to regression analysis and deserve to be treated accordingly. Let X and Y be two random variables with joint probability density function $f(x,y)$. The conditional expectation of Y given that $X = x$ is defined as

$$E(Y|X = x) = \int y\, f(y|x)\, dy = \int y \frac{f(x,y)}{f_X(x)}\, dy = m(x),$$

where $f(y|x)$ is the conditional probability density function (conditional pdf) of Y given $X = x$, and $f_X(x)$ is the marginal pdf of X. The mean function might be quite nonlinear even for simple-looking densities.

Example 4.2.
Consider the roof distribution with joint pdf

$$f(x,y) = x + y \qquad \text{for} \qquad 0 \le x \le 1 \quad \text{and} \quad 0 \le y \le 1,$$

with $f(x,y) = 0$ elsewhere, and marginal pdf

$$f_X(x) = \int_0^1 f(x,y)\, dy = x + \frac{1}{2} \qquad \text{for} \quad 0 \le x \le 1,$$

with $f_X(x) = 0$ elsewhere. Hence we get

$$E(Y|X=x) = \int y \frac{f(x,y)}{f_X(x)}\,dy = \int_0^1 y \frac{x+y}{x+\frac{1}{2}}\,dy = \frac{\frac{1}{2}x+\frac{1}{3}}{x+\frac{1}{2}} = m(x)$$

which is an obviously nonlinear function. □

Note that $E(Y|X=x)$ is a function of x alone. Consequently, we may abbreviate this term as $m(x)$. If we vary x we get a set of conditional expectations. This mapping from x to $m(x)$ is called the *conditional expectation function* and is often denoted as $E(Y|X)$. This tells us how Y and X are related "on average". Therefore, it is of immediate interest to estimate $m(\bullet)$.

Fixed and Random Design

We started the discussion in the preceeding section by assuming that both X and Y are random variables with joint pdf $f(x,y)$. The natural sampling scheme in this setup is to draw a random sample from the bivariate distribution that is characterized by $f(x,y)$. That is, we randomly draw observations of the form $\{X_i, Y_i\}$, $i=1,\ldots,n$. Before the sample is drawn, we can view the n pairs $\{X_i, Y_i\}$ as identically and independently distributed pairs of random variables. This sampling scheme will be referred to as the *random design*.

We will concentrate on random design in the following derivations. However, there are applications (especially in the natural sciences) where the researcher is able to control the values of the predictor variable X and Y is the sole random variable. As an example, imagine an experiment that is supposed to provide evidence for the link between a person's beer consumption (X) and his or her reaction time (Y) in a traffic incident. Here the researcher will be able to specify the amount of beer the testee is given before the experiment is conducted. Hence X will no longer be a random variable, while Y still will be. This setup is usually referred to as the *fixed design*. In repeated sampling, in the fixed design case the density $f_X(x)$ is known (it is induced by the researcher). This additional knowledge (relative to the random design case, where $f_X(x)$ is unknown) will simplify the estimation of $m(\bullet)$, as well as deriving statistical properties of the estimator used, as we shall see below. A special case of the fixed design model is the e.g. equispaced sequence $x_i = i/n, i = 0,\ldots,n$, on $[0,1]$.

4.1.2 Kernel Regression

As we just mentioned, kernel regression estimators depend on the type of the design.

Random Design

The derivation of the estimator in the random design case starts with the definition of conditional expectation:

$$m(x) = E(Y|X = x) = \int y \frac{f(x,y)}{f_X(x)} \, dy = \frac{\int y \, f(x,y) \, dy}{f_X(x)}. \tag{4.3}$$

Given that we have observations of the form $\{X_i, Y_i\}$, $i = 1, \ldots, n$, the only unknown quantities on the right hand side of (4.3) are $f(x,y)$ and $f_X(x)$. From our discussion of kernel density estimation we know how to estimate probability density functions. Consequently, we plug in kernel estimates for $f_X(x)$ and $f(x,y)$ in (4.3). Estimating $f_X(x)$ is straightforward. To estimate $f(x,y)$ we employ the multiplicative kernel density estimator (with product kernel) of Section 3.6

$$\widehat{f}_{h,g}(x,y) = \frac{1}{n} \sum_{i=1}^{n} K_h\left(\frac{x - X_i}{h}\right) K_g\left(\frac{y - Y_i}{g}\right). \tag{4.4}$$

Hence, for the numerator of (4.3) we get

$$\begin{aligned} \int y \, \widehat{f}_{h,g}(x,y) dy &= \frac{1}{n} \sum_{i=1}^{n} \frac{1}{h} K\left(\frac{x - X_i}{h}\right) \int \frac{y}{g} K\left(\frac{y - Y_i}{g}\right) dy \qquad (4.5) \\ &= \frac{1}{n} \sum_{i=1}^{n} K_h(x - X_i) \int (sg + Y_i) K(s) \, ds \\ &= \frac{1}{n} \sum_{i=1}^{n} K_h(x - X_i) Y_i, \end{aligned}$$

where we used the facts that kernel functions integrate to 1 and are symmetric around zero. Plugging in leads to the Nadaraya-Watson estimator introduced by Nadaraya (1964) and Watson (1964)

$$\widehat{m}_h(x) = \frac{n^{-1} \sum_{i=1}^{n} K_h(x - X_i) Y_i}{n^{-1} \sum_{j=1}^{n} K_h(x - X_j)}, \tag{4.6}$$

which is the natural extension of kernel estimation to the problem of estimating an unknown conditional expectation function. Several points are noteworthy:

- Rewriting (4.6) as

$$\widehat{m}_h(x) = \frac{1}{n} \sum_{i=1}^{n} \left(\frac{K_h(x - X_i)}{n^{-1} \sum_{j=1}^{n} K_h(x - X_j)} \right) Y_i = \frac{1}{n} \sum_{i=1}^{n} W_{hi}(x) Y_i \tag{4.7}$$

reveals that the Nadaraya-Watson estimator can be seen as a weighted (local) average of the response variables Y_i (note $\frac{1}{n}\sum_{i=1}^{n} W_{hi}(x) = 1$). In fact, the Nadaraya-Watson estimator shares this weighted local average property with several other smoothing techniques, e.g. k-nearest-neighbor and spline smoothing, see Subsections 4.2.1 and 4.2.3.

- Note that just as in kernel density estimation the bandwidth h determines the degree of smoothness of $\widehat{m}_h$, see Figure 4.2. To motivate this, let h go to either extreme. If $h \to 0$, then $W_{hi}(x) \to n$ if $x = X_i$ and is not defined elsewhere. Hence, at an observation X_i, $\widehat{m}_h(X_i)$ converges to Y_i, i.e. we get an interpolation of the data. On the other hand if $h \to \infty$ then $W_{hi}(x) \to 1$ for all values of x, and $\widehat{m}_h(X_i) \to \overline{Y}$, i.e. the estimator is a constant function that assigns the sample mean of Y to each x. Choosing h so that a good compromise between over- and undersmoothing is achieved, is once again a crucial problem.
- You may wonder what happens if the denominator of $W_{hi}(x)$ is equal to zero. In this case, the numerator is also equal to zero, and the estimate is not defined. This can happen in regions of sparse data.

Fixed Design

In the fixed design model, $f_X(x)$ is assumed to be known and a possible kernel estimator for this sampling scheme employs weights of the form

$$W_{hi}^{FD}(x) = \frac{K_h(x - x_i)}{f_X(x)}. \tag{4.8}$$

Thus, estimators for the fixed design case are of simpler structure and are easier to analyze in their statistical properties.

Since our main interest is the random design case, we will only mention a very particular fixed design kernel regression estimator: For the case of ordered design points $x_{(i)}, i = 1, \ldots, n$, from some interval $[a, b]$ Gasser & Müller (1984) suggested the following weight sequence

$$W_{hi}^{GM}(x) = n \int_{s_{i-1}}^{s_i} K_h(x - u)du, \tag{4.9}$$

where $s_i = \left(x_{(i)} + x_{(i+1)}\right)/2$, $s_0 = a$, $s_{n+1} = b$. Note that as for the Nadaraya-Watson estimator, the weights $W_{hi}^{GM}(x)$ sum to 1.

To show how the weights (4.9) are related to the intuitively appealing formula (4.8) note that by the mean value theorem

$$(s_i - s_{i-1})\, K_h(x - \xi) = \int_{s_{i-1}}^{s_i} K_h(x - u)\, du \tag{4.10}$$

for some ξ between s_i and s_{i-1}. Moreover,

$$n(s_i - s_{i-1}) \approx \frac{1}{f_X(x)}. \tag{4.11}$$

Plugging in (4.10) and (4.11) into (4.8) gives

$$W_{hi}^{FD}(x) = \frac{K_h(x - x_{(i)})}{f_X(x)} \approx n \int_{s_{i-1}}^{s_i} K_h(x-u)\,du = W_{hi}^{GM}(x).$$

We will meet the Gasser-Müller estimator $\frac{1}{n}\sum_{i=1}^{n} W_{hi}^{GM}(x) Y_i$ again in the following section where the statistical properties of kernel regression estimators are discussed.

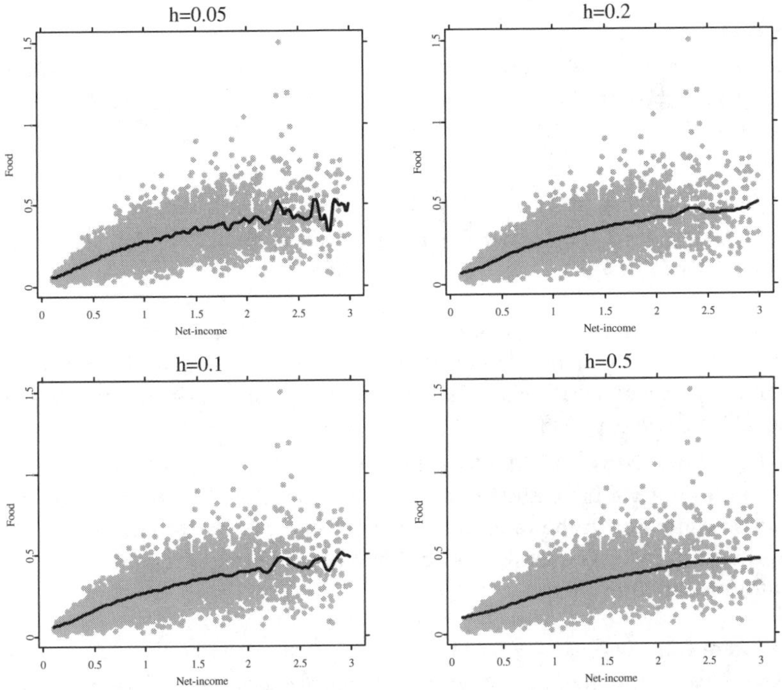

Figure 4.2. Four kernel regression estimates for the 1973 U.K. Family Expenditure data with bandwidths $h = 0.05$, $h = 0.1$, $h = 0.2$, and $h = 0.5$ `SPMregress`

Statistical Properties

Are kernel regression estimators consistent? In the previous chapters we showed that an estimator is consistent in deriving its mean squared error (MSE), showing that the MSE converges, and appealing to the fact that convergence in mean square implies convergence in probability (the latter being the condition stated in the definition of consistency).

Moreover, the MSE helped in assessing the speed with which convergence is attained. In the random design case it is very difficult to derive the MSE of the Nadaraya-Watson estimator since it is the ratio (and not the sum) of two estimators. It turns out that one can show that the Nadaraya-Watson estimator is consistent in the random design case without explicit recurrence to the MSE of this estimator. The conditions under which this result holds are summarized in the following theorem:

Theorem 4.1.
Assume the univariate random design model and the regularity conditions $\int |K(u)|du < \infty$, $u\ K(u) \to 0$ *for* $|u| \to \infty$, $EY^2 < \infty$. *Suppose also* $h \to 0$, $nh \to \infty$, *then*

$$\frac{1}{n}\sum_{i=1}^{n} W_{hi}(x)Y_i = \widehat{m}_h(x) \xrightarrow{P} m(x)$$

where for x *holds* $f_X(x) > 0$ *and* x *is a point of continuity of* $m(x)$, $f_X(x)$, *and* $\sigma^2(x) = \mathrm{Var}(Y|X = x)$.

The proof involves showing that — considered separately — both the numerator and the denominator of $\widehat{m}_h(x)$ converge. Then, as a consequence of Slutsky's theorem, it can be shown that $\widehat{m}_h(x)$ converges. For more details see Härdle (1990, p. 39ff).

Certainly, we would like to know the speed with which the estimator converges but we have already pointed out that the MSE of the Nadaraya-Watson estimator in the random design case is very hard to derive. For the fixed design case, Gasser & Müller (1984) have derived the MSE of the estimator named after them:

Theorem 4.2.
Assume the univariate fixed design model and the conditions: $K(\bullet)$ *has support* $[-1,1]$ *with* $K(-1) = K(1) = 0$, $m(\bullet)$ *is twice continuously differentiable,* $\max_i |x_i - x_{i-1}| = O(n^{-1})$. *Assume* $\mathrm{Var}(\varepsilon_i) = \sigma^2$, $i = 1,\ldots,n$. *Then, under* $h \to 0$, $nh \to \infty$ *it holds*

$$\mathrm{MSE}\left\{\frac{1}{n}\sum_{i=1}^{n} W_{hi}^{GM}(x)Y_i\right\} \approx \frac{1}{nh}\sigma^2\|K\|_2^2 + \frac{h^4}{4}\mu_2^2(K)\{m''(x)\}^2.$$

As usual, the (asymptotic) MSE has two components, the variance term $\sigma^2\|K\|_2^2/(nh)$ and the squared bias term $h^4\mu_2^2(K)\{m''(x)\}^2/4$. Hence, if we increase the bandwidth h we face the familiar trade-off between decreasing the variance while increasing the squared bias.

To get a similar result for the random design case, we linearize the Nadaraya-Watson estimator as follows

$$\widehat{m}_h(x) = \frac{\widehat{r}_h(x)}{\widehat{f}_h(x)},$$

thus

$$\begin{aligned}\widehat{m}_h(x) - m(x) &= \left\{\frac{\widehat{r}_h(x)}{\widehat{f}_h(x)} - m(x)\right\}\left[\frac{\widehat{f}_h(x)}{f_X(x)} + \left\{1 - \frac{\widehat{f}_h(x)}{f_X(x)}\right\}\right] \\ &= \frac{\widehat{r}_h(x) - m(x)\widehat{f}_h(x)}{f_X(x)} \\ &\quad + \{\widehat{m}_h(x) - m(x)\}\frac{f_X(x) - \widehat{f}_h(x)}{f_X(x)}. \end{aligned} \tag{4.12}$$

It can be shown that of the two terms on the right hand side, the first term is the leading term in the distribution of $\widehat{m}_h(x) - m(x)$, whereas the second term can be neglected. Hence, the MSE of $\widehat{m}_h(x)$ can be approximated by calculating

$$E\left\{\frac{\widehat{r}_h(x) - m(x)\widehat{f}_h(x)}{f_X(x)}\right\}^2.$$

The following theorem can be derived this way:

Theorem 4.3.
Assume the univariate random design model and the conditions $\int |K(u)|du < \infty$, $\lim u\,K(u) = 0$ for $|u| \to \infty$ and $EY^2 < \infty$ hold. Suppose $h \to 0$, $nh \to \infty$, then

$$\mathrm{MSE}\{\widehat{m}_h(x)\} \approx \underbrace{\frac{1}{nh}\frac{\sigma^2(x)}{f_X(x)}\|K\|_2^2}_{\textit{variance part}} + \underbrace{\frac{h^4}{4}\left\{m''(x) + 2\frac{m'(x)f_X'(x)}{f_X(x)}\right\}^2 \mu_2^2(K)}_{\textit{bias part}}, \tag{4.13}$$

where for x holds $f_X(x) > 0$ and x is a point of continuity of $m'(x)$, $m''(x)$, $f_X(x)$, $f_X'(x)$, and $\sigma^2(x) = \mathrm{Var}(Y|X = x)$.

Let AMSE denote the asymptotic MSE. Most components of this formula are constants w.r.t. n and h, and we may write denoting constant terms by C_1 and C_2, respectively

$$\text{AMSE}(n,h) = \frac{1}{nh}C_1 + h^4 C_2.$$

Minimizing this expression with respect to h gives the optimal bandwidth $h_{opt} \sim n^{-1/5}$. If you plug a bandwidth $h \sim n^{-1/5}$ into (4.13), you will find that the AMSE is of order $O(n^{-4/5})$, a rate of convergence that is slower than the rate obtained by the LS estimator in linear regression but is the same as for estimating a density function (cf. Section 3.2).

As in the density estimation case, AMSE depends on unknown quantities like $\sigma^2(x)$ or $m''(x)$. Once more, we are faced with the problem of finding a bandwidth-selection rule that has desirable theoretical properties and is applicable in practice. We have displayed Nadaraya-Watson kernel regression estimates with different bandwidths in Figure 4.2. The issue of bandwidth selection will be discussed later on in Section 4.3.

4.1.3 Local Polynomial Regression and Derivative Estimation

The Nadaraya-Watson estimator can be seen as a special case of a larger class of kernel regression estimators: Nadaraya-Watson regression corresponds to a local constant least squares fit. To motivate local linear and higher order local polynomial fits, let us first consider a Taylor expansion of the unknown conditional expectation function $m(\bullet)$:

$$m(t) \approx m(x) + m'(x)(t-x) + \cdots + m^{(p)}(x)(t-x)^p \frac{1}{p!} \tag{4.14}$$

for t in a neighborhood of the point x. This suggests *local polynomial* regression, namely to fit a polynomial in a neighborhood of x. The neighborhood is realized by including kernel weights into the minimization problem

$$\min_{\beta} \sum_{i=1}^{n} \left\{ Y_i - \beta_0 - \beta_1(X_i - x) - \ldots - \beta_p(X_i - x)^p \right\}^2 K_h(x - X_i), \tag{4.15}$$

where β denotes the vector of coefficients $(\beta_0, \beta_1, \ldots, \beta_p)^\top$. The result is therefore a weighted least squares estimator with weights $K_h(x - X_i)$. Using the notations

$$\mathbf{X} = \begin{pmatrix} 1 & X_1 - x & (X_1 - x)^2 & \ldots & (X_1 - x)^p \\ 1 & X_2 - x & (X_2 - x)^2 & \ldots & (X_2 - x)^p \\ \vdots & \vdots & \vdots & \ddots & \vdots \\ 1 & X_n - x & (X_n - x)^2 & \ldots & (X_n - x)^p \end{pmatrix}, \quad \boldsymbol{Y} = \begin{pmatrix} Y_1 \\ Y_2 \\ \vdots \\ Y_n \end{pmatrix},$$

$$\mathbf{W} = \begin{pmatrix} K_h(x - X_1) & 0 & \dots & 0 \\ 0 & K_h(x - X_2) & \dots & 0 \\ \vdots & \vdots & \ddots & \vdots \\ 0 & 0 & \dots & K_h(x - X_n) \end{pmatrix},$$

we can compute $\widehat{\beta}$ which minimizes (4.15) by the usual formula for a weighted least squares estimator

$$\widehat{\beta}(x) = \left(\mathbf{X}^\top \mathbf{W} \mathbf{X}\right)^{-1} \mathbf{X}^\top \mathbf{W} \mathbf{Y}. \tag{4.16}$$

It is important to note that — in contrast to parametric least squares — this estimator varies with x. Hence, this is really a local regression at the point x. Denote the components of $\widehat{\beta}(x)$ by $\widehat{\beta}_0(x), \dots, \widehat{\beta}_p(x)$. The local polynomial estimator of the regression function m is

$$\widehat{m}_{p,h}(x) = \widehat{\beta}_0(x) \tag{4.17}$$

due to the fact that we have $m(x) \approx \beta_0(x)$ by comparing (4.14) and (4.15). The whole curve $\widehat{m}_{p,h}(\bullet)$ is obtained by running the above local polynomial regression with varying x. We have included the parameter h in the notation since the final estimator depends obviously on the bandwidth parameter h as it does the Nadaraya-Watson estimator.

Let us gain some more insight into this by computing the estimators for special values of p. For $p = 0$ $\widehat{\beta}$ reduces to $\widehat{\beta}_0$, which means that the *local constant* estimator is nothing else as our well known Nadaraya-Watson estimator, i.e.

$$\widehat{m}_{0,h}(x) = \widehat{m}_h(x) = \frac{\sum_{i=1}^n K_h(x - X_i) Y_i}{\sum_{i=1}^n K_h(x - X_i)}.$$

Now turn to $p = 1$. Denote

$$S_{h,j}(x) = \sum_{i=1}^n K_h(x - X_i)(X_i - x)^j,$$

$$T_{h,j}(x) = \sum_{i=1}^n K_h(x - X_i)(X_i - x)^j Y_i,$$

then we can write

$$\widehat{\beta}(x) = \begin{pmatrix} S_{h,0}(x) & S_{h,1}(x) \\ S_{h,1}(x) & S_{h,2}(x) \end{pmatrix}^{-1} \begin{pmatrix} T_{h,0}(x) \\ T_{h,1}(x) \end{pmatrix} \tag{4.18}$$

which yields the *local linear* estimator

$$\widehat{m}_{1,h}(x) = \widehat{\beta}_0(x) = \frac{T_{h,0}(x)\, S_{h,2}(x) - T_{h,1}(x)\, S_{h,1}(x)}{S_{h,0}(x)\, S_{h,2}(x) - S_{h,1}^2(x)}. \tag{4.19}$$

Here we used the usual matrix inversion formula for 2×2 matrices. Of course, (4.18) can be generalized for arbitrary large p. The general formula is

$$\widehat{\beta}(x) = \begin{pmatrix} S_{h,0}(x) & S_{h,1}(x) & \dots & S_{h,p}(x) \\ S_{h,1}(x) & S_{h,2}(x) & \dots & S_{h,p+1}(x) \\ \vdots & \vdots & \ddots & \vdots \\ S_{h,p}(x) & S_{h,p+1}(x) & \dots & S_{h,2p}(x) \end{pmatrix}^{-1} \begin{pmatrix} T_{h,0}(x) \\ T_{h,1}(x) \\ \vdots \\ T_{h,p}(x) \end{pmatrix}. \tag{4.20}$$

Introducing the notation $e_0 = (1, 0, \dots, 0)^\top$ for the first unit vector in $\mathbb{R}^{p+1}$, we can write the local linear estimator as

$$\widehat{m}_{1,h}(x) = e_0^\top \left(\mathbf{X}^\top \mathbf{W} \mathbf{X}\right)^{-1} \mathbf{X}^\top \mathbf{W} \mathbf{Y}.$$

Note that the Nadaraya-Watson estimator could also be written as

$$\widehat{m}_h(x) = \frac{T_{h,0}(x)}{S_{h,0}(x)}.$$

Example 4.3.
The local linear estimator $\widehat{m}_{1,h}$ for our running example is displayed in Figure 4.3. What can we conclude from comparing this fit with the Nadaraya-Watson fit in Figure 4.1? The main difference to see is that the local linear fit reacts more sensitively on the boundaries of the fit.

Another graphical difference will appear, when we compare local linear and Nadaraya-Watson estimates with optimized bandwidths (see Section 4.3). Then we will see that the local linear fit will be influenced less by outliers like those which cause the "bump" in the right part of both Engel curves. □

Here we can discuss this effect by looking at the asymptotic MSE of the local linear regression estimator:

$$\mathrm{AMSE}\{\widehat{m}_{1,h}(x)\} = \frac{1}{nh} \frac{\sigma^2(x)}{f_X(x)} \|K\|_2^2 + \frac{h^4}{4} \left\{m''(x)\right\}^2 \mu_2^2(K). \tag{4.21}$$

This formula is dealt with in more detail when we come to multivariate regression, see Section 4.5. The AMSE in the local linear case differs from that for the Nadaraya-Watson estimator (4.13) only with regard to the bias. It is easy to see that the bias of the local linear fit is design-independent and disappears when $m(\bullet)$ is linear. Thus, a local linear fit can improve the function estimation in regions with sparse observations, for instance in the high net-income region in our Engel curve example. Let us also mention that the bias

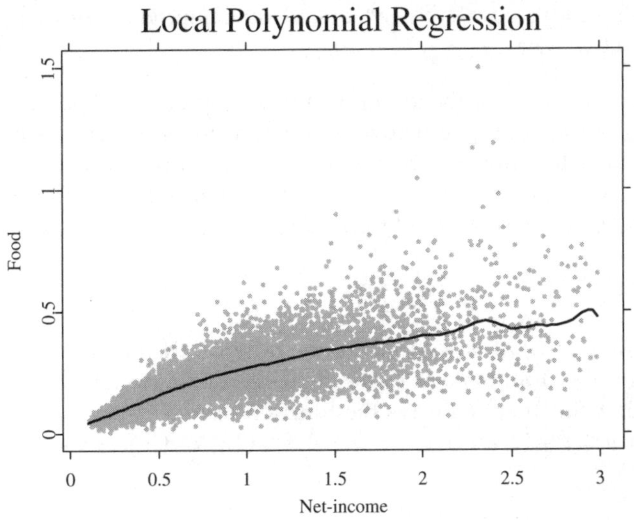

Figure 4.3. Local polynomial regression, $p = 1, h = 0.2$, U.K. Family Expenditure Survey 1973 `SPMlocpolyreg`

of the local linear estimator has the same form as that of the Gasser-Müller estimator, i.e. the bias in the fixed design case.

The local linear estimator achieves further improvement in the boundary regions. In the case of Nadaraya-Watson estimates we typically observe problems due to the one-sided neighborhoods at the boundaries. The reason is that in local constant modeling, more or less the same points are used to estimate the curve near the boundary. Local polynomial regression overcomes this by fitting a higher degree polynomial here.

For estimating regression functions, the order p is usually taken to be one (local linear) or three (local cubic regression). As we have seen, the local linear fit performs (asymptotically) better than the Nadaraya-Watson estimator (local constant). This holds generally: Odd order fits outperform even order fits. Some additional remarks should be made in summary:

- As the Nadaraya-Watson estimator, the local polynomial estimator is a weighted (local) average of the response variables Y_i.
- As for all other kernel methods the bandwidth h determines the degree of smoothness of $\widehat{m}_{p,h}$. For $h \to 0$, we observe the same result as for the Nadaraya-Watson estimator, namely at an observation X_i, $\widehat{m}_{p,h}(X_i)$ con-

verges to Y_i. The behavior is different for $h \to \infty$. An infinitely large h makes all weights equal, thus we obtain a parametric pth order polynomial fit in that case.

A further advantage of the local polynomial approach is that it provides an easy way of estimating derivatives of the function $m(\bullet)$. The natural approach would be to estimate m by $\widehat{m}$ and then to compute the derivative $\widehat{m}'$. But an alternative and more efficient method is obtained by comparing (4.14) and (4.15) again. From this we get the *local polynomial derivative estimator*

$$\widehat{m}_{p,h}^{(\nu)}(x) = \nu!\,\widehat{\beta}_\nu(x) \tag{4.22}$$

for the νth derivative of $m(\bullet)$. Usually the order of the polynomial is $p = \nu + 1$ or $p = \nu + 3$ in analogy to the regression case (recall that the zero derivative of a function is always the function itself). Also in analogy, the "odd" order $p = \nu + 2\ell + 1$ outperforms the "even" order $p = \nu + 2\ell$.

Example 4.4.
To estimate the first ($\nu = 1$) derivative of our Engel curve we could take $p = 2$ (*local quadratic* derivative estimator). This is done to get Figure 4.4. Note that we have used a rule-of-thumb bandwidth here, see Fan & Müller (1995, p. 92) and Fan & Gijbels (1996, p. 111). □

4.2 Other Smoothers

This section intends to give a short overview of nonparametric smoothing techniques that differ from the kernel method. For further references on these and other nonparametric smoothers see the bibliographic notes at the end of this chapter.

4.2.1 Nearest-Neighbor Estimator

As we have seen above, kernel regression estimation can be viewed as a method of computing weighted averages of the response variables in a fixed neighborhood around x, the width of this neighborhood being governed by the bandwidth h. The k-nearest-neighbor (k-NN) estimator can also be viewed as a weighted average of the response variables in a neighborhood around x, with the important difference that the neighborhood width is not fixed but variable. To be more specific, the values of Y used in computing the average, are those which belong to the k observed values of X that are

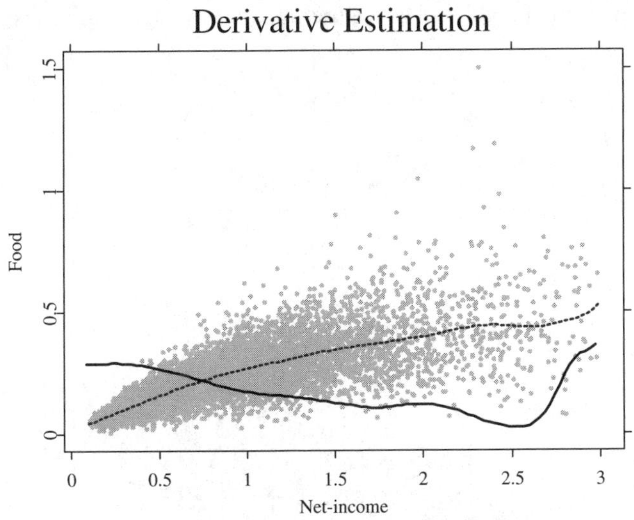

Figure 4.4. Local polynomial regression ($p = 1$) and derivative estimation, ($p = 2$), h by rule of thumb, U.K. Family Expenditure Survey 1973 `SPMderivest`

nearest to the point x, at which we would like to estimate $m(x)$. Formally, the k-NN estimator can be written as

$$\widehat{m}_k(x) = \frac{1}{n} \sum_{i=1}^{n} W_{ki}(x) Y_i, \tag{4.23}$$

where the weights $\{W_{ki}(x)\}_{i=1}^n$ are defined as

$$W_{ki}(x) = \begin{cases} n/k \text{ if } & i \in J_x \\ 0 & \text{otherwise} \end{cases} \tag{4.24}$$

with the set of indices

$$J_x = \{i : X_i \text{ is one of the } k \text{ nearest observations to } x\}.$$

If we estimate $m(\bullet)$ at a point x where the data are sparse then it might happen that the k nearest neighbors are rather far away from x (and each other), thus consequently we end up with a wide neighborhood around x for which an average of the corresponding values of Y is computed. Note that k is the smoothing parameter of this estimator. Increasing k makes the estimate smoother.

Example 4.5.
A k-NN estimation of the Engel curve (net-income vs. food) is shown in Figure 4.5. □

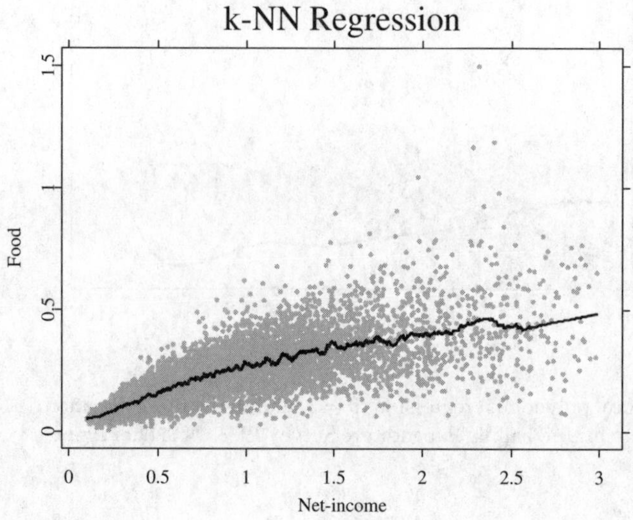

Figure 4.5. k-Nearest-neighbor regression, $k = 101$, U.K. Family Expenditure Survey 1973 `SPMknnreg`

The k-NN estimator can be viewed as a kernel estimator with uniform kernel $K(u) = \frac{1}{2}\mathrm{I}(|u| \leq 1)$ and variable bandwidth $R = R(k)$, with $R(k)$ being the distance between x and its furthest k-nearest neighbor:

$$\widehat{m}_k(x) = \frac{\sum_{i=1}^{n} K_R(x - X_i)Y_i}{\sum_{i=1}^{n} K_R(x - X_i)}. \tag{4.25}$$

The k-NN estimator can be generalized in this sense by considering kernels other than the uniform kernel. Bias and variance of this more general k-NN estimator are given in the following theorem by Mack (1981).

Theorem 4.4.
Let $k \to \infty$, $k/n \to 0$ and $n \to \infty$. Then

$$E\{\widehat{m}_k(x)\} - m(x) \approx \frac{\mu_2(K)}{8f_X(x)^2}\left\{m''(x) + 2\,\frac{m'(x)f'_X(x)}{f_X(x)}\right\}\left(\frac{k}{n}\right)^2 \tag{4.26}$$

$$Var\{\widehat{m}_k(x)\} \approx 2\|K\|_2^2\,\frac{\sigma^2(x)}{k}. \tag{4.27}$$

Obviously, unlike the variance of the Nadaraya-Watson kernel regression estimator, the variance of the k-NN regression estimator does not depend on $f_X(x)$, which makes sense since the k-NN estimator always averages over k observations, regardless of how dense the data is in the neighborhood of the point x where we estimate $m(\bullet)$. Consequently, $Var\{\widehat{m}_k(x)\} \sim \frac{1}{k}$. By choosing

$$k = 2nhf_X(x), \tag{4.28}$$

we obtain a k-NN estimator that is approximately identical to a kernel estimator with bandwidth h in the leading terms of the MSE.

4.2.2 Median Smoothing

Median smoothing may be described as the nearest-neighbor technique to solve the problem of estimating the *conditional median function*, rather than the conditional expectation function, which has been our target so far. The conditional median $\text{med}(Y|X = x)$ is more robust to outliers than the conditional expectation $E(Y|X = x)$. Moreover, median smoothing allows us to model discontinuities in the regression curve $\text{med}(Y|X)$. Formally, the median smoother is defined as

$$\widehat{m}(x) = \text{med}\{Y_i : i \in J_x\}, \tag{4.29}$$

where

$$J_x = \{i : X_i \quad \text{is one of the } k \text{ nearest neighbors of } x\}.$$

That is, the median of those Y_is is computed, for which the corresponding X_i is one of the k nearest neighbors of x.

Example 4.6.
We display such a median smoother for our running Engle curve example in Figure 4.6. Note that in contrast to the k-NN estimator, extreme values of food expenditures do no longer affect the estimator. □

4.2.3 Spline Smoothing

Spline smoothing can be motivated by considering the residual sum of squares (RSS) as a criterion for the goodness of fit of a function $m(\bullet)$ to the data. The residual sum of squares is defined as

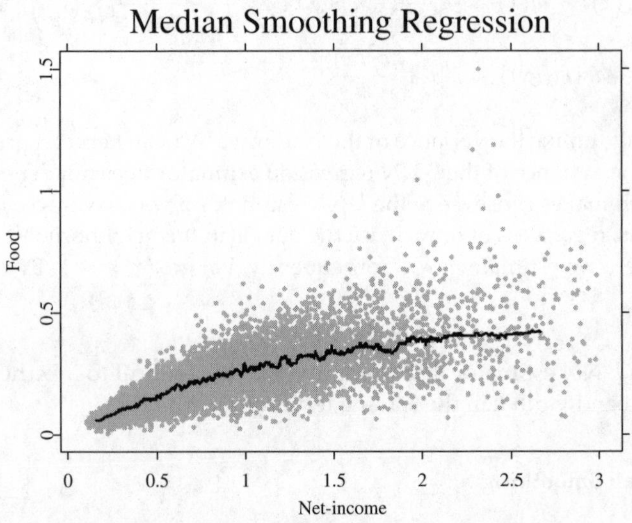

Figure 4.6. Median smoothing regression, $k = 101$, U.K. Family Expenditure Survey 1973 `SPMmesmooreg`

$$\text{RSS} = \sum_{i=1}^{n} \{Y_i - m(X_i)\}^2 .$$

Yet, one can define the function $m(X_i) = Y_i,\ i = 1, \ldots, n$ that is minimizing the RSS but is merely interpolating the data, without exploiting any structure that might be present in the data. Spline smoothing solves this problem by adding a stabilizer that penalizes non-smoothness of $m(\bullet)$. One possible stabilizer is given by

$$\|m''\|_2^2 = \int \{m''(x)\}^2 \ dx. \tag{4.30}$$

The use of m'' can be motivated by the fact that the curvature of $m(x)$ increases with $|m''(x)|$. Using the penalty term (4.30) we may restate the minimization problem as

$$\widehat{m}_\lambda = \arg\min_m S_\lambda(m) \tag{4.31}$$

with

$$S_\lambda(m) = \sum_{i=1}^{n} \{Y_i - m(X_i)\}^2 + \lambda \|m''\|_2^2. \tag{4.32}$$

If we consider the class of all twice differentiable functions on the interval $[a, b] = [X_{(1)}, X_{(n)}]$ (where $X_{(i)}$ denotes ith order statistic) then the (unique) minimizer of (4.32) is given by the cubic spline estimator $\widehat{m}_\lambda(x)$, which consists of cubic polynomials

$$p_i(x) = \alpha_i + \beta_i x + \gamma_i x^2 + \delta_i x^3, \quad i = 1, \ldots, n-1,$$

between adjacent $X_{(i)}, X_{(i+1)}$-values.

The parameter λ controls the weight given to the stabilizer in the minimization. The higher λ is, the more weight is given to $\|m''\|_2^2$ and the smoother the estimate. As $\lambda \to 0$, $\widehat{m}_\lambda(\bullet)$ is merely an interpolation of the observations of Y. If $\lambda \to \infty$, $\widehat{m}_\lambda(\bullet)$ tends to a linear function.

Let us now consider the spline estimator in more detail. For the estimator to be twice continuously differentiable we have to make sure that there are no jumps in the function, as well in its first and second derivative if evaluated at $X_{(i)}$. Formally, we require

$$p_i\left(X_{(i)}\right) = p_{i-1}\left(X_{(i)}\right), \; p_i'\left(X_{(i)}\right) = p_{i-1}'\left(X_{(i)}\right), \; p_i''\left(X_{(i)}\right) = p_{i-1}''\left(X_{(i)}\right).$$

Additionally a boundary condition has to be fulfilled. Typically this is

$$p_1''\left(X_{(1)}\right) = p_{n-1}''\left(X_{(n)}\right) = 0.$$

These restrictions, along with the conditions for minimizing $S_\lambda(m)$ w.r.t. the coefficients of p_i, define a system of linear equations which can be solved in only $O(n)$ calculations.

To illustrate this, we present some details on the computational algorithm introduced by Reinsch, see Green & Silverman (1994). Observe that the residual sum of squares

$$\text{RSS} = \sum_{i=1}^{n} \{Y_{(i)} - m(X_{(i)})\}^2 = (\boldsymbol{Y} - \boldsymbol{m})^\top (\boldsymbol{Y} - \boldsymbol{m}) \tag{4.33}$$

where $\boldsymbol{Y} = (Y_{(1)}, \ldots, Y_{(n)})^\top$ with $Y_{(i)}$ the corresponding value to $X_{(i)}$ and

$$\boldsymbol{m} = \left(m(X_{(1)}), \ldots, m(X_{(n)})\right)^\top.$$

If $m(\bullet)$ were indeed a piecewise cubic polynomial on intervals $[X_{(i)}, X_{(i+1)}]$ then the penalty term could be expressed as a quadratic form in $\boldsymbol{m}$

$$\int \{m''(x)\}^2 \, dx = \boldsymbol{m}^\top \mathbf{K} \boldsymbol{m} \tag{4.34}$$

with a matrix $\mathbf{K}$ that can be decomposed to

$$\mathbf{K} = \mathbf{Q}\mathbf{R}^{-1}\mathbf{Q}^\top.$$

Here, **Q** and **R** are band matrices and functions of $h_i = X_{(i+1)} - X_{(i)}$. More precisely, **Q** is a $n \times (n-1)$ matrix with elements

$$q_{j-1,j} = -\frac{1}{h_{j-1}},\ q_{jj} = -\frac{1}{h_{j-1}} - \frac{1}{h_{j-1}},\ q_{j+1,j} = -\frac{1}{h_j}$$

and $q_{ij} = 0$ for $|i-j| > 1$. **R** is a symmetric $(n-1) \times (n-1)$ matrix with elements

$$r_{jj} = \frac{1}{3}\left(h_{j-1} + h_j\right),\ r_{j,j+1} = r_{j+1,j} = \frac{1}{6}h_j,$$

and $r_{ij} = 0$ for $|i-j| > 1$. From (4.33) and (4.34) it follows that the smoothing spline is obtained by

$$\widehat{\boldsymbol{m}}_\lambda = (\mathbf{I} + \lambda\mathbf{K})^{-1}\boldsymbol{Y}, \tag{4.35}$$

with **I** denoting the n-dimensional identity matrix. Because of the band structure of **Q** and **R** (4.35) can be solved indeed in $O(n)$ steps using a Cholesky decomposition.

Example 4.7.
In Figure 4.7 we illustrate the resulting cubic spline estimate for our running Engel curve example. □

From (4.35) we see that the spline smoother is a linear estimator in Y_i, i.e. weights $W_{\lambda i}(x)$ exist, such that

$$\widehat{m}_\lambda(x) = \frac{1}{n}\sum_{i=1}^{n} W_{\lambda i}(x) Y_i.$$

It can be shown that under certain conditions the spline smoother is asymptotically equivalent to a kernel smoother that employs the so-called *spline kernel*

$$K_S(u) = \frac{1}{2}\exp\left(-\frac{|u|}{\sqrt{2}}\right)\sin\left(\frac{|u|}{\sqrt{2}} + \frac{\pi}{4}\right),$$

with local bandwidth $h(X_i) = \lambda^{1/4} n^{-1/4} f(X_i)^{-1/4}$ (Silverman, 1984).

4.2.4 Orthogonal Series

Under regularity conditions, functions can be represented as a series of basis functions (e.g. a Fourier series). Suppose that $m(\bullet)$ can be represented by such a Fourier series. That is, suppose that

$$m(x) = \sum_{j=0}^{\infty} \beta_j \varphi_j(x), \tag{4.36}$$

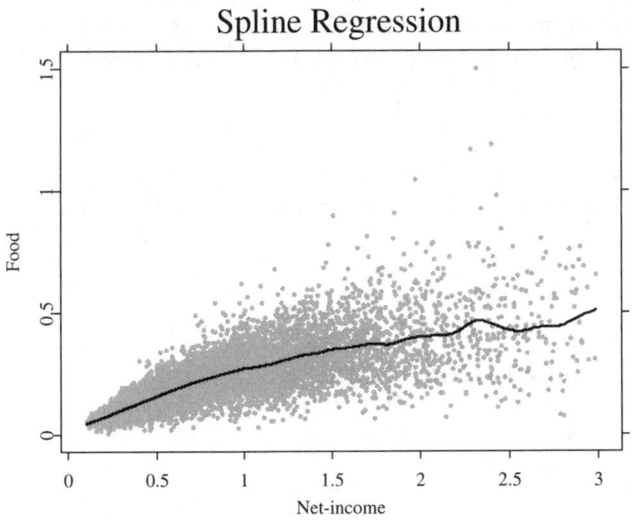

Figure 4.7. Spline regression, $\lambda = 0.005$, U.K. Family Expenditure Survey 1973 SPMspline

where $\{\varphi_j\}_{j=0}^{\infty}$ is a known basis of functions and $\{\beta_j\}_{j=0}^{\infty}$ are the unknown Fourier coefficients. Our goal is to estimate the unknown Fourier coefficients. Note that we indeed have an infinite sum in (4.36) if there are infinitely many non-zero β_js.

Obviously, an infinite number of coefficients cannot be estimated from a finite number of observations. Hence, one has to choose the number of terms N (which, as indicated, is a function of the sample size n) that will be included in the Fourier series representation. Thus, in principle, series estimation proceeds in three steps:

(a) select a basis of functions,
(b) select N, where N is an integer less than n, and
(c) estimate the N unknown coefficients by a suitable method.

N is the smoothing parameter of series estimation. The larger N is the more terms are included in the Fourier series and the estimate tends toward interpolating the data. On the contrary, small values of N will produce relatively smooth estimates.

Regarding the estimation of the coefficients $\{\beta_j\}_{j=0}^{N}$ there are basically two methods: One method involves looking at the finite version (i.e. the sum up to $N = N$) of (4.36) as a regression equation and estimating the coefficients by regressing the Y_i on $\varphi_0(X_i), \ldots, \varphi_N(X_i)$.

Example 4.8.
We have applied this for the Engel curve again. As functions φ_j we used the Legendre polynomials (orthogonalized polynomials, see Example 4.9 below) in Figure 4.8. □

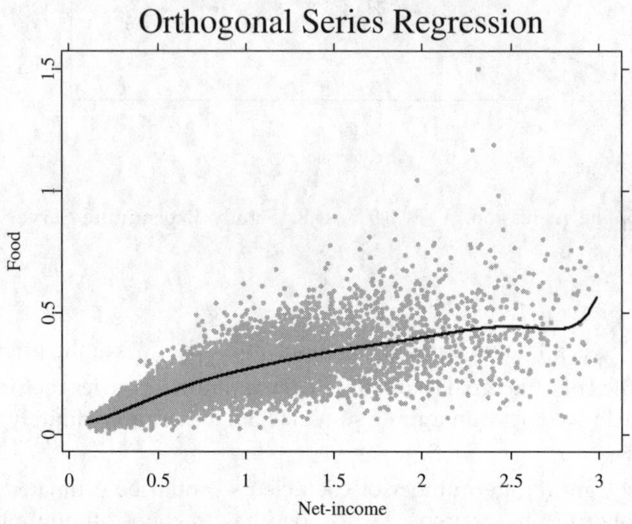

Figure 4.8. Orthogonal series regression using Legendre polynomials, $N = 9$, U.K. Family Expenditure Survey 1973 SPMorthogon

An alternative approach is concerned with choosing the basis of functions $\{\varphi_j\}_{j=0}^{\infty}$ to be orthonormal. The orthonormality requirement can be formalized as

$$\int \varphi_j(x)\varphi_k(x)dx = \delta_{jk} = \begin{cases} 0, \text{ if } & j \neq k; \\ 1, \text{ if } & j = k. \end{cases}$$

The following two examples show such orthonormal bases.

Example 4.9.
Consider the Legendre polynomials on $[-1, 1]$

$$p_0(x) = \frac{1}{\sqrt{2}},\ p_1(x) = \frac{\sqrt{3}x}{\sqrt{2}},\ p_2(x) = \frac{\sqrt{2}}{2\sqrt{5}}(3x^2 - 1),\ \ldots$$

Higher order Legendre polynomials can be computed by $(m+1)p_{m+1}(x) = (2m+1)\,x\,p_m(x) - m\,p_{m-1}(x)$. □

Example 4.10.
Consider the wavelet basis $\{\psi_{jk}\}$ on $\mathbb{R}$ generated from a mother wavelet $\psi(\bullet)$. A wavelet ψ_{jk} can be computed by

$$\psi_{jk}(x) = 2^{j/2}\psi(2^j x - k),$$

where 2^j is a scale factor and k a location shift. A simple example of a mother wavelet is the Haar wavelet

$$\psi(x) = \begin{cases} -1 & x \in [0, 1/2] \\ 1 & x \in (1/2, 1] \\ 0 & \text{otherwise.} \end{cases}$$

which is a simple step function. □

The coefficients β_j can be calculated from

$$\beta_j = \sum_{k=0}^{\infty} \beta_k \delta_{jk} = \sum_{k=0}^{\infty} \beta_k \int \varphi_k(x)\varphi_j(x)\,dx = \int m(x)\varphi_j(x)dx \tag{4.37}$$

If we find a way to estimate the unknown function $m(x)$ in (4.37) then we will end up with an estimate of β_j. For Fourier series expansions and wavelets the β_j can be approximated using the fast Fourier and fast wavelet transform (FFT and FWT), respectively. See Härdle, Kerkyacharian, Picard & Tsybakov (1998) for more details.

Example 4.11.
Wavelets are particularly suited to fit regression functions that feature varying frequencies and jumps. Figure 4.9 shows the wavelet fit (Haar basis) for simulated data from a regression curve that combines a sine part with varying frequency and a constant part. To apply the fast wavelet transform, $n = 2^8 = 256$ data points were generated. □

4.3 Smoothing Parameter Selection

As we pointed out in the preceeding sections, for some nonparametric estimators at least an asymptotic connection can be made to kernel regression

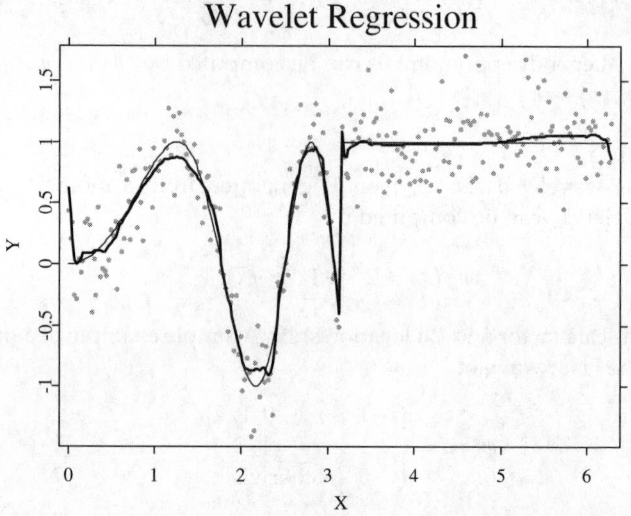

Figure 4.9. Wavelet regression and original curve for simulated data, $n = 256$ SPMwavereg

estimators. Hence, in this section we will be focusing on finding a good way of choosing the smoothing parameter of kernel regression estimators, namely the bandwidth h.

What conditions do we require for a bandwidth selection rule to be "good"? First of all it should have theoretically desirable properties. Secondly, it has to be applicable in practice. Regarding the first condition, there have been a number of criteria proposed that measure in one way or another how close the estimate is to the true curve. It will be instructive to go through these measures one by one:

- We are already familiar with the *mean squared error*

$$\mathrm{MSE}(x,h) = \mathrm{MSE}\{\widehat{m}_h(x)\} = E[\{\widehat{m}_h(x) - m(x)\}^2]. \tag{4.38}$$

  $m(x)$ is just an unknown constant, but the estimator $\widehat{m}_h(x)$ is a random variable. Hence, the expectation in (4.38) is taken with respect to the distribution of this random variable. Since $\widehat{m}_h(x)$ is a function of the random variables $\{X_i, Y_i\}_{i=1}^n$ it follows that

$$\mathrm{MSE}(x,h) = \int \cdots \int \{\widehat{m}_h(x) - m(x)\}^2 \cdot f(x_1,\ldots,x_n,y_1,\ldots,y_n)\,dx_1\cdots dx_n\,dy_1\cdots dy_n.$$

The MSE measures the squared deviation of the estimate $\widehat{m}_h$ from m at a single point x. If we are interested in how well we estimate an entire m then we should really use a global measure of closeness of the estimate to the true curve. Moreover, in the previous section, where we derived an approximate formula for the MSE, we have already mentioned that the AMSE-optimal bandwidth is a complicated function of unknown quantities like $m''(x)$, or $\sigma^2(x)$. One might argue that these unknowns may be replaced by consistent estimates but to get such an estimate for $m''(x)$, the choice of a bandwidth h is required, which is the very problem we are trying to solve.

- The *integrated squared error* ISE

$$\text{ISE}(h) = \text{ISE}\{\widehat{m}_h\} = \int_{-\infty}^{\infty} \{\widehat{m}_h(x) - m(x)\}^2 w(x) f_X(x)\, dx \tag{4.39}$$

is a global discrepancy measure. But it is still a random variable as different samples will produce different values of $\widehat{m}_h(x)$, and thereby different values of ISE. The weight function $w(\bullet)$ may be used to assign less weight to observations in regions of sparse data (to reduce the variance in this region) or at the tail of the distribution of X (to trim away boundary effects).

- The *mean integrated squared error* MISE

$$\begin{aligned} \text{MISE}(h) &= \text{MISE}\{\widehat{m}_h\} = E\{\text{ISE}(h)\} \\ &= \int \cdots \int \left[\int_{-\infty}^{\infty} \{\widehat{m}_h(x) - m(x)\}^2 w(x) f_X(x) dx \right] \\ &\qquad \cdot f(x_1, \ldots, x_n, y_1, \ldots, y_n)\, dx_1 \cdots dx_n\, dy_1 \cdots dy_n \end{aligned}$$

is not a random variable. It is the expected value of the random variable ISE with the expectation being taken with respect to all possible samples of X and Y.

- The *averaged squared error* ASE

$$\text{ASE}(h) = \text{ASE}\{\widehat{m}_h\} = \frac{1}{n} \sum_{j=1}^{n} \{\widehat{m}_h(X_j) - m(X_j)\}^2 w(X_j) \tag{4.40}$$

is a discrete approximation to ISE, and just like the ISE it is both a random variable and a global measure of discrepancy.

- The *mean averaged squared error* MASE

$$\text{MASE}(h) = \text{MASE}\{\widehat{m}_h\} = E\{\text{ASE}(h) | X_1 = x_1, \ldots, X_n = x_n\} \tag{4.41}$$

is the conditional expectation of ASE, where the expectation is taken w.r.t. the joint distribution of $Y_1, Y_2, \ldots, Y_n$. If we view the $X_1, \ldots, X_n$ as random variables then MASE is a random variable.

Which discrepancy measure should be used to derive a rule for choosing h? A natural choice would be MISE or its asymptotic version AMISE since we have some experience of its optimization from the density case. The AMISE in the regression case, however, involves more unknown quantities than the AMISE in the density estimator. As a result, plug-in approaches are mainly used for the local linear estimator due to its simpler bias formula. See for instance Wand & Jones (1995, pp. 138–139) for some examples.

We will discuss two approaches of rather general applicability: cross-validation and penalty terms. For the sake of simplicity, we restrict ourselves to bandwidth selection for the Nadaraya-Watson estimator here. For that estimator is has been shown (Marron & Härdle, 1986) that ASE, ISE and MISE lead asymptotically to the same level of smoothing. Hence, we can use the criterion which is the easiest to calculate and manipulate: the discrete $\text{ASE}(h)$.

4.3.1 A Closer Look at the Averaged Squared Error

We want to find the bandwidth h that minimizes $\text{ASE}(h)$. For easy reference, let us write down $\text{ASE}(h)$ in more detail:

$$\begin{aligned}\text{ASE}(h) = & \frac{1}{n}\sum_{i=1}^{n} m^2(X_i)w(X_i) + \frac{1}{n}\sum_{i=1}^{n} \widehat{m}_h^2(X_i)w(X_i) \\ & -2\frac{1}{n}\sum_{i=1}^{n} m(X_i)\widehat{m}_h(X_i)w(X_i).\end{aligned} \tag{4.42}$$

We already pointed out that ASE is a random variable. Its conditional expectation, MASE, is given by

$$\begin{aligned}\text{MASE}(h) &= E\{\text{ASE}(h)|X_1 = x_1,\ldots,X_n = x_n\} \qquad (4.43)\\ &= \frac{1}{n}\sum_{i=1}^{n} E[\{\widehat{m}_h(X_i) - m(X_i)\}^2\,|X_1 = x_1,\ldots,X_n = x_n]\,w(X_i)\\ &= \frac{1}{n}\sum_{i=1}^{n}\Big[\underbrace{Var\{\widehat{m}_h(X_i)|X_1 = x_1,\ldots,X_n = x_n\}}_{v(h)}\\ &\qquad + \underbrace{\text{Bias}^2\{\widehat{m}_h(X_i)|X_1 = x_1,\ldots,X_n = x_n\}}_{b^2(h)}\Big]\,w(X_i),\end{aligned}$$

with squared bias

$$b^2(h) = \frac{1}{n}\sum_{i=1}^{n}\left\{\frac{1}{n}\sum_{j=1}^{n}\frac{K_h(X_i - X_j)}{\widehat{f}_h(X_i)}m(X_j) - m(X_i)\right\}^2 w(X_i) \tag{4.44}$$

and variance

$$v(h) = \frac{1}{n}\sum_{i=1}^{n}\left[\frac{1}{n^2}\sum_{j=1}^{n}\left\{\frac{K_h(X_i - X_j)}{\widehat{f}_h(X_i)}\right\}^2 \sigma(X_j)^2\right] w(X_i). \qquad (4.45)$$

The following example shows the dependence of squared bias, variance and its sum MSE on the bandwidth h.

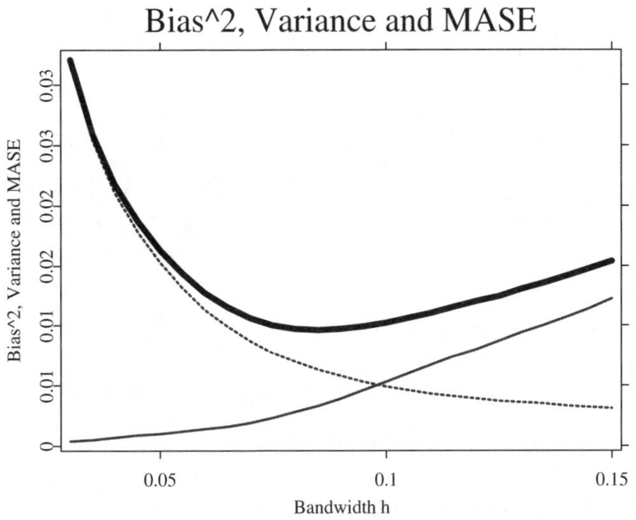

Figure 4.10. MASE (thick line), squared bias (thin solid line) and variance part (thin dashed line) for simulated data, weights $w(x) = I(x \in [0.05, 0.95])$ SPMsimulmase

Example 4.12.
The squared bias is increasing in h as can be seen in Figure 4.10 where $b^2(h)$ is plotted along with the decreasing $v(h)$ and their sum MASE (thick line). Apparently, there is the familiar trade-off that increasing h will reduce the variance but increase the squared bias. The minimum MASE is achieved at $h = 0.085$.

You may wonder how we are able to compute these quantities since they involve the unknown $m(\bullet)$. The answer is simple: We have generated the data ourselves, determining the regression function

$$m(x) = \{\sin(2\pi x^3)\}^3$$

beforehand. The data have been generated according to

$$Y_i = m(X_i) + \varepsilon_i, \quad X_i \sim U[0,1], \quad \varepsilon_i \sim N(0, 0.01),$$

see Figure 4.11. □

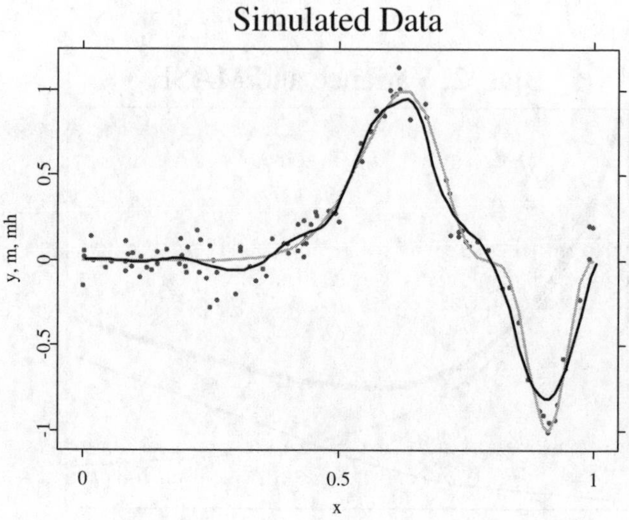

Figure 4.11. Simulated data with true and estimated curve SPMsimulmase

What is true for MASE is also true for ASE(h): it involves $m(\bullet)$, the function we want to estimate. Therefore, we have to replace ASE(h) with an approximation that can be computed from the data. A naive way of replacing $m(\bullet)$ would be to use the observations of Y instead, i.e.

$$p(h) = \frac{1}{n} \sum_{i=1}^{n} \{Y_i - \widehat{m}_h(X_i)\}^2 \, w(X_i), \tag{4.46}$$

which is called the *resubstitution estimate* and is essentially a weighted residual sum of squares (RSS). However, there is a problem with this approach since Y_i is used in $\widehat{m}_h(X_i)$ to predict itself. As a consequence, $p(h)$ can be made arbitrarily small by letting $h \to 0$ (in which case $\widehat{m}(\bullet)$ is an interpolation of the Y_is).

To gain more insight into this matter let us expand $p(h)$ by adding and subtracting $m(X_i)$:

$$
\begin{aligned}
p(h) &= \frac{1}{n}\sum_{i=1}^{n} [\{Y_i - m(X_i)\} + \{m(X_i) - \widehat{m}_h(X_i)\}]^2 \, w(X_i) \\
&= \frac{1}{n}\sum_{i=1}^{n} \varepsilon_i^2 w(X_i) + \mathrm{ASE}(h) \\
&\qquad - \frac{2}{n}\sum_{i=1}^{n} \varepsilon_i \{\widehat{m}_h(X_i) - m(X_i)\} w(X_i),
\end{aligned} \tag{4.47}
$$

where $\varepsilon_i = Y_i - m(X_i)$. Note that the first term $\frac{1}{n}\sum_{i=1}^{n} \varepsilon_i^2 w(X_i)$ of (4.47) does not depend on h, and the second term is $\mathrm{ASE}(h)$. Hence, minimizing $p(h)$ would surely lead to the same result as minimizing $\mathrm{ASE}(h)$ if it weren't for the third term $\sum_{i=1}^{n} \varepsilon_i \{\widehat{m}_h(X_i) - m(X_i)\} w(X_i)$. In fact, if we calculate the conditional expectation of $p(h)$

$$
\begin{aligned}
& E\{p(h) | X_1 = x_1, \ldots, X_n = x_n\} \\
&= \frac{1}{n}\sum_{i=1}^{n} \sigma^2(x_i) w(x_i) + E\{\mathrm{ASE}(h) | \, X_1 = x_1, \ldots, X_n = x_n\} \\
&\qquad - \frac{2}{n^2}\sum_{i=1}^{n} W_{hi}(x_i) \sigma^2(x_i) w(x_i)
\end{aligned} \tag{4.48}
$$

we observe that the third term (recall the definition of W_{hi} in (4.7)), which is the conditional expectation of

$$
-\frac{2}{n}\sum_{i=1}^{n} \varepsilon_i \{\widehat{m}_h(X_i) - m(X_i)\} w(X_i),
$$

tends to zero at the same rate as the variance $v(h)$ in (4.45) and has a negative sign. Therefore, $p(h)$ is downwardly biased as an estimate of $\mathrm{ASE}(h)$, just as the bandwidth minimizing $p(h)$ is downwardly biased as an estimate of the bandwidth minimizing $\mathrm{ASE}(h)$.

In the following two sections we will examine two ways out of this dilemma. The method of *cross-validation* replaces $\widehat{m}_h(X_i)$ in (4.46) with the leave-one-out-estimator $\widehat{m}_{h,-i}(X_i)$. In a different approach $p(h)$ is multiplied by a penalizing function which corrects for the downward bias of the resubstitution estimate.

4.3.2 Cross-Validation

We already familiarized ourselves with know cross-validation in the context of bandwidth selection in kernel density estimation. This time around, we will use it as a remedy for the problem that in

$$
p(h) = \frac{1}{n}\sum_{i=1}^{n} \{Y_i - \widehat{m}_h(X_i)\}^2 \, w(X_i) \tag{4.49}
$$

Y_i is used in $\widehat{m}_h(X_i)$ to predict itself. Cross-validation solves this problem by employing the *leave-one-out*-estimator

$$\widehat{m}_{h,-i}(X_i) = \frac{\sum_{j \neq i} K_h(X_i - X_j)Y_j}{\sum_{j \neq i} K_h(X_i - X_j)}. \tag{4.50}$$

That is, in estimating $\widehat{m}_h(\bullet)$ at X_i the ith observation is left out (as reflected in the subscript "$-i$"). This leads to the *cross-validation function*

$$CV(h) = \frac{1}{n}\sum_{i=1}^{n}\left\{Y_i - \widehat{m}_{h,-i}(X_i)\right\}^2 w(X_i). \tag{4.51}$$

In terms of the analysis of the previous section, it can be shown that the conditional expectation of the third term of (4.47), is equal to zero if we use $\widehat{m}_{h,-i}(X_i)$ instead of $\widehat{m}_h(X_i)$, i.e.

$$E\left[-\frac{2}{n}\sum_{i=1}^{n}\varepsilon_i\{\widehat{m}_{h,-i}(X_i) - m(X_i)\}w(X_i)\Big| X_1 = x_1, \ldots, X_n = x_n\right] = 0.$$

This means minimizing $CV(h)$ is (on average) equivalent to minimizing ASE(h) since the first term in (4.47) is independent of h. We can conclude that with the bandwidth selection rule "choose $\widehat{h}$ to minimize $CV(h)$" we have found a rule that is both theoretically desirable and applicable in practice.

Example 4.13.
Let us apply the cross-validation method to the Engel curve example now. Figure 4.12 shows the Nadaraya-Watson kernel regression curve (recall that we always used the Quartic kernel for the figures) with the bandwidth chosen by minimizing the cross-validation criterion $CV(h)$.

For comparison purposes, let us consider bandwidth selection for a different nonparametric smoothing method. You can easily see that applying the cross-validation approach to local polynomial regression presents no problem. This is what we have done in Figure 4.13. Here we show the local linear estimate with cross-validated bandwidth for the same data. As we already pointed out in Subsection 4.1.3 the estimate shows more stable behavior in the high net-income region (regions with small number of observations) and outperforms the Nadaraya-Watson estimate at the boundaries. □

4.3.3 Penalizing Functions

Recall the formula (4.48)for the conditional expectation of $p(h)$. That is,

$$E\left\{p(h)|X_1 = x_1, \ldots, X_n = x_n\right\} \neq E\{\text{ASE}(h)|X_1 = x_1, \ldots, X_n = x_n\}.$$

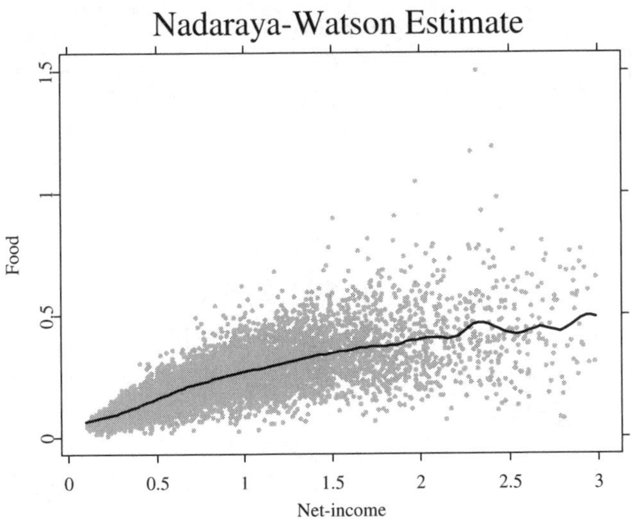

Figure 4.12. Nadaraya-Watson kernel regression with cross-validated bandwidth $\widehat{h}_{CV} = 0.15$, U.K. Family Expenditure Survey 1973 SPMnadwaest

You might argue that this inequality is not all that important as long as the bandwidth minimizing $E\{p(h)|X_1 = x_1, \ldots, X_n = x_n\}$ is equal to the bandwidth minimizing $E\{\text{ASE}(h)|X_1 = x_1, \ldots, X_n = x_n\}$. Unfortunately, one of the two terms causing the inequality, the last term of (4.48), depends on h and is causing the downward bias. The *penalizing function* approach corrects for the downward bias by multiplying $p(h)$ by a correction factor that penalizes too small h. The "corrected version" of $p(h)$ can be written as

$$G(h) = \frac{1}{n}\sum_{i=1}^{n}\{Y_i - \widehat{m}_h(X_i)\}^2 \, \Xi\left(\frac{1}{n}W_{hi}(X_i)\right) w(X_i), \tag{4.52}$$

with a correction function Ξ. As we will see in a moment, a penalizing function $\Xi(u)$ with first-order Taylor expansion $\Xi(u) = 1 + 2u + O(u^2)$ for $u \to 0$, will work well. Using this Taylor expansion we can write (4.52) as

$$\begin{aligned} G(h) = \frac{1}{n}\sum_{i=1}^{n}&\left[\varepsilon_i^2 + \{m(X_i) - \widehat{m}_h(X_i)\}^2 - 2\varepsilon_i\{m(X_i) - \widehat{m}_h(X_i)\}\right] \\ &\cdot\left\{1 + \frac{2}{n}W_{hi}(X_i)\right\} w(X_i) + O\left((nh)^{-2}\right). \end{aligned} \tag{4.53}$$

Multiplying out and ignoring terms of higher order, we get

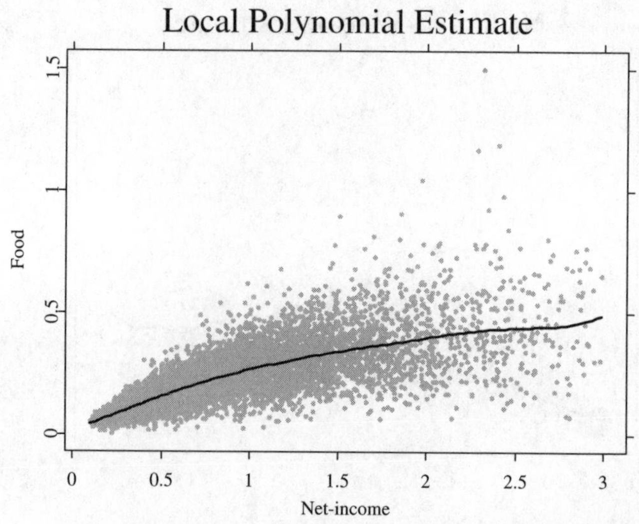

Figure 4.13. Local polynomial regression ($p = 1$) with cross-validated bandwidth $\widehat{h}_{CV} = 0.56$, U.K. Family Expenditure Survey 1973 `SPMlocpolyest`

$$G(h) \approx \frac{1}{n}\sum_{i=1}^{n} \varepsilon_i^2 w(X_i) + \text{ASE}(h) - \frac{2}{n}\sum_{i=1}^{n} \varepsilon_i \left\{\widehat{m}(X_i) - m_h(X_i)\right\} w(X_i)$$
$$+ \frac{2}{n^2}\sum_{i=1}^{n} \varepsilon_i^2 W_{hi}(X_i) w(X_i). \tag{4.54}$$

The first term in (4.54) does not depend on h. The expectation of the third term, conditional on $X_1 = x_1, \ldots, X_n = x_n$, is equal to the negative value of the last term of (4.48). But this is just the conditional expectation of the last term in (4.54), with a negative sign in front. Hence, the last two terms cancel each other out asymptotically and $G(h)$ is roughly equal to $\text{ASE}(h)$.

The following list presents a number of penalizing functions that satisfy the expansion $\Xi(u) = 1 + 2u + O(u^2), \quad u \to 0$:

(1) *Shibata's model selector* (Shibata, 1981),

$$\Xi_S(u) = 1 + 2u;$$

(2) *Generalized cross-validation* (Craven and Wahba, 1979; Li, 1985),

$$\Xi_{GCV}(u) = (1 - u)^{-2};$$

(3) *Akaike's Information Criterion* (Akaike, 1970),

$$\Xi_{AIC}(u) = \exp(2u);$$

(4) *Finite Prediction Error* (Akaike, 1974),

$$\Xi_{FPE}(u) = (1+u)/(1-u);$$

(5) *Rice's T* (Rice, 1984),

$$\Xi_T(u) = (1-2u)^{-1}.$$

To see how these various functions differ in the degree of penalizing small values of h, consider Figure 4.14.

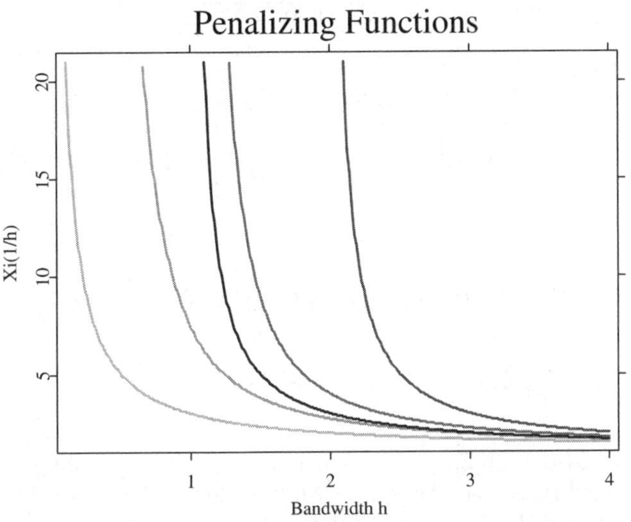

Figure 4.14. Penalizing functions $\Xi(h^{-1})$ as a function of h (from left to right: S, AIC, FPE, GCV, T) `SPMpenalize`

The functions differ in the relative weight they give to variance and bias of $\widehat{m}_h(x)$. Rice's T gives the most weight to variance reduction while Shibata's model selector stresses bias reduction the most. The differences displayed in the graph are not substantial, however. If we denote the bandwidth minimizing $G(h)$ with $\widehat{h}$ and the minimizer of ASE(h) with $\widehat{h}_0$ then for $n \to \infty$

$$\frac{\text{ASE}(\widehat{h})}{\text{ASE}(\widehat{h}_0)} \xrightarrow{P} 1 \quad \text{and} \quad \frac{\widehat{h}}{\widehat{h}_0} \xrightarrow{P} 1.$$

Thus, regardless of which specific penalizing function we use, we can assume that with an increasing number of observations $\widehat{h}$ approximates the ASE minimizing bandwidth $\widehat{h}_0$. Hence, choosing the bandwidth minimizing $G(h)$ is another "good" rule for bandwidth-selection in kernel regression estimation.

Note that

$$\begin{aligned} CV(h) &= \frac{1}{n}\sum_{i=1}^{n}\{Y_i - \widehat{m}_{h,-i}(X_i)\}^2 w(X_i) \\ &= \frac{1}{n}\sum_{i=1}^{n}\{Y_i - \widehat{m}_h(X_i)\}^2 \left\{\frac{Y_i - \widehat{m}_{h,-i}(X_i)}{Y_i - \widehat{m}_h(X_i)}\right\}^2 w(X_i) \end{aligned}$$

and

$$\begin{aligned} \frac{Y_i - \widehat{m}_h(X_i)}{Y_i - \widehat{m}_{h,-i}(X_i)} &= \frac{\sum_j K_h(X_i - X_j)Y_j - Y_i \sum_j K_h(X_i - X_j)}{\sum_{j\neq i} K_h(X_i - X_j)Y_j - Y_i \sum_{j\neq i} K_h(X_i - X_j)} \\ &\quad \cdot \frac{\sum_{j\neq i} K_h(X_i - X_j)}{\sum_j K_h(X_i - X_j)} \\ &= 1 \cdot \left\{1 - \frac{K_h(0)}{\sum_j K_h(X_i - X_j)}\right\} = 1 - \frac{1}{n} W_{hi}(X_i). \end{aligned}$$

Hence

$$\begin{aligned} CV(h) &= \frac{1}{n}\sum_{i=1}^{n}\{Y_i - \widehat{m}_h(X_i)\}^2 \left\{1 - \frac{1}{n}W_{hi}(X_i)\right\}^{-2} w(X_i) \\ &= \frac{1}{n}\sum_{i=1}^{n}\{Y_i - \widehat{m}_h(X_i)\}^2\, \Xi_{GCV}\left(\frac{1}{n}W_{hi}(X_i)\right) w(X_i) \end{aligned}$$

i.e. $CV(h) = G(h)$ with Ξ_{GCV}. An analogous result is possible for local polynomial regression, see Härdle & Müller (2000). Therefore the cross-validation approach is equivalent to the penalizing functions concept and has the same asymptotic properties. (Note, that this equivalence does not hold in general for other smoothing approaches.)

4.4 Confidence Regions and Tests

As in the case of density estimation, confidence intervals and bands can be based on the asymptotic normal distribution of the regression estimator. We will restrict ourselves to the Nadaraya-Watson case in order to show the essential concepts. In the latter part of this section we address the related topic of specification tests, which test the hypothesis of a parametric against the alternative nonparametric regression function.

4.4.1 Pointwise Confidence Intervals

Now that you have become familiar with nonparametric regression, you may want to know: How close *is* the smoothed curve to the true curve? Recall that we asked the same question when we introduced the method of kernel *density* estimation. There, we made use of (pointwise) confidence intervals and (global) confidence bands. But to construct this measure, we first had to derive the (asymptotic) sampling distribution.

The following theorem establishes the asymptotic distribution of the Nadaraya-Watson kernel estimator for one-dimensional predictor variables.

Theorem 4.5.
Suppose that m and f_X are twice differentiable, $\int |K(u)|^{2+\kappa}\,du < \infty$ for some $\kappa > 0$, x is a continuity point of $\sigma^2(x)$ and $E(|Y|^{2+\kappa}|X = x)$ and $f_X(x) > 0$. Take $h = cn^{-1/5}$. Then

$$n^{2/5}\left\{\widehat{m}_h(x) - m(x)\right\} \xrightarrow{L} N\left(b_x, v_x^2\right)$$

with

$$b_x = c^2\mu_2(K)\left\{\frac{m''(x)}{2} + \frac{m'(x)f_X'(x)}{f_X(x)}\right\},\ v_x^2 = \frac{\sigma^2(x)\|K\|_2^2}{cf_X(x)}.$$

The asymptotic bias b_x is proportional to the second moment of the kernel and a measure of local curvature of m. This measure of local curvature is not a function of m alone but also of the marginal density. At maxima or minima, the bias is a multiple of $m''(x)$ alone; at inflection points it is just a multiple of $\{m'(x)f_X'(x)\}/f_X(x)$ only.

We now use this result to define confidence intervals. Suppose that the bias is of negligible magnitude compared to the variance, e.g. if the bandwidth h is sufficiently small. Then we can compute *approximate confidence intervals* with the following formula:

$$\left[\widehat{m}_h(x) - z_{1-\frac{\alpha}{2}}\sqrt{\frac{\|K\|_2\widehat{\sigma}^2(x)}{nh\widehat{f}_h(x)}},\ \widehat{m}_h(x) + z_{1-\frac{\alpha}{2}}\sqrt{\frac{\|K\|_2\widehat{\sigma}^2(x)}{nh\widehat{f}_h(x)}}\right] \tag{4.55}$$

where $z_{1-\frac{\alpha}{2}}$ is the $(1-\frac{\alpha}{2})$-quantile of the standard normal distribution and the estimate of the variance $\sigma^2(x)$ is given by

$$\widehat{\sigma}^2(x) = \frac{1}{n}\sum_{i=1}^{n} W_{hi}(x)\{Y_i - \widehat{m}_h(x)\}^2,$$

with W_{hi} the weights from The Nadaraya-Watson estimator.

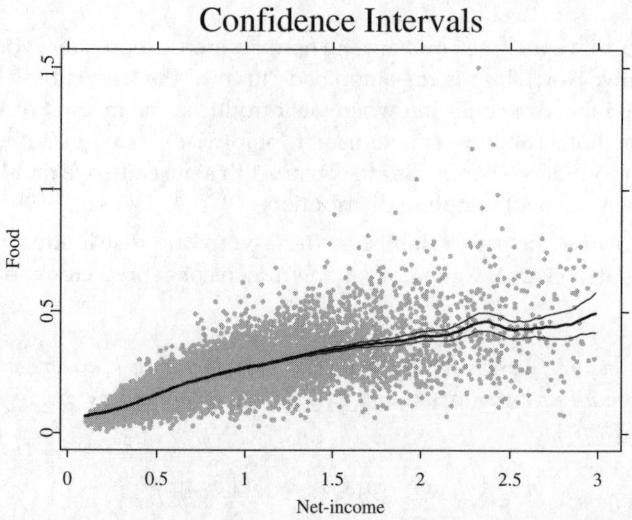

Figure 4.15. Nadaraya-Watson kernel regression and 95% confidence intervals, $h = 0.2$, U.K. Family Expenditure Survey 1973 SPMengelconf

Example 4.14.
Figure 4.15 shows the Engel curve from the 1973 U.K. net-income versus food example with confidence intervals. As we can see, the bump in the right part of the regression curve is not significant at 5% level. □

4.4.2 Confidence Bands

As we have seen in the density case, *uniform confidence bands* for $m(\bullet)$ need rather restrictive assumptions. The derivation of uniform confidence bands is again based on Bickel & Rosenblatt (1973).

Theorem 4.6.
Suppose that the support of X is $[0,1]$, $f_X(x) > 0$ on $[0,1]$, and that $m(\bullet)$, $f_X(\bullet)$ and $\sigma(\bullet)$ are twice differentiable. Moreover, assume that K is differentiable with support $[-1,1]$ with $K(-1) = K(1) = 0$, $E(|Y|^k|X = x)$ is bounded for all k. Then for $h_n = n^{-\kappa}$, $\kappa \in (\frac{1}{5}, \frac{1}{2})$

$$P\left(\text{for all } x \in [0,1] : \widehat{m}_h(x) - z_{n,\alpha} \sqrt{\frac{\widehat{\sigma}_h^2(x)\|K\|_2^2}{nh\widehat{f}_h(x)}} \right.$$
$$\left. \le m(x) \le \widehat{m}_h(x) + z_{n,\alpha} \sqrt{\frac{\widehat{\sigma}_h^2(x)\|K\|_2^2}{nh\widehat{f}_h(x)}} \right) \longrightarrow 1-\alpha,$$

where

$$z_{n,\alpha} = \left\{ \frac{-\log\{-\frac{1}{2}\log(1-\alpha)\}}{(2\kappa \log n)^{1/2}} + d_n \right\}^{1/2},$$

$$d_n = (2\kappa \log n)^{1/2} + (2\kappa \log n)^{-1/2} \log \left(\frac{1}{2\pi} \frac{\|K'\|_2}{\|K\|_2} \right)^{1/2}.$$

In practice, the data $X_1, \ldots, X_n$ are transformed to the interval $[0,1]$, then the confidence bands are computed and rescaled to the original scale of $X_1, \ldots, X_n$.

The following comprehensive example covers local polynomial kernel regression as well as optimal smoothing parameter selection and confidence bands.

Example 4.15.
The behavior of foreign exchange (FX) rates has been the subject of many recent investigations. A correct understanding of the foreign exchange rate dynamics has important implications for international asset pricing theories, the pricing of contingent claims and policy-oriented questions.

In the past, one of the most important exchange rates was that of Deutsche Mark (DM) to US Dollar (USD). The data that we consider here are from Olsen & Associates, Zürich. They contains the following numbers of quotes during the period Oct 1 1992 and Sept 30 1993. The data have been transformed as described in Bossaerts, Hafner & Härdle (1996).

We present now the regression smoothing approach with local linear estimation of the conditional mean (*mean function*) and the conditional variance (*variance function*) of the FX returns

$$Y_t = \log(S_t / S_{t-1}),$$

with S_t being the FX rates. An extension of the *autoregressive conditional heteroscedasticity* model (ARCH model) is the *conditional heteroscedastic autoregressive nonlinear* model (CHARN model)

$$Y_t = m(Y_{t-1}) + \sigma(Y_{t-1})\xi_t. \tag{4.56}$$

The task is to estimate the mean function $m(x) = E(Y_t | Y_{t-1} = x)$ and the variance function $\sigma^2(x) = E(Y_t^2 | Y_{t-1} = x) - E^2(Y_t | Y_{t-1} = x)$. As already

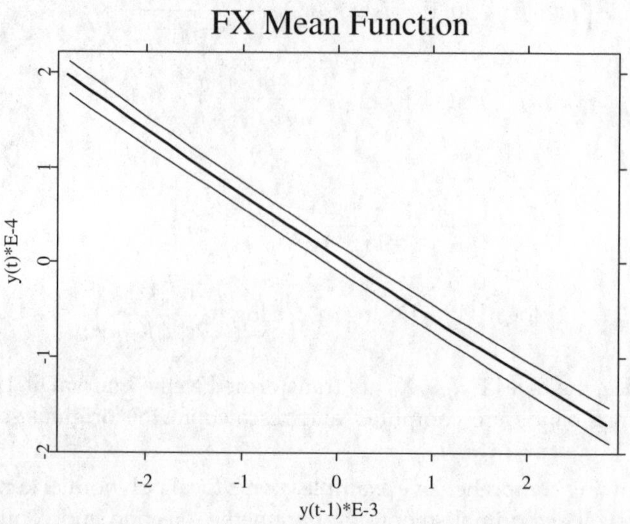

Figure 4.16. The estimated mean function for DM/USD with uniform confidence bands, shown is only the truncated range $(-0.003, 0.003)$ SPMfxmean

mentioned we use local linear estimation here. For details of assumptions and asymptotics of the local polynomial procedure in time series see Härdle & Tsybakov (1997). Here, local linear estimation means to compute the following weighted least squares problems

$$\widehat{\beta}(x) = \arg\min_{\beta} \sum_{t=1}^{n} \{Y_t - \beta_0 - \beta_1(Y_{t-1} - x)\}^2 K_h(Y_{t-1} - x)$$

$$\widehat{\gamma}(x) = \arg\min_{\gamma} \sum_{t=1}^{n} \{Y_t^2 - \gamma_0 - \gamma_1(Y_{t-1} - x)\}^2 K_h(Y_{t-1} - x).$$

Denoting the true regression function of $E(Y^2|Y_{t-1})$

$$s(x) = E(Y_t^2|Y_{t-1} = x),$$

then the estimators of $m(x)$ and $s(x)$ are the first elements of the vectors $\widehat{\beta}(x)$ and $\widehat{\gamma}(x)$, respectively. Consequently, a possible variance estimate is

$$\widehat{\sigma}^2(x) = \widehat{s}_{1,h}(x) - \widehat{m}_{1,h}^2(x),$$

with

$$\widehat{m}_{1,h}(x) = e_0^\top \widehat{\beta}(x), \quad \widehat{s}_{1,h}(x) = e_0^\top \widehat{\gamma}(x)$$

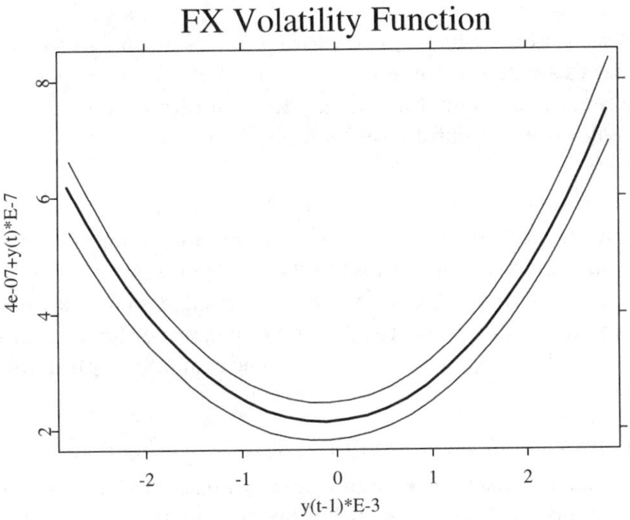

Figure 4.17. The estimated variance function for DM/USD with uniform confidence bands, shown is only the truncated range $(-0.003, 0.003)$ SPMfxvolatility

and $e_0 = (1, 0)^\top$ the first unit vector in $\mathbb{R}^2$.

The estimated functions are plotted together with approximate 95% confidence bands, which can be obtained from the asymptotic normal distribution of the local polynomial estimator. The cross-validation optimal bandwidth $h = 0.00914$ is used for the local linear estimation of the mean function in Figure 4.16. As indicated by the 95% confidence bands, the estimation is not very robust at the boundaries. Therefore, Figure 4.16 covers a truncated range. Analogously, the variance estimate is shown in Figure 4.17, using the cross-validation optimal bandwidth $h = 0.00756$.

The basic results are the mean reversion and the "smiling" shape of the conditional variance. Conditional heteroscedasticity appears to be very distinct. For DM/USD a "reverted leverage effect" can be observed, meaning that the conditional variance is higher for positive lagged returns than for negative ones of the same size. But note that the difference is still within the 95% confidence bands. □

4.4.3 Hypothesis Testing

In this book we will treat the topic of testing not as a topic of its own, being aware that this would be an enormous task. Instead, we concentrate on cases where regression estimators have a direct application in specification testing. We will only concentrate on methodology and skip any discussion about efficiency.

As this is the first section where we deal with testing, let us start with some brief, but general considerations about non- and semiparametric testing. Firstly, you should free your mind of the facts that you know about testing in the parametric world. No parameter is estimated so far, consequently it cannot be the target of interest to test for significance or linear restrictions of the parameters. Looking at our nonparametric estimates typical questions that may arise are:

- Is there indeed an impact of X on Y?
- Is the estimated function $m(\bullet)$ significantly different from the traditional parameterization (e.g. the linear or log-linear model)?

Secondly, in contrast to parametric regression, with non- and semiparametrics the problems of estimation and testing are not equivalent anymore. We speak here of equivalence in the sense that, in the parametric world, interval estimation corresponds to parameter testing. It turns out that the optimal rates of convergence are different for nonparametric estimation and nonparametric testing. As a consequence, the choice of smoothing parameter is an issue to be discussed separately in both cases. Moreover, the optimality discussion for nonparametric testing is in general quite a controversial one and far from being obvious. This unfortunately concerns all aspects of nonparametric testing. For instance, the construction of confidence bands around a nonparametric function estimate to decide whether it is significantly different from being linear, can lead to a much too conservative and thus inefficient testing procedure.

Let us now turn to the fundamentals of nonparametric testing. Indeed, the appropriateness of a parametric model may be judged by comparing the parametric fit with a nonparametric estimator. This can be done in various ways, e.g. you may use a (weighted) squared deviation between the two models. A simple (but in many situations inefficient) approach would be to use critical values from the asymptotic distribution of this statistic. Better results are usually obtained by approximating the distribution of the test statistics using a resampling method.

Before introducing a specific test statistic we have to specify the hypothesis H_0 and the alternative H_1. To make it easy let us start with a nonparametric regression $m(x) = E(Y|X = x)$. Our first null hypothesis is that X has no impact on Y. If we assume $EY = 0$ (otherwise take $\widetilde{Y}_i = Y_i - \overline{Y}$), then we may be interested to test

$$H_0 : m(x) \equiv 0 \text{ vs. } H_1 : m(x) \neq 0.$$

As throughout this chapter we do not want to make any assumptions about the function $m(\bullet)$ other than smoothness conditions. Having an estimate of $m(x)$ at hand, e.g. the Nadaraya-Watson estimate $\widehat{m}(\bullet) = \widehat{m}_h(\bullet)$ from (4.6), a natural measure for the deviation from zero is

$$T_1 = n\sqrt{h} \int \{\widehat{m}(x) - 0\}^2 \, \widetilde{w}(x) \, dx, \tag{4.57}$$

where $\widetilde{w}(x)$ denotes a weight function (typically chosen by the empirical researcher). This weight function often serves to trim the boundaries or regions of sparse data. If the weight function is equal to $f_X(x)w(x)$, i.e.

$$\widetilde{w}(x) = f_X(x)w(x)$$

with $f_X(x)$ being the density of X and $w(x)$ another weight function, one could take the empirical version of (4.57)

$$T_2 = \sqrt{h} \sum_{i=1}^{n} \{\widehat{m}(x) - 0\}^2 \, w(x) \tag{4.58}$$

as a test statistic.

It is clear that under H_0 both test statistics T_1 and T_2 must converge to zero, whereas under H_1 the condition $n \to \infty$ also lets the statistic increase to infinity. Note that under H_0 our estimate $\widehat{m}(\bullet)$ does not have any bias (cf. Theorem 4.3) that could matter in the squared deviation $\{\widehat{m}(x) - 0\}^2$. Actually, with the same assumptions we needed for the kernel estimator $\widehat{m}(\bullet)$, we find that under the null hypothesis T_1 and T_2 converge to a $N(0, V)$ distribution with

$$V = 2 \int \frac{\sigma^4(x)\widetilde{w}^2(x)}{f_X^2(x)} \, dx \int (K \star K)^2(x) \, dx. \tag{4.59}$$

As used previously, $\sigma^2(x)$ denotes the conditional variance $\text{Var}(Y|X = x)$.

Let us now consider the more general null hypothesis. Suppose we are interested in a specific parametric model given by $E(Y|X = x) = m_\theta(\bullet)$ and $m_\theta(\bullet)$ is a (parametric) function, known up to the parameter θ. This means

$$H_0 : m(x) \equiv m_\theta(x) \text{ vs. } H_1 : m(x) \neq m_\theta(x).$$

A consistent estimator $\widehat{\theta}$ for θ is usually easy to obtain (by least squares, maximum likelihood, or as a moment estimator, for example). The analog to statistic (4.58) is then obtained by using the deviation from $m_{\widehat{\theta}}$, i.e.

$$\sqrt{h}\sum_{i=1}^{n}\left\{\widehat{m}(x) - m_{\widehat{\theta}}(x)\right\}^2 w(x). \tag{4.60}$$

However, this test statistic involves the following problem: Whereas $m_{\widehat{\theta}}(\bullet)$ is (asymptotically) unbiased and converging at rate $\sqrt{n}$, our nonparametric estimate $\widehat{m}(x)$ has a "kernel smoothing" bias and converges at rate $\sqrt{nh}$. For that reason, Härdle & Mammen (1993) propose to introduce an artificial bias by replacing $m_{\widehat{\theta}}(\bullet)$ with

$$\widehat{m}_{\widehat{\theta}}(x) = \frac{\sum_{i=1}^{n} K_h(x - X_i)\, m_{\widehat{\theta}}(X_i)}{\sum_{i=1}^{n} K_h(x - X_i)} \tag{4.61}$$

in statistic (4.60). More specifically, we use

$$T = \sqrt{h}\sum_{i=1}^{n}\left\{\widehat{m}(x) - \widehat{m}_{\widehat{\theta}}(x)\right\}^2 w(x). \tag{4.62}$$

As a result of this, under H_0 the bias of $\widehat{m}(\bullet)$ cancels out that of $\widehat{m}_{\widehat{\theta}_0}$ and the convergence rates are also the same.

Example 4.16.
Consider the expected wage (Y) as a function of years of professional experience (X). The common parameterization for this relationship is

$$H_0 : E(Y|X = x) = m(x) = \beta_0 + \beta_1 x + \beta_2 x^2,$$

and we are interested in verifying this quadratic form. So, we firstly estimate $\theta = (\beta_0, \beta_1, \beta_2)^\top$ by least squares and set $m_{\widehat{\theta}}(x) = \widehat{\beta}_0 + \widehat{\beta}_1 x + \widehat{\beta}_2 x^2$. Secondly, we calculate the kernel estimates $\widehat{m}(\bullet)$ and $\widehat{m}_{\widehat{\theta}}(\bullet)$, as in (4.6) and (4.61). Finally, we apply test statistic (4.62) to these two smoothers. If the statistic is "large", we reject H_0. □

The remaining question of the example is: How to find the critical value for "large"? The typical approach in parametric statistics is to obtain the critical value from the asymptotic distribution. This is principally possible in our nonparametric problem as well:

Theorem 4.7.
Assume the conditions of Theorem 4.3, further that $\widehat{\theta}$ is a $\sqrt{n}$-consistent estimator for θ and $h = O(n^{-1/5})$. Then it holds

$$T - \frac{1}{\sqrt{h}}\|K\|_2^2 \int \sigma^2(x) w(x)\, dx$$
$$\xrightarrow{L} N\left(0, 2\int \sigma^4(x) w^2(x)\, dx \int (K \star K)^2(x)\, dx\right)$$

As in the parametric case, we have to estimate the variance expression in the normal distribution. However, with an appropriate estimate for $\sigma^2(x)$ this is no obstacle. The main practical problem here is the very slow convergence of T towards the normal distribution.

For that reason, approximations of the critical values corresponding to the finite sample distribution are used. The most popular way to approximate this finite sample distribution is via a resampling scheme: simulate the distribution of your test statistic under the hypothesis (i.e. "resample") and determine the critical values based on that simulated distribution. This method is called Monte Carlo method or bootstrap, depending on how the distribution of the test statistic can be simulated. Depending on the context, different resampling procedures have to be applied. Later on, for each particular case we will introduce not only the test statistic but also an appropriate resampling method.

For our current testing problem the possibly most popular resampling method is the so-called *wild bootstrap* introduced by Wu (1986). One of its advantages is that it allows for a heterogeneous variance in the residuals. Härdle & Mammen (1993) introduced wild bootstrap into the context of nonparametric hypothesis testing as considered here. The principal idea is to resample from the residuals $\widehat{\varepsilon}_i$, $i = 1, \ldots, n$, that we got under the null hypothesis. Each bootstrap residual ε_i^* is drawn from a distribution that coincides with the distribution of $\widehat{\varepsilon}_i$ up to the first three moments. The testing procedure then consists of the following steps:

(a) Estimate the regression function $m_{\widehat{\theta}}(\bullet)$ under the null hypothesis and construct the residuals $\widehat{\varepsilon}_i = Y_i - m_{\widehat{\theta}}(X_i)$.

(b) For each X_i, draw a bootstrap residual ε_i^* so that

$$E(\varepsilon_i^*) = 0\,, \quad E(\varepsilon_i^{*2}) = \widehat{\varepsilon}_i^2, \quad \text{and} \quad E(\varepsilon_i^{*3}) = \widehat{\varepsilon}_i^3\,.$$

(c) Generate a bootstrap sample $\{(Y_i^*, X_i)\}_{i=1,\ldots,n}$ by setting

$$Y_i^* = m_{\widehat{\theta}}(X_i) + \varepsilon_i^*\,.$$

(d) From this sample, calculate the bootstrap test statistic T^* in the same way as the original T is calculated.

(e) Repeat steps (b) to (d) n_{boot} times (n_{boot} being several hundred or thousand) and use the n_{boot} generated test statistics T^* to determine the quantiles of the test statistic under the null hypothesis. This gives you approximative values for the critical values for your test statistic T.

One famous method which fulfills the conditions in step (c) is the so-called *golden cut method*. Here we draw ε_i^* from the two-point distribution with probability mass at

$$a = \frac{1-\sqrt{5}}{2}\,\widehat{\varepsilon}_i, \quad b = \frac{1+\sqrt{5}}{2}\,\widehat{\varepsilon}_i,$$

occurring with probabilities $q = (5+\sqrt{5})/10$ and $1-q$, respectively. In the second part of this book you will see more examples of Monte Carlo and bootstrap methods.

Let us mention that besides the type of test statistics that we introduced here, other distance measures are plausible. However, all test statistics can be considered as estimates of one of the following expressions:

$$E\left[w(X)\{m(X) - m_\theta(X)\}^2\right], \tag{4.63}$$
$$E\left[w(X)\{m(X) - m_\theta(X)\}\varepsilon_X\right], \tag{4.64}$$
$$E\left[w(X)\varepsilon_X\, E[\varepsilon_X|X]\right], \tag{4.65}$$
$$E\left[w(X)\{\sigma^2(X) - \sigma^2_\theta(X)\}\right], \tag{4.66}$$

ε_X being the residuum under H_0 at point x, and $w(x)$ the weight function as above. Furthermore, $\sigma^2_\theta(X)$ is the error variance under the hypothesis, and $\sigma^2(X)$ the one under the alternative. Obviously, our test statistics (4.58) and (4.62) are estimates of expression (4.63).

The question "Which is the best test statistic?" has no simple answer. An optimal test should keep the nominal significance level under the hypothesis and provide the highest power under the alternative. However, in practice it turns out that the behavior of a specific test may depend on the model, the error distribution, the design density and the weight function. This leads to an increasing number of proposals for the considered testing problem. We refer to the bibliographic notes for additional references to (4.63)–(4.66) and for further test approaches.

4.5 Multivariate Kernel Regression

In the previous section several techniques for estimating the conditional expectation function m of the bivariate distribution of the random variables Y and X were presented. Recall that the conditional expectation function is an interesting target for estimation since it tells us how Y and X are related on average. In practice, however, we will mostly be interested in specifying how the response variable Y depends on a *vector* of exogenous variables, denoted by $\mathbf{X}$. This means we aim to estimate the conditional expectation

$$E(Y|\mathbf{X}) = E\left(Y|X_1,\ldots,X_d\right) = m(\mathbf{X}), \tag{4.67}$$

where $\mathbf{X} = (X_1,\ldots,X_d)^\top$. Consider the relation

$$E(Y|\mathbf{X}) = \int y f(y|x)\, dy = \frac{\int y f(y,x)\, dy}{f_{\mathbf{X}}(x)}.$$

If we replace the multivariate density $f(y,x)$ by its kernel density estimate

$$\widehat{f}_{h,\mathbf{H}}(y,x) = \frac{1}{n}\sum_{i=1}^{n} K_h(Y_i - y)\, \mathcal{K}_{\mathbf{H}}(\mathbf{X}_i - x)$$

and $f_{\mathbf{X}}(x)$ by (3.60) we arrive at the multivariate generalization of the Nadaraya-Watson estimator:

$$\widehat{m}_{\mathbf{H}}(x) = \frac{\sum_{i=1}^{n} \mathcal{K}_{\mathbf{H}}\left(\mathbf{X}_i - x\right) Y_i}{\sum_{i=1}^{n} \mathcal{K}_{\mathbf{H}}\left(\mathbf{X}_i - x\right)}. \tag{4.68}$$

Hence, the multivariate kernel regression estimator is again a weighted sum of the observed responses Y_i. Depending on the choice of the kernel, $\widehat{m}_{\mathbf{H}}(x)$ is a weighted average of those Y_i where $\mathbf{X}_i$ lies in a ball or cube around x.

Note also, that the multivariate Nadaraya-Watson estimator is a local constant estimator. The definition of local polynomial kernel regression is a straightforward generalization of the univariate case. Let us illustrate this with the example of a local linear regression estimate. The minimization problem here is

$$\min_{\beta_0,\beta_1} \sum_{i=1}^{n} \left\{ Y_i - \beta_0 - \beta_1^\top (\mathbf{X}_i - x) \right\}^2 \mathcal{K}_{\mathbf{H}}(\mathbf{X}_i - x).$$

The solution to the problem can hence be equivalently written as

$$\widehat{\beta} = (\widehat{\beta}_0, \widehat{\beta}_1^\top)^\top = \left(\mathbf{X}^\top \mathbf{W} \mathbf{X}\right)^{-1} \mathbf{X}^\top \mathbf{W} \mathbf{Y} \tag{4.69}$$

using the notations

$$\mathbf{X} = \begin{pmatrix} 1 & (\mathbf{X}_1 - x)^\top \\ \vdots & \vdots \\ 1 & (\mathbf{X}_n - x)^\top \end{pmatrix}, \quad \mathbf{Y} = \begin{pmatrix} Y_1 \\ \vdots \\ Y_n \end{pmatrix},$$

and $\mathbf{W} = \operatorname{diag}\left(\mathcal{K}_{\mathbf{H}}(\mathbf{X}_1 - x), \ldots, \mathcal{K}_{\mathbf{H}}(\mathbf{X}_n - x)\right)$. In (4.69) $\widehat{\beta}_0$ estimates the regression function itself, whereas $\widehat{\beta}_1$ estimates the partial derivatives w.r.t. the components x. In the following we denote the multivariate local linear estimator as

$$\widehat{m}_{1,\mathbf{H}}(x) = \widehat{\beta}_0(x). \tag{4.70}$$

4.5.1 Statistical Properties

The asymptotic conditional variances of the Nadaraya-Watson estimator $\widehat{m}_{\mathbf{H}}$ and the local linear $\widehat{m}_{1,\mathbf{H}}$ are identical and their derivation can be found in detail in Ruppert & Wand (1994):

$$\operatorname{Var}\left\{\widehat{m}_{\mathbf{H}}(\boldsymbol{x})|\boldsymbol{X}_1,\ldots,\boldsymbol{X}_n\right\} = \frac{1}{n\det(\mathbf{H})}\,\|\mathcal{K}\|_2^2\,\frac{\sigma(\boldsymbol{x})}{f_{\boldsymbol{X}}(\boldsymbol{x})}\left\{1+o_p(1)\right\}, \tag{4.71}$$

with $\sigma(\boldsymbol{x})$ denoting $\operatorname{Var}(Y|\boldsymbol{X}=\boldsymbol{x})$.

In the following we will sketch the derivation of the asymptotic conditional bias. We have seen this remarkable difference between both estimators already in the univariate case. Denote $\boldsymbol{M}$ the second order Taylor expansion of $(m(\boldsymbol{X}_1),\ldots,m(\boldsymbol{X}_1))^\top$, i.e.

$$\boldsymbol{M} \approx m(\boldsymbol{x})\mathbf{1}_n + \boldsymbol{L}(\boldsymbol{x}) + \frac{1}{2}\boldsymbol{Q}(\boldsymbol{x}) = \mathbf{X}\begin{pmatrix} m(\boldsymbol{x}) \\ \nabla_m(\boldsymbol{x}) \end{pmatrix} + \frac{1}{2}\boldsymbol{Q}(\boldsymbol{x}), \tag{4.72}$$

where

$$\boldsymbol{L}(\boldsymbol{x}) = \begin{pmatrix} (\boldsymbol{X}_1-\boldsymbol{x})^\top\nabla_m(\boldsymbol{x}) \\ \vdots \\ (\boldsymbol{X}_n-\boldsymbol{x})^\top\nabla_m(\boldsymbol{x}) \end{pmatrix}, \quad \boldsymbol{Q}(\boldsymbol{x}) = \begin{pmatrix} (\boldsymbol{X}_1-\boldsymbol{x})^\top\mathcal{H}_m(\boldsymbol{x})(\boldsymbol{X}_1-\boldsymbol{x}) \\ \vdots \\ (\boldsymbol{X}_n-\boldsymbol{x})^\top\mathcal{H}_m(\boldsymbol{x})(\boldsymbol{X}_1-\boldsymbol{x}) \end{pmatrix}$$

and ∇ and $\mathcal{H}$ being the gradient and the Hessian, respectively. Additionally to (3.62) it holds

$$\frac{1}{n}\sum_{i=1}^{n}\mathcal{K}_{\mathbf{H}}(\boldsymbol{X}_i-\boldsymbol{x})\,(\boldsymbol{X}_i-\boldsymbol{x}) = \mu_2(\mathcal{K})\mathbf{H}\mathbf{H}^\top\,\nabla_f(\boldsymbol{x}) + o_p(\mathbf{H}\mathbf{H}^\top\mathbf{1}_d),$$

$$\frac{1}{n}\sum_{i=1}^{n}\mathcal{K}_{\mathbf{H}}(\boldsymbol{X}_i-\boldsymbol{x})\,(\boldsymbol{X}_i-\boldsymbol{x})(\boldsymbol{X}_i-\boldsymbol{x})^\top = \mu_2(\mathcal{K})f_{\boldsymbol{X}}(\boldsymbol{x})\mathbf{H}\mathbf{H}^\top\,\nabla_f(\boldsymbol{x}) + o_p(\mathbf{H}\mathbf{H}^\top),$$

see Ruppert & Wand (1994). Therefore the denominator of the conditional asymptotic expectation of the Nadaraya-Watson estimator $\widehat{m}_{\mathbf{H}}$ is approximately $f_{\boldsymbol{X}}(\boldsymbol{x})$. Using $E(Y|\boldsymbol{X}_1,\ldots,\boldsymbol{X}_n) = \boldsymbol{M}$ and the Taylor expansion for $\boldsymbol{M}$ we have

$$\begin{aligned} E\left\{\widehat{m}_{\mathbf{H}}|\boldsymbol{X}_1,\ldots,\boldsymbol{X}_n\right\} \approx \left\{f_{\boldsymbol{X}}(\boldsymbol{x})+o_p(1)\right\}^{-1}\Bigg\{&\frac{1}{n}\sum_{i=1}^{n}\mathcal{K}_{\mathbf{H}}(\boldsymbol{X}_i-\boldsymbol{x})m(\boldsymbol{x}) \\ &+\sum_{i=1}^{n}\mathcal{K}_{\mathbf{H}}(\boldsymbol{X}_i-\boldsymbol{x})(\boldsymbol{X}_i-\boldsymbol{x})^\top\nabla_m(\boldsymbol{x}) \\ &+\sum_{i=1}^{n}\mathcal{K}_{\mathbf{H}}(\boldsymbol{X}_i-\boldsymbol{x})(\boldsymbol{X}_i-\boldsymbol{x})^\top\mathcal{H}_m(\boldsymbol{x})(\boldsymbol{X}_i-\boldsymbol{x})\Bigg\}. \end{aligned}$$

Hence

$$E\{\widehat{m}_{\mathbf{H}}|\mathbf{X}_1,\ldots,\mathbf{X}_n\} \approx \{f_{\mathbf{X}}(\boldsymbol{x})\}^{-1}\Big[f_{\mathbf{X}}(\boldsymbol{x})m(\boldsymbol{x}) + \mu_2(\mathcal{K})\,\nabla_m\,\mathbf{H}\mathbf{H}^\top\,\nabla_f + \frac{1}{2}\mu_2(\mathcal{K})f_{\mathbf{X}}(\boldsymbol{x})\,\mathrm{tr}\{\mathbf{H}^\top\mathcal{H}_m(\boldsymbol{x})\mathbf{H}\}\Big],$$

such that we obtain the following theorem.

Theorem 4.8.
The conditional asymptotic bias and variance of the multivariate Nadaraya-Watson kernel regression estimator are

$$\mathrm{Bias}\,\{\widehat{m}_{\mathbf{H}}|\mathbf{X}_1,\ldots,\mathbf{X}_n\} \approx \mu_2(\mathcal{K})\frac{\nabla_m(\boldsymbol{x})^\top\mathbf{H}\mathbf{H}^\top\,\nabla_f(\boldsymbol{x})}{f_{\mathbf{X}}(\boldsymbol{x})} + \frac{1}{2}\mu_2(\mathcal{K})\,tr\{\mathbf{H}^\top\mathcal{H}_m(\boldsymbol{x})\mathbf{H}\}$$

$$\mathrm{Var}\,\{\widehat{m}_{\mathbf{H}}|\mathbf{X}_1,\ldots,\mathbf{X}_n\} \approx \frac{1}{n\det(\mathbf{H})}\,\|\mathcal{K}\|_2^2\,\frac{\sigma(\boldsymbol{x})}{f_{\mathbf{X}}(\boldsymbol{x})}$$

in the interior of the support of $f_{\mathbf{X}}$.

Let us now turn to the local linear case. Recall that we use the notation $\boldsymbol{e}_0 = (1,0,\ldots,0)^\top$ for the first unit vector in $\mathbb{R}^d$. Then we can write the local linear estimator as

$$\widehat{m}_{1,\mathbf{H}}(\boldsymbol{x}) = \boldsymbol{e}_0^\top\left(\mathbf{X}^\top\mathbf{W}\mathbf{X}\right)^{-1}\mathbf{X}^\top\mathbf{W}\mathbf{Y}.$$

Now we have using (4.69) and (4.72),

$$\begin{aligned} &E\{\widehat{m}_{1,\mathbf{H}}|\mathbf{X}_1,\ldots,\mathbf{X}_n\} - m(\boldsymbol{x}) \\ &= \boldsymbol{e}_0^\top\left(\mathbf{X}^\top\mathbf{W}\mathbf{X}\right)^{-1}\mathbf{X}^\top\mathbf{W}\mathbf{X}\left\{\begin{pmatrix} m(\boldsymbol{x}) \\ \nabla_m(\boldsymbol{x})\end{pmatrix} + \frac{1}{2}\mathbf{Q}(\boldsymbol{x})\right\} - m(\boldsymbol{x}) \\ &= \frac{1}{2}\boldsymbol{e}_0^\top\left(\mathbf{X}^\top\mathbf{W}\mathbf{X}\right)^{-1}\mathbf{X}^\top\mathbf{W}\mathbf{Q}(\boldsymbol{x}) \end{aligned}$$

since $\boldsymbol{e}_0^\top[m(\boldsymbol{x}),\nabla_m(\boldsymbol{x})^\top] = m(\boldsymbol{x})$. Hence, the numerator of the asymptotic conditional bias only depends on the quadratic term. This is one of the key points in asymptotics for local polynomial estimators. If we were to use local polynomials of order p and expand (4.72) up to order $p+1$, then only the term of order $p+1$ would appear in the numerator of the asymptotic conditional bias. Of course this is to be paid with a more complicated structure of the denominator. The following theorem summarizes bias and variance for the estimator $\widehat{m}_{1,\mathbf{H}}$.

Theorem 4.9.
The conditional asymptotic bias and variance of the multivariate local linear regression estimator are

$$\text{Bias}\,\{\widehat{m}_{1,\mathbf{H}}|\mathbf{X}_1,\ldots,\mathbf{X}_n\} \approx \frac{1}{2}\mu_2(\mathcal{K})\, tr\{\mathbf{H}^\top \mathcal{H}_m(\boldsymbol{x})\mathbf{H}\}$$
$$\text{Var}\,\{\widehat{m}_{1,\mathbf{H}}|\mathbf{X}_1,\ldots,\mathbf{X}_n\} \approx \frac{1}{n\det(\mathbf{H})}\,\|\mathcal{K}\|_2^2\,\frac{\sigma(\boldsymbol{x})}{f_{\mathbf{X}}(\boldsymbol{x})}$$

in the interior of the support of $f_{\mathbf{X}}$.

For all omitted details we refer again to Ruppert & Wand (1994). They also point out that the local linear estimate has the same order conditional bias in the interior as well as in the boundary of the support of $f_{\mathbf{X}}$.

4.5.2 Practical Aspects

The computation of local polynomial estimators can be done by any statistical package that is able to run weighted least squares regression. However, since we estimate a function, this weighted least squares regression has to be performed in all observation points or on a grid of points in $\mathbb{R}^d$. Therefore, explicit formulae, which can be derived at least for lower dimensions d are useful.

Example 4.17.
Consider $d = 2$ and for fixed $\boldsymbol{x}$ the sums

$$S_{jk} = S_{jk}(\boldsymbol{x}) = \sum_{i=1}^{n} \mathcal{K}_{\mathbf{H}}(\boldsymbol{x}-\mathbf{X}_i)(X_{1i}-x_1)^j(X_{2i}-x_2)^k,$$
$$T_{jk} = T_{jk}(\boldsymbol{x}) = \sum_{i=1}^{n} \mathcal{K}_{\mathbf{H}}(\boldsymbol{x}-\mathbf{X}_i)(X_{1i}-x_1)^j(X_{2i}-x_2)^k Y_i.$$

Then for the local linear estimate we can write

$$\widehat{\beta} = \begin{pmatrix} S_{00} & S_{10} & S_{01} \\ S_{10} & S_{20} & S_{11} \\ S_{01} & S_{11} & S_{02} \end{pmatrix}^{-1} \begin{pmatrix} T_{00} \\ T_{10} \\ T_{01} \end{pmatrix}. \tag{4.73}$$

Here it is still possible to fit the explicit formula for the estimated regression function on one line:

$$\widehat{m}_{1,\mathbf{H}}(\boldsymbol{x}) = \frac{(S_{20}S_{02}-S_{11}^2)T_{00} + (S_{10}S_{11}-S_{01}S_{20})T_{01} + (S_{01}S_{11}-S_{02}S_{10})T_{10}}{2S_{01}S_{10}S_{11} - S_{02}S_{10}^2 - S_{00}S_{11}^2 - S_{01}^2S_{20} + S_{00}S_{02}S_{20}}. \tag{4.74}$$

To estimate the regression plane we have to apply (4.74) on a two-dimensional grid of points.

Figure 4.18 shows the Nadaraya-Watson and the local linear two-dimensional estimate for simulated data. We use 500 design points uniformly distributed in $[0,1] \times [0,1]$ and the regression function

$$m(\boldsymbol{x}) = \sin(2\pi x_1) + x_2.$$

The error term is $N(0, \frac{1}{4})$. The bandwidth is chosen as $h_1 = h_2 = 0.3$ for both estimators. □

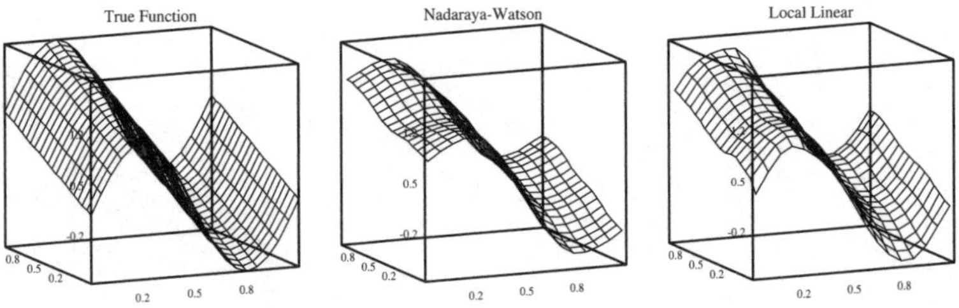

Figure 4.18. Two-dimensional local linear estimate `SPMtruenadloc`

Nonparametric kernel regression function estimation is not limited to bivariate distributions. Everything can be generalized to higher dimensions but unfortunately some problems arise. A practical problem is the graphical display for higher dimensional multivariate functions. This problem has already been considered in Chapter 3 when we discussed the graphical representation of multivariate density estimates. The corresponding remarks for plotting functions of up to three-dimensional arguments apply here again.

A general problem in multivariate nonparametric estimation is the so-called *curse of dimensionality*. Recall that the nonparametric regression estimators are based on the idea of local (weighted) averaging. In higher dimensions the observations are sparsely distributed even for large sample sizes, and consequently estimators based on local averaging perform unsatisfactorily in this situation. Technically, one can explain this effect by looking at the AMSE again. Consider a multivariate regression estimator with the same bandwidth h for all components, e.g. a Nadaraya-Watson or local linear estimator with bandwidth matrix $\mathbf{H} = h \cdot \mathbf{I}$. Here the asymptotic MSE will also depend on d:

$$\text{AMSE}(n, h) = \frac{1}{nh^d} C_1 + h^4 C_2.$$

where C_1 and C_2 are constants that neither depend on n nor h. If we derive the optimal bandwidth we find that $h_{opt} \sim n^{-1/(4+d)}$ and hence the rate of convergence for AMSE is $n^{-4/(4+d)}$. You can clearly see that the speed of convergence decreases dramatically for higher dimensions d.

Bibliographic Notes

An introduction to kernel regression methods can be found in the monographs of Silverman (1986), Härdle (1990), Bowman & Azzalini (1997), Simonoff (1996) and Pagan & Ullah (1999). The books of Scott (1992) and Wand & Jones (1995) deal particularly with the multivariate case. For detailed derivations of the asymptotic properties we refer to Collomb (1985) and Gasser & Müller (1984). The latter reference also considers boundary kernels to reduce the bias at the boundary regions of the explanatory variables.

Locally weighted least squares were originally studied by Stone (1977), Cleveland (1979) and Lejeune (1985). Technical details of asymptotic expansions for bias and variance can be found in Ruppert & Wand (1994). Monographs concerning local polynomial fitting are Wand & Jones (1995), Fan & Gijbels (1996) and Simonoff (1996). Computational aspects, in particular the WARPing technique (binning) for kernel and local polynomial regression, are discussed in Härdle & Scott (1992) and Fan & Marron (1994). The monograph of Loader (1999) discusses local regression in combination with likelihood-based estimation.

For comprehensive works on spline smoothing see Eilers & Marx (1996), Wahba (1990) and and Green & Silverman (1994). Good resources for wavelets are Daubechies (1992), Donoho & Johnstone (1994), Donoho & Johnstone (1995) and Donoho, Johnstone, Kerkyacharian & Picard (1995). The books of Eubank (1999) and Schimek (2000b) provide extensive overviews on a variety of different smoothing methods.

For a monograph on testing in nonparametric models see Hart (1997). The concepts presented by equations (4.63)–(4.66) are in particular studied by the following articles: González Manteiga & Cao (1993) and Härdle & Mammen (1993) introduced (4.63), Gozalo & Linton (2001) studied (4.64) motivated by Lagrange multiplier tests. Equation (4.65) was originally introduced by Zheng (1996) and independently discussed by Fan & Li (1996). Finally, (4.66) was proposed by Dette (1999) in the context of testing for parametric structures in the regression function. For an introductory presentation see Yatchew (2003). In general, all test approaches are also possible for multivariate regressors.

More sophisticated is the *minimax approach* for testing nonparametric alternatives studied by Ingster (1993). This approach tries to maximize the power for the worst alternative case, i.e. the one that is closest to the hypothesis but could still be detected. Rather different approaches have been introduced by Bierens (1990) and Bierens & Ploberger (1997), who consider (integrated) conditional moment tests or by Stute (1997) and Stute, González Manteiga & Presedo-Quindimi (1998), who verify, via bootstrap, whether the residuals of the hypothesis integrated over the empirical distribution of the regressor variable X converge to a centered Gaussian process. There is further literature about *adaptive testing* which tries to find the smoothing parameter that maximizes the power when holding the type one error, see for example Ledwina (1994), Kallenberg & Ledwina (1995), Spokoiny (1996) and Spokoiny (1998).

Exercises

Exercise 4.1. Consider X, Y with a bivariate normal distribution, i.e.

$$\begin{pmatrix} X \\ Y \end{pmatrix} \sim N\left(\begin{pmatrix} \mu \\ \eta \end{pmatrix}, \begin{pmatrix} \sigma^2 & \rho\sigma\tau \\ \rho\sigma\tau & \tau^2 \end{pmatrix}\right)$$

with density

$$f(x,y) = \frac{1}{2\pi\sigma\tau\sqrt{1-\rho^2}} \cdot \exp\left\{\frac{-\left(\frac{x-\mu}{\sigma}\right)^2 - \left(\frac{y-\eta}{\tau}\right)^2 + 2\rho\left(\frac{x-\mu}{\sigma}\right)\left(\frac{y-\eta}{\tau}\right)}{2\left(1-\rho^2\right)}\right\}.$$

This means $X \sim N\left(\mu, \sigma^2\right), Y \sim N\left(\eta, \tau^2\right)$ and $corr\,(X,Y) = \rho$. Show that the regression function $m(x) = E(Y|X = x)$ is linear, more exactly

$$E(Y|X = x) = \alpha + \beta x,$$

with $\alpha = \eta - \mu\rho\tau/\sigma$ and $\beta = \rho\tau/\sigma$.

Exercise 4.2. Explain the effects of

$$2\frac{m'(x)f_X'(x)}{f_X(x)}$$

in the bias part of the Nadaraya-Watson MSE, see equation (4.13). When will the bias be large/small?

Exercise 4.3. Calculate the pointwise optimal bandwidth $h_{opt}(x)$ from the Nadaraya-Watson AMSE, see equation (4.13).

Exercise 4.4. Show that $\widehat{m}_{0,h}(x) = \widehat{m}_h(x)$, i.e. the local constant estimator (4.17) equals the Nadaraya-Watson estimator.

Exercise 4.5. Show that from (4.33) and (4.34) indeed the spline formula $\widehat{\boldsymbol{m}}_\lambda = (\mathbf{I} + \lambda\mathbf{K})^{-1}\mathbf{Y}$ in equation (4.35) follows.

Exercise 4.6. Compare the kernel, the k-NN, the spline and a linear regression fit for a data set.

Exercise 4.7. Show that the Legendre polynomials p_0 and p_1 are indeed orthogonal functions.

Exercise 4.8. Compute and plot the confidence intervals for the Nadaraya-Watson kernel estimate and the local polynomial estimate for a data set.

Exercise 4.9. Prove that the solution of the minimization problem (4.31) is a piecewise cubic polynomial function which is twice continuously differentiable (cubic spline).

Exercise 4.10. Discuss the behavior of the smoothing parameters h in kernel regression, k in nearest-neighbor estimation, λ in spline smoothing and N in orthogonal series estimation.

Summary

- ⋆ The regression function $m(\bullet)$ which relates an independent variable X and a dependent variable Y is the conditional expectation

$$m(x) = E(Y|X = x).$$

- ⋆ A natural kernel regression estimate for a random design can be obtained by replacing the unknown densities in $E(Y|X = x)$ by kernel density estimates. This yields the Nadaraya-Watson estimator

$$\widehat{m}_h(x) = \frac{1}{n} \sum_{i=1} W_{hi}(x) Y_i$$

with weights

$$W_{hi}(x) = \frac{n\, K_h(x - X_i)}{\sum_{j=1}^{n} K_h(x - X_j)}.$$

For a fixed design we can use the Gasser-Müller estimator with

$$W_{hi}^{GM}(x) = n \int_{s_{i-1}}^{s_i} K_h(x - u) du.$$

- ⋆ The asymptotic MSE of the Nadaraya-Watson estimator is

$$\begin{aligned} \text{AMSE}\{\widehat{m}_h(x)\} = {} & \frac{1}{nh} \frac{\sigma^2(x)}{f_X(x)} \|K\|_2^2 \\ & + \frac{h^4}{4} \left\{ m''(x) + 2\frac{m'(x) f_X'(x)}{f_X(x)} \right\}^2 \mu_2^2(K), \end{aligned}$$

the asymptotic MSE of the Gasser-Müller estimator is identical up to the $2\frac{m'(x)f_X'(x)}{f_X(x)}$ term.

⋆ The Nadaraya-Watson estimator is a local constant least squares estimator. Extending the local constant approach to local polynomials of degree p yields the minimization problem:

$$\min_{\beta} \sum_{i=1}^{n} \left\{Y_i - \beta_0 - \beta_1(X_i - x) - \ldots - \beta_p(X_i - x)^p\right\}^2 K_h(x - X_i),$$

where $\widehat{\beta}_0$ is the estimator of the regression curve and the $\widehat{\beta}_\nu$ are proportional to the estimates for the derivatives.

⋆ For the regression problem, odd order local polynomial estimators outperform even order regression fits. In particular, the asymptotic MSE for the local linear kernel regression estimator is

$$\text{AMSE}\{\widehat{m}_{1,h}(x)\} = \frac{1}{nh}\frac{\sigma^2(x)}{f_X(x)}\|K\|_2^2 + \frac{h^4}{4}\left\{m''(x)\right\}^2 \mu_2^2(K).$$

The bias does not depend on f_X, f_X' and m' which makes the the local linear estimator more design adaptive and improves the behavior in boundary regions.

⋆ The k-NN estimator has in its simplest form the representation

$$\widehat{m}_k(x) = \frac{1}{k}\sum_{i=n}^{n} Y_i \,\mathrm{I}\{X_i \text{ is among } k \text{ nearest neighbors of } x\}.$$

This can be refined using kernel weights instead of uniform weighting of all observations nearest to x. The variance of the k-NN estimator does not depend on $f_X(x)$.

⋆ Median smoothing is a version of k-NN which estimates the conditional median rather than the conditional mean:

$$\widehat{m}_k(x) = \text{med}\{Y_i : X_i \text{ is among of } k \text{ nearest neighbors of } x\}.$$

It is more robust to outliers.

- ⋆ The smoothing spline minimizes a penalized residual sum of squares over all twice differentiable functions $m(\bullet)$:

$$S_\lambda(m) = \sum_{i=1}^{n} \{Y_i - m(X_i)\}^2 + \lambda \|m''\|_2^2,$$

the solution consists of piecewise cubic polynomials in $[X_{(i)}, X_{(i+1)}]$. Under regularity conditions the spline smoother is equivalent to a kernel estimator with higher order kernel and design dependent bandwidth.

- ⋆ Orthogonal series regression (Fourier regression, wavelet regression) uses the fact that under certain conditions, functions can be represented by a series of basis functions. The coefficients of the basis functions have to be estimated. The smoothing parameter N is the number of terms in the series.

- ⋆ Bandwidth selection in regression is usually done by cross-validation or the penalized residual sum of squares.

- ⋆ Pointwise confidence intervals and uniform confidence bands can be constructed analogously to the density estimation case.

- ⋆ Nonparametric regression estimators for univariate data can be easily generalized to the multivariate case. A general problem is the curse of dimensionality.

Part II

Semiparametric Models

5

Semiparametric and Generalized Regression Models

In the previous part of this book we found the curse of dimensionality to be one of the major problems that arises when using *nonparametric multivariate* regression techniques. For the practitioner, a further problem is that for more than two regressors, graphical illustration or interpretation of the results is hardly ever possible. Truly multivariate regression models are often far too flexible and general for making detailed inference.

5.1 Dimension Reduction

Researchers have looked for possible remedies, and a lot of effort has been allocated to developing methods which reduce the complexity of high dimensional regression problems. This refers to the reduction of dimensionality as well as allowance for partly parametric modeling. Not surprisingly, one follows the other. The resulting models can be grouped together as so-called *semiparametric* models.

All models that we will study in the following chapters can be motivated as generalizations of well-known parametric models, mainly of the linear model

$$E(Y|\boldsymbol{X}) = m(\boldsymbol{X}) = \boldsymbol{X}^\top \boldsymbol{\beta}$$

or its generalized version

$$E(Y|\boldsymbol{X}) = m(\boldsymbol{X}) = G\{\boldsymbol{X}^\top \boldsymbol{\beta}\}\,. \tag{5.1}$$

Here G denotes a known function, $\boldsymbol{X}$ is the d-dimensional vector of regressors and $\boldsymbol{\beta}$ is a coefficient vector that is to be estimated from observations for Y and $\boldsymbol{X}$.

Let us take a closer look at model (5.1). This model is known as the generalized linear model. Its use and estimation are extensively treated in McCullagh & Nelder (1989). Here we give only some selected motivating examples.

What is the reason for introducing this functional G, called the link? (Note that other authors call its inverse G^{-1} the link.) Clearly, if G is the identity we are back in the classical linear model. As a first alternative let us consider a quite common approach for investigating growth models. Here, the model is often assumed to be multiplicative instead of additive, i.e.

$$Y = \prod_{j=1}^{d} X_j^{\beta_j} \cdot \varepsilon, \quad E\log(\varepsilon) = 0 \tag{5.2}$$

in contrast to

$$Y = \prod_{j=1}^{d} X_j^{\beta_j} + \xi, \quad E\xi = 0. \tag{5.3}$$

Depending on whether we have multiplicative errors ε or additive errors ξ, we can transform model (5.2) to

$$E\{\log(Y)|\boldsymbol{X}\} = \sum_{j=1}^{d} \beta_j \log(X_j) \tag{5.4}$$

and model (5.3) to

$$E(Y|\boldsymbol{X}) = \exp\left\{\sum_{j=1}^{d} \beta_j \log(X_j)\right\}. \tag{5.5}$$

Considering now $\log(\boldsymbol{X})$ as the regressor instead of $\boldsymbol{X}$, equation (5.5) is equivalent to (5.1) with $G(\bullet) = \exp(\bullet)$. Equation (5.4), however, is a transformed model, see the bibliographic notes for references on this model family.

The most common cases in which link functions are used are binary responses ($Y \in \{0,1\}$) or multicategorical ($Y \in \{0,1,\ldots,J\}$) responses and count data ($Y \sim$ Poisson). For the binary case, let us introduce an example that we will study in more detail in Chapters 7 and 9.

Example 5.1.
Imagine we are interested in possible determinants of the migration decision of East Germans to leave the East for West Germany. Think of Y^* as being the net-utility from migrating from the eastern part of Germany to the western part. Utility itself is not observable but we can observe characteristics of the decision makers and the alternatives that affect utility. As Y^* is not observable it is called a latent variable. Let the observable characteristics

Table 5.1. Descriptive statistics for migration data, $n = 3235$

		Yes	No	(in %)	
Y	MIGRATION INTENTION	38.5	61.5		
X_1	FAMILY/FRIENDS IN WEST	85.6	11.2		
X_2	UNEMPLOYED/JOB LOSS CERTAIN	19.7	78.9		
X_3	CITY SIZE 10,000–100,000	29.3	64.2		
X_4	FEMALE	51.1	49.8		
		Min	Max	Mean	S.D.
X_5	AGE (in years)	18	65	39.84	12.61
X_6	HOUSEHOLD INCOME (in DM)	200	4000	2194.30	752.45

be summarized in a vector $\boldsymbol{X}$. This vector $\boldsymbol{X}$ may contain variables such as education, age, sex and other individual characteristics. A selection of such characteristics is shown in Table 5.1. □

In Example 5.1, we hope that the vector of regressors $\boldsymbol{X}$ captures the variables that systematically affect each person's utility whereas unobserved or random influences are absorbed by the term ε. Suppose further, that the components of $\boldsymbol{X}$ influence net-utility through a multivariate function $v(\bullet)$ and that the error term is additive. Then the latent-variable model is given by

$$Y^* = v(\boldsymbol{X}) - \varepsilon \quad \text{and} \quad Y = \begin{cases} 1 & \text{if } Y^* > 0, \\ 0 & \text{otherwise.} \end{cases} \tag{5.6}$$

Hence, what we really observe is the binary variable Y that takes on the value 1 if net-utility is positive (person intends to migrate) and 0 otherwise (person intends to stay). Then some calculations lead to

$$P(Y = 1 \mid \boldsymbol{X} = \boldsymbol{x}) = E(Y \mid \boldsymbol{X} = \boldsymbol{x}) = G_{\varepsilon|\boldsymbol{x}}\{v(\boldsymbol{x})\} \tag{5.7}$$

with $G_{\varepsilon|\boldsymbol{x}}$ being the cdf of ε conditional on $\boldsymbol{x}$.

Recall that standard parametric models assume that ε is independently distributed of $\boldsymbol{X}$ with known distribution function $G_{\varepsilon|\boldsymbol{x}} = G$, and that the index $v(\bullet)$ has the following simple form:

$$v(\boldsymbol{x}) = \beta_0 + \boldsymbol{x}^\top \boldsymbol{\beta}. \tag{5.8}$$

The most popular distribution assumptions regarding the error are the normal and the logistic ones, leading to the so-called *probit* or *logit* models with $G(\bullet) = \Phi(\bullet)$ (Gaussian cdf), respectively $G(\bullet) = \exp(\bullet)/\{1 + \exp(\bullet)\}$. We will learn how to estimate the coefficients β_0 and $\boldsymbol{\beta}$ in Section 5.2.

The binary choice model can be easily extended to the multicategorical case, which is usually called *discrete choice model*. We will not discuss extensions for multicategorical responses here. Some references for these models are mentioned in the bibliographic notes.

Several approaches have been proposed to reduce dimensionality or to generalize parametric regression models in order to allow for nonparametric relationships. Here, we state three different approaches:

- variable selection in nonparametric regression,
- generalization of (5.1) to a nonparametric link function,
- generalization of (5.1) to a semi- or nonparametric index,

which are discussed in more detail.

5.1.1 Variable Selection in Nonparametric Regression

The intention of variable selection is to choose an appropriate subset of variables, $\boldsymbol{X}_r = (X_{j_1}, \ldots, X_{j_r})^\top \in \boldsymbol{X} = (X_1, \ldots, X_d)^\top$, from the set of all variables that could potentially enter the regression. Of course, the selection of the variables could be determined by the particular problem at hand, i.e. we choose the variables according to insights provided by some underlying economic theory. This approach, however, does not really solve the statistical side of our modeling process. The curse of dimensionality could lead us to keep the number of variables as low as possible. On the other hand, fewer variables could in turn reduce the explanatory power of the model. Thus, after having chosen a set of variables on theoretical grounds in a first step, we still do not know how many and, more importantly, which of these variables will lead to optimal regression results. Therefore, a variable selection method is needed that uses a statistical selection criterion.

Vieu (1994) has proposed to use the integrated square error ISE to measure the quality of a given subset of variables. In theory, a subset of variables is defined to be an optimal subset if it minimizes the integrated squared error:

$$\mathrm{ISE}(\boldsymbol{X}_r^{opt}) = \min_{\boldsymbol{X}_r} \mathrm{ISE}(\boldsymbol{X}_r)$$

where $\boldsymbol{X}_r \subset \boldsymbol{X}$. In practice, the ISE is replaced by its sample analog, the multivariate analog of the cross validation function (3.38). After the variables have been selected, the conditional expectation of Y on $\boldsymbol{X}_r$ is calculated by some kind of standard nonparametric multivariate regression technique such as the kernel regression estimator.

5.1.2 Nonparametric Link Function

Index models play an important role in econometrics. An *index* is a summary of different variables into one number, e.g. the price index, the growth index, or the cost-of-living index. It is clear that by summarizing all the information contained in the variables $X_1, \ldots, X_d$ into one "single index" term we will greatly reduce the dimensionality of a problem. Models based on such an index are known as *single index models* (SIM). In particular we will discuss single index models of the following form:

$$E(Y|\boldsymbol{X}) = m(\boldsymbol{X}) = g\left\{v_{\boldsymbol{\beta}}(\boldsymbol{X})\right\}, \tag{5.9}$$

where $g(\bullet)$ is an *unknown link* function and $v_{\boldsymbol{\beta}}(\bullet)$ an up to $\boldsymbol{\beta}$ specified index function. The estimation can be carried out in two steps. First, we estimate $\boldsymbol{\beta}$. Then, using the index values for our observations, we can estimate g by nonparametric regression. Note that estimating $g(\bullet)$ by regressing the Y on $v_{\widehat{\boldsymbol{\beta}}}(\boldsymbol{X})$ is only a one-dimensional regression problem.

Obviously, (5.9) generalizes (5.7) in that we do not assume the link function G to be known. For that purpose we replaced G by g to emphasize that the link function needs to be estimated. Notice, that often the general index function $v_{\boldsymbol{\beta}}(\boldsymbol{X})$ is replaced by the linear index $\boldsymbol{X}^\top\boldsymbol{\beta}$. Equations (5.5) and (5.6) together with (5.8) give examples for such linear index functions.

5.1.3 Semi- or Nonparametric Index

In many applications a canonical partitioning of the explanatory variables exists. In particular, if there are categorical or discrete explanatory variables we may want to keep them separate from the other design variables. Note that only the continuous variables in the nonparametric part of the model cause the curse of dimensionality (Delgado & Mora, 1995). In the following chapters we will study the following models:

- *Additive Model* (AM)
  The standard additive model is a generalization of the multiple linear regression model by introducing one-dimensional nonparametric functions in the place of the linear components. Here, the conditional expectation of Y given $\boldsymbol{X} = (X_1, \ldots, X_d)^\top$ is assumed to be the sum of unknown functions of the explanatory variables plus an intercept term:

$$E(Y|\boldsymbol{X}) = c + \sum_{j=1}^{d} g_j(X_j) \tag{5.10}$$

  Observe how reduction is achieved in this model: Instead of estimating one function of several variables, as we do in completely nonparametric regression, we merely have to estimate d functions of one-dimensional variables X_j.
- *Partial Linear Model* (PLM)
  Suppose we only want to model parts of the index linearly. This could

be for analytical reasons or for reasons going back to economic theory. For instance, the impact of a dummy variable $X_1 \in \{0,1\}$ might be sufficiently explained by estimating the coefficient β_1.

For the sake of clarity, let us now separate the d-dimensional vector of explanatory variables into $\boldsymbol{U} = (U_1,\ldots,U_p)^\top$ and $\boldsymbol{T} = (T_1,\ldots,T_q)^\top$. The regression of Y on $\boldsymbol{X} = (\boldsymbol{U},\boldsymbol{T})$ is assumed to have the form:

$$E(Y|\boldsymbol{U},\boldsymbol{T}) = \boldsymbol{U}^\top \boldsymbol{\beta} + m(\boldsymbol{T}) \tag{5.11}$$

where $m(\bullet)$ is an unknown multivariate function of the vector $\boldsymbol{T}$. Thus, a partial linear model can be interpreted as a sum of a purely parametric part, $\boldsymbol{U}^\top \boldsymbol{\beta}$, and a purely nonparametric part, $m(\boldsymbol{T})$. Not surprisingly, estimating $\boldsymbol{\beta}$ and $m(\bullet)$ involves the combination of both parametric and nonparametric regression techniques.

- *Generalized Additive Model* (GAM)
  Just like the (standard) additive model, generalized additive models are based on the sum of d nonparametric functions of the d variables $\boldsymbol{X}$ (plus an intercept term). In addition, they allow for a known parametric link function, $G(\bullet)$, that relates the sum of functions to the dependent variable:

$$E(Y|\boldsymbol{X}) = G\left\{c + \sum_{j=1}^{d} g_j(X_j)\right\}. \tag{5.12}$$

- *Generalized Partial Linear Model* (GPLM)
  Introducing a link $G(\bullet)$ for a partial linear model $\boldsymbol{U}^\top \boldsymbol{\beta} + m(\boldsymbol{T})$ yields the *generalized partial linear model* (GPLM):

$$E(Y|\boldsymbol{U},\boldsymbol{T}) = G\left\{\boldsymbol{U}^\top \boldsymbol{\beta} + m(\boldsymbol{T})\right\}.$$

  G denotes a known link function as in the GAM. In contrast to the GAM, $m(\bullet)$ is possibly a multivariate nonparametric function of the variable $\boldsymbol{T}$.
- *Generalized Partial Linear Partial Additive Model* (GAPLM)
  In high dimensions of $\boldsymbol{T}$ the estimate of the nonparametric function $m(\bullet)$ in the GPLM faces the same problems as the fully nonparametric multidimensional regression function estimates: the curse of dimensionality and the practical problem of interpretability. Hence, it is useful to think about a lower dimensional modeling of the nonparametric part. This leads to the GAPLM with an additive structure in the nonparametric component:

$$E(Y|\boldsymbol{U},\boldsymbol{T}) = G\left\{\boldsymbol{U}^\top \boldsymbol{\beta} + \sum_{j=1}^{q} g_j(T_j)\right\}.$$

  Here, the $g_j(\bullet)$ will be univariate nonparametric functions of the variables T_j. In the case of an identity function G we speak of an additive partial linear model (APLM)

More discussion and motivation is given in the following chapters where the different models are discussed in detail and the specific estimation procedures are presented. Before proceeding with this task, however, we will first introduce some facts about the parametric generalized linear model (GLM). The following section is intended to give more insight into this model since its concept and the technical details of its estimation will be necessary for its semiparametric modification in Chapters 6 to 9.

5.2 Generalized Linear Models

Generalized linear models (GLM) extend the concept of the widely used linear regression model. The linear model assumes that the response Y (the dependent variable) is equal to a linear combination $\mathbf{X}^\top \boldsymbol{\beta}$ and a normally distributed error term:

$$Y = \mathbf{X}^\top \boldsymbol{\beta} + \varepsilon.$$

The least squares estimator $\widehat{\boldsymbol{\beta}}$ is adapted to these assumptions. However, the restriction of linearity is far too strict for a variety of practical situations. For example, a continuous distribution of the error term implies that the response Y has a continuous distribution as well. Hence, this standard linear regression model fails, for example, when dealing with binary data (Bernoulli Y) or with count data (Poisson Y).

Nelder & Wedderburn (1972) introduced the term *generalized linear models* (GLM). A good resource of material on this model is the monograph of McCullagh & Nelder (1989). The essential feature of the GLM is that the regression function, i.e. the expectation $\mu = E(Y|\mathbf{X})$ of Y is a monotone function of the index $\eta = \mathbf{X}^\top \boldsymbol{\beta}$. We denote the function which relates μ and η by G:

$$E(Y|\mathbf{X}) = G(\mathbf{X}^\top \boldsymbol{\beta}) \quad \Longleftrightarrow \quad \mu = G(\eta).$$

This function G is called the *link function*. (We remark that Nelder & Wedderburn (1972), McCullagh & Nelder (1989) actually denote G^{-1} as the link function.)

5.2.1 Exponential Families

In the GLM framework we assume that the distribution of Y is a member of the *exponential family*. The exponential family covers a broad range of distributions, for example discrete as the Bernoulli or Poisson distribution and continuous as the Gaussian (normal) or Gamma distribution.

A distribution is said to be a member of the exponential family if its probability function (if Y discrete) or its density function (if Y continuous) has the structure

$$f(y,\theta,\psi) = \exp\left\{\frac{y\theta - b(\theta)}{a(\psi)} + c(y,\psi)\right\} \tag{5.13}$$

with some specific functions $a(\bullet)$, $b(\bullet)$ and $c(\bullet)$. These functions differ for the distinct Y distributions. Generally speaking, we are only interested in estimating the parameter θ. The additional parameter ψ is — as the variance σ^2 in the linear regression — a nuisance parameter. McCullagh & Nelder (1989) call θ the *canonical parameter*.

Example 5.2.
Suppose Y is normally distributed, i.e. $Y \sim N(\mu,\sigma^2)$. Hence we can write its density as

$$\varphi(y) = \frac{1}{\sqrt{2\pi}\sigma}\exp\left\{\frac{-1}{2\sigma^2}(y-\mu)^2\right\} = \exp\left\{y\frac{\mu}{\sigma^2} - \frac{\mu^2}{2\sigma^2} - \frac{y^2}{2\sigma^2} - \log(\sqrt{2\pi}\sigma)\right\}$$

and we see that the normal distribution is a member of the exponential family with

$$a(\psi) = \sigma^2,\ b(\theta) = \frac{\mu^2}{2},\ c(y,\psi) = -\frac{y^2}{2\sigma^2} - \log(\sqrt{2\pi}\sigma),$$

where we set $\psi = \sigma$ and $\theta = \mu$. □

Example 5.3.
Suppose now Y is Bernoulli distributed, i.e. its probability function is

$$P(Y=y) = \mu^y(1-\mu)^{1-y} = \begin{cases} \mu & \text{if } y = 1, \\ 1-\mu & \text{if } y = 0. \end{cases}$$

This can be transformed into

$$P(Y=y) = \left(\frac{\mu}{1-\mu}\right)^y (1-\mu) = \exp\left\{y\log\left(\frac{p}{1-\mu}\right)\right\}(1-\mu)$$

using the *logit*

$$\theta = \log\left(\frac{\mu}{1-\mu}\right) \iff \mu = \frac{e^\theta}{1+e^\theta}.$$

Thus we have an exponential family with

$$a(\psi) = 1,\ b(\theta) = -\log(1-\mu) = \log(1+e^\theta),\ c(y,\psi) \equiv 0.$$

This is a distribution without an additional nuisance parameter ψ. □

It is known that the least squares estimator $\widehat{\beta}$ in the classical linear model is also the maximum-likelihood estimator for normally distributed errors. By imposing that the distribution of Y belongs to the exponential family it is possible to stay in the framework of maximum-likelihood for the GLM. Moreover, the use of the general concept of exponential families has the advantage that we can derive properties of different distributions at the same time.

To derive the maximum-likelihood algorithm in detail, we need to present some more properties of the probability function or density function $f(\bullet)$. First of all, f is a density (w.r.t. the Lebesgue measure in the continuous and w.r.t. the counting measure in the discrete case). This allows us to write

$$\int f(y, \theta, \psi)\, dy = 1.$$

Under some suitable regularity conditions (it is possible to exchange differentiation and integration) this yields

$$\begin{aligned} 0 &= \frac{\partial}{\partial\theta} \int f(y, \theta, \psi)\, dy = \int \frac{\partial}{\partial\theta} f(y, \theta, \psi)\, dy \\ &= \int \left\{ \frac{\partial}{\partial\theta} \log f(y, \theta, \psi) \right\} f(y, \theta, \psi)\, dy = E\left\{ \frac{\partial}{\partial\theta} \ell(y, \theta, \psi) \right\}, \end{aligned}$$

where $\ell(y, \theta, \psi)$ denotes the *log-likelihood*, i.e.

$$\ell(y, \theta, \psi) = \log f(y, \theta, \psi). \tag{5.14}$$

The function $\frac{\partial}{\partial\theta}\ell(y, \theta, \psi)$ is typically called *score* and it is known that

$$E\left\{ \frac{\partial^2}{\partial\theta^2} \ell(y, \theta, \psi) \right\} = -E\left\{ \frac{\partial}{\partial\theta} \ell(y, \theta, \psi) \right\}^2.$$

This and taking first and second derivatives of (5.13) gives now

$$0 = E\left\{ \frac{Y - b'(\theta)}{a(\psi)} \right\}, \quad \text{and} \quad E\left\{ \frac{-b''(\theta)}{a(\psi)} \right\} = -E\left\{ \frac{Y - b'(\theta)}{a(\psi)} \right\}^2.$$

We can conclude

$$\begin{aligned} E(Y) &= \mu = b'(\theta), \\ \operatorname{Var}(Y) &= V(\mu)a(\psi) = b''(\theta)a(\psi). \end{aligned}$$

We observe that the expectation of Y only depends on θ whereas the variance of Y depends on the parameter of interest θ and the nuisance parameter ψ. Typically one assumes that the factor $a(\psi)$ is identical over all observations.

5.2.2 Link Functions

Apart from the distribution of Y, the link function G is another important part of the GLM. Recall the notation

$$\eta = \boldsymbol{X}^\top \boldsymbol{\beta},\ \mu = G(\eta).$$

In the case that

$$\boldsymbol{X}^\top \boldsymbol{\beta} = \eta = \theta$$

the link function is called *canonical link* function. For models with a canonical link, some theoretical and practical problems are easier to solve. Table 5.2 summarizes characteristics for some exponential functions together with canonical parameters θ and their canonical link functions. Note that for the binomial and the negative binomial distribution we assume the parameter k to be known. The case of binary Y is a special case of the binomial distribution ($k = 1$).

What link functions can we choose apart from the canonical? For most of the models a number of special link functions exist. For binomial Y for example, the logistic or Gaussian link functions are often used. Recall that a binomial model with the canonical logit link is called *logit* model. If the binomial distribution is combined with the Gaussian link, it is called *probit* model. A further alternative for binomial Y is the complementary log-log link

$$\eta = \log\{-\log(1-\mu)\}.$$

A very flexible class of link functions is the class of power functions which are also called Box-Cox transformations (Box & Cox, 1964). They can be defined for all models for which we have observations with positive mean. This family is usually specified as

$$\eta = \begin{cases} \mu^\lambda & \text{if } \lambda \neq 0, \\ \log \mu & \text{if } \lambda = 0. \end{cases}$$

5.2.3 Iteratively Reweighted Least Squares Algorithm

As already pointed out, the estimation method of choice for a GLM is maximizing the likelihood function with respect to $\boldsymbol{\beta}$. Suppose that we have the vector of observations $\boldsymbol{Y} = (Y_1, \ldots, Y_n)^\top$ and denote their expectations (given $\boldsymbol{X}_i = \boldsymbol{x}_i$) by the vector $\boldsymbol{\mu} = (\mu_1, \ldots, \mu_n)^\top$. More precisely, we have

$$\mu_i = G(\boldsymbol{x}_i^\top \boldsymbol{\beta}).$$

The log-likelihood of the vector $\boldsymbol{Y}$ is then

$$\ell(\boldsymbol{Y}, \boldsymbol{\mu}, \psi) = \sum_{i=1}^{n} \ell(Y_i, \theta_i, \psi), \tag{5.15}$$

where $\theta_i = \theta(\eta_i) = \theta(\boldsymbol{x}_i^\top \boldsymbol{\beta})$ and $\ell(\bullet)$ on the right hand side of (5.15) denotes the individual log-likelihood contribution for each observation i.

Table 5.2. Characteristics of some GLM distributions

Notation	Range of y	$b(\theta)$	$\mu(\theta)$	Canonical link $\theta(\mu)$	Variance $V(\mu)$	$a(\psi)$
Bernoulli $B(\mu)$	$\{0,1\}$	$\log(1+e^\theta)$	$\frac{e^\theta}{1+e^\theta}$	logit	$\mu(1-\mu)$	1
Binomial $B(k,\mu)$	$[0,k]$ integer	$k\log(1+e^\theta)$	$\frac{ke^\theta}{1+e^\theta}$	$\log\left(\frac{\mu}{k-\mu}\right)$	$\mu\left(1-\frac{\mu}{k}\right)$	1
Poisson $P(\mu)$	$[0,\infty)$ integer	$\exp(\theta)$	$\exp(\theta)$	log	μ	1
Negative Binomial $NB(\mu,k)$	$[0,\infty)$ integer	$-k\log(1-e^\theta)$	$\frac{ke^\theta}{1-e^\theta}$	$\log\left(\frac{\mu}{k+\mu}\right)$	$\mu+\frac{\mu^2}{k}$	1
Normal $N(\mu,\sigma^2)$	$(-\infty,\infty)$	$\theta^2/2$	θ	identity	1	σ^2
Gamma $G(\mu,\nu)$	$(0,\infty)$	$-\log(-\theta)$	$-1/\theta$	reciprocal	μ^2	$1/\nu$
Inverse Gaussian $IG(\mu,\sigma^2)$	$(0,\infty)$	$-(-2\theta)^{1/2}$	$\frac{-1}{\sqrt{(-2\theta)}}$	squared reciprocal	μ^3	σ^2

Example 5.4.
For $Y_i \sim N(\mu_i,\sigma^2)$ we have

$$\ell(Y_i,\theta_i,\psi) = \log\left(\frac{1}{\sqrt{2\pi}\sigma}\right) - \frac{1}{2\sigma^2}(Y_i-\mu_i)^2.$$

This gives the sample log-likelihood

$$\ell(\mathbf{Y},\boldsymbol{\mu},\sigma) = n\log\left(\frac{1}{\sqrt{2\pi}\sigma}\right) - \frac{1}{2\sigma^2}\sum_{i=1}^{n}(Y_i-\mu_i)^2. \tag{5.16}$$

Obviously, maximizing the log-likelihood for $\boldsymbol{\beta}$ under normal Y is equivalent to minimizing the least squares criterion as the objective function. □

Example 5.5.
The calculation in Example 5.3 shows that the individual log-likelihood for the binary responses Y_i equals $\ell(Y_i, \theta_i, \psi) = Y_i \log(\mu_i) + (1 - Y_i) \log(1 - \mu_i)$. This leads to the sample version

$$\ell(\boldsymbol{Y}, \boldsymbol{\mu}, \psi) = \sum_{i=1}^{n} \{Y_i \log(\mu_i) + (1 - Y_i) \log(1 - \mu_i)\}. \tag{5.17}$$

Note that one typically defines $0 \cdot \log(0) = 0$. □

Let us remark that in the case where the distribution of Y itself is unknown, but its two first moments can be specified, then the quasi-likelihood may replace the log-likelihood (5.14). This means we assume that

$$E(Y) = \mu,$$
$$\text{Var}(Y) = a(\psi)\, V(\mu).$$

The quasi-likelihood is defined by

$$\ell(y, \theta, \psi) = \frac{1}{a(\psi)} \int_{\mu(\theta)}^{y} \frac{(s - y)}{V(s)}\, ds\,, \tag{5.18}$$

cf. Nelder & Wedderburn (1972). If Y comes from an exponential family then the derivatives of (5.14) and (5.18) coincide. Thus, (5.18) establishes in fact a generalization of the likelihood approach.

Alternatively to the log-likelihood the *deviance* is used often. The deviance function is defined as

$$D(\boldsymbol{Y}, \boldsymbol{\mu}, \psi) = 2\,\{\ell(\boldsymbol{Y}, \boldsymbol{\mu}^{max}, \psi) - \ell(\boldsymbol{Y}, \boldsymbol{\mu}, \psi)\}\,, \tag{5.19}$$

where $\boldsymbol{\mu}^{max}$ (typically $\boldsymbol{Y}$) is the non-restricted vector maximizing $\ell(\boldsymbol{Y}, \bullet, \psi)$. The deviance (up to the factor $a(\psi)$) is the GLM analog of the *residual sum of squares* (RSS) in linear regression and compares the log-likelihood ℓ for the "model" μ with the maximal achievable value of ℓ. Since the first term in (5.19) is not dependent on the model and therefore not on $\boldsymbol{\beta}$, minimization of the deviance corresponds exactly to maximization of the log-likelihood.

Before deriving the algorithm to determine $\boldsymbol{\beta}$, let us have a look at (5.15) again. From $\ell(Y_i, \theta_i, \psi) = \log f(Y_i, \theta_i, \psi)$ and (5.13) we see

$$\ell(\boldsymbol{Y}, \boldsymbol{\mu}, \psi) = \sum_{i=1}^{n} \left\{ \frac{Y_i \theta_i - b(\theta_i)}{a(\psi)} - c(Y_i, \psi) \right\}. \tag{5.20}$$

Obviously, neither $a(\psi)$ nor $c(Y_i, \psi)$ have an influence on the maximization, hence it is sufficient to consider

$$\widetilde{\ell}(\boldsymbol{Y}, \boldsymbol{\mu}) = \sum_{i=1}^{n} \{Y_i \theta_i - b(\theta_i)\}. \tag{5.21}$$

We will now maximize (5.21) w.r.t. $\boldsymbol{\beta}$. For that purpose take the first derivative of (5.21). This yields the gradient

$$\mathcal{D}(\boldsymbol{\beta}) = \frac{\partial}{\partial \boldsymbol{\beta}} \widetilde{\ell}(\boldsymbol{Y}, \boldsymbol{\mu}) = \sum_{i=1}^{n} \{Y_i - b'(\theta_i)\} \frac{\partial}{\partial \boldsymbol{\beta}} \theta_i \tag{5.22}$$

and our optimization problem is to solve

$$\mathcal{D}(\boldsymbol{\beta}) = 0,$$

a (in general) nonlinear system of equations in $\boldsymbol{\beta}$. For that reason, an iterative method is needed. One possible solution is the *Newton-Raphson algorithm*, a generalization of the Newton algorithm for the multidimensional parameter. Denote $\mathcal{H}(\boldsymbol{\beta})$ the *Hessian* of the log-likelihood, i.e. the matrix of second derivatives with respect to all components of $\boldsymbol{\beta}$. Then, one Newton-Raphson iteration step for $\boldsymbol{\beta}$ is

$$\widehat{\boldsymbol{\beta}}^{new} = \widehat{\boldsymbol{\beta}}^{old} - \left\{ \mathcal{H}(\widehat{\boldsymbol{\beta}}^{old}) \right\}^{-1} \mathcal{D}(\widehat{\boldsymbol{\beta}}^{old}).$$

A variant of the Newton-Raphson is the *Fisher scoring algorithm* which replaces the Hessian by its expectation (w.r.t. the observations Y_i)

$$\widehat{\boldsymbol{\beta}}^{new} = \widehat{\boldsymbol{\beta}}^{old} - \left\{ E\mathcal{H}(\widehat{\boldsymbol{\beta}}^{old}) \right\}^{-1} \mathcal{D}(\widehat{\boldsymbol{\beta}}^{old}).$$

To present both algorithms in a more detailed way, we need again some additional notation. Recall that we have $\mu_i = G(\boldsymbol{x}_i^\top \boldsymbol{\beta}) = b'(\theta_i)$, $\eta_i = \boldsymbol{x}_i^\top \boldsymbol{\beta}$ and $b'(\theta_i) = \mu_i = G(\eta_i)$. For the first and second derivatives of θ_i we obtain (after some calculation)

$$\frac{\partial}{\partial \boldsymbol{\beta}} \theta_i = \frac{G'(\eta_i)}{V(\mu_i)} \boldsymbol{x}_i$$
$$\frac{\partial^2}{\partial \boldsymbol{\beta} \boldsymbol{\beta}^\top} \theta_i = \frac{G''(\eta_i)V(\mu_i) - G'(\eta_i)^2 V'(\mu_i)}{V(\mu_i)^2} \boldsymbol{x}_i \boldsymbol{x}_i^\top.$$

Using this, we can express the gradient of the log-likelihood as

$$\mathcal{D}(\boldsymbol{\beta}) = \sum_{i=1}^{n} \{Y_i - \mu_i\} \frac{G'(\eta_i)}{V(\mu_i)} \boldsymbol{x}_i.$$

For the Hessian we get

$$\begin{aligned}\mathcal{H}(\boldsymbol{\beta}) &= \sum_{i=1}^{n}\left\{-b''(\theta_i)\left(\frac{\partial}{\partial\boldsymbol{\beta}}\theta_i\right)\left(\frac{\partial}{\partial\boldsymbol{\beta}}\theta_i\right)^\top - \{Y_i - b'(\theta_i)\}\frac{\partial^2}{\partial\boldsymbol{\beta}\boldsymbol{\beta}^\top}\theta_i\right\}\\ &= \sum_{i=1}^{n}\left\{\frac{G'(\eta_i)^2}{V(\mu_i)} - \{Y_i - \mu_i\}\frac{G''(\eta_i)V(\mu_i) - G'(\eta_i)^2V'(\mu_i)}{V(\mu_i)^2}\right\}\boldsymbol{x}_i\boldsymbol{x}_i^\top .\end{aligned}$$

Since $EY_i = \mu_i$ it turns out that the Fisher scoring algorithm is easier: We replace $\mathcal{H}(\boldsymbol{\beta})$ by

$$E\mathcal{H}(\boldsymbol{\beta}) = \sum_{i=1}^{n}\left\{\frac{G'(\eta_i)^2}{V(\mu_i)}\right\}\boldsymbol{x}_i\boldsymbol{x}_i^\top .$$

For the sake of simplicity let us concentrate on the Fisher scoring for the moment. Define the weight matrix

$$\mathbf{W} = \mathrm{diag}\left(\frac{G'(\eta_1)^2}{V(\mu_1)},\ldots,\frac{G'(\eta_n)^2}{V(\mu_n)}\right).$$

Additionally, define

$$\widetilde{\boldsymbol{Y}} = \left(\frac{Y_1-\mu_1}{G'(\eta_1)},\ldots,\frac{Y_n-\mu_n}{G'(\eta_n)}\right)^\top$$

and the design matrix

$$\mathbf{X} = \begin{pmatrix}\boldsymbol{x}_1^\top\\ \vdots\\ \boldsymbol{x}_1^\top\end{pmatrix}.$$

Then one iteration step for $\boldsymbol{\beta}$ can be rewritten as

$$\begin{aligned}\boldsymbol{\beta}^{new} &= \boldsymbol{\beta}^{old} + (\mathbf{X}^\top\mathbf{W}\mathbf{X})^{-1}\mathbf{X}^\top\mathbf{W}\widetilde{\boldsymbol{Y}}\\ &= (\mathbf{X}^\top\mathbf{W}\mathbf{X})^{-1}\mathbf{X}^\top\mathbf{W}\mathbf{Z}\end{aligned} \tag{5.23}$$

where $\mathbf{Z} = (Z_1,\ldots,Z_n)^\top$ is the vector of *adjusted dependent variables*

$$Z_i = \boldsymbol{x}_i^\top\boldsymbol{\beta}^{old} + (Y_i - \mu_i)\{G'(\eta_i)\}^{-1}. \tag{5.24}$$

The iteration stops when the parameter estimate or the log-likelihood (or both) do not change significantly any more. We denote the resulting parameter estimate by $\widehat{\boldsymbol{\beta}}$.

We see that each iteration step (5.23) is the result of a weighted least squares regression on the adjusted variables Z_i on $\boldsymbol{x}_i$. Hence, a GLM can be estimated by *iteratively reweighted least squares* (IRLS). Note further that in the linear regression model, where we have $G' \equiv 1$ and $\mu_i = \eta_i = \boldsymbol{x}_i^\top\boldsymbol{\beta}$, no iteration is necessary. The Newton-Raphson algorithm can be given in a similar way (with the more complicated weights and a different formula for the adjusted variables). There are several remarks on the algorithm:

- In the case of a canonical link function, the Newton-Raphson and the Fisher scoring algorithm coincide. Here the second derivative of θ_i is zero. Additionally we have

$$b'(\theta_i) = G(\theta_i) \quad \Longrightarrow \quad b''(\theta_i) = G'(\theta_i) = V(\mu_i).$$

This also simplifies the weight matrix $\mathbf{W}$.

- We still have to address the problem of starting values. A naive way would be just to start with some arbitrary $\boldsymbol{\beta}_0$, as e.g. $\boldsymbol{\beta}_0 = 0$. It turns out that we do not in fact need a starting value for $\boldsymbol{\beta}$ since the adjusted dependent variable can be equivalently initialized by appropriate $\eta_{i,0}$ and $\mu_{i,0}$. Typically the following choices are made, we refer here to McCullagh & Nelder (1989).
  - For all but binomial models:
    $\mu_{i,0} = Y_i$ and $\eta_{i,0} = G^{-1}(\mu_{i,0})$
  - For binomial models:
    $\mu_{i,0} = (Y_i + \frac{1}{2})/(k+1)$ and $\eta_{i,0} = G^{-1}(\mu_{i,0})$.
    (k denotes the binomial weights, i.e. $k = 1$ in the Bernoulli case.)
- An estimate $\widehat{\psi}$ for the dispersion parameter ψ can be obtained from

$$\widehat{a}(\psi) = \frac{1}{n} \sum_{i=1}^{n} \frac{(Y_i - \widehat{\mu}_i)^2}{V(\widehat{\mu}_i)}, \tag{5.25}$$

when $\widehat{\mu}_i$ denotes the estimated regression function for the ith observation.

The resulting estimator $\widehat{\boldsymbol{\beta}}$ has an asymptotic normal distribution, except of course for the standard linear regression case with normal errors where $\widehat{\boldsymbol{\beta}}$ has an exact normal distribution.

Theorem 5.1.
Under regularity conditions and as $n \to \infty$ we have for the estimated coefficient vector

$$\sqrt{n}(\widehat{\boldsymbol{\beta}} - \boldsymbol{\beta}) \xrightarrow{L} N(0, \boldsymbol{\Sigma}).$$

Denote further by $\widehat{\boldsymbol{\mu}}$ the estimator of $\boldsymbol{\mu}$. Then, for deviance and log-likelihood it holds approximately: $D(\mathbf{Y}, \widehat{\boldsymbol{\mu}}, \psi) \sim \chi^2_{n-d}$ and $2\{\ell(\mathbf{Y}, \widehat{\boldsymbol{\mu}}, \psi) - \ell(\mathbf{Y}, \boldsymbol{\mu}, \psi)\} \sim \chi^2_d$.

The asymptotic covariance of the coefficient $\widehat{\boldsymbol{\beta}}$ can be estimated by

$$\widehat{\boldsymbol{\Sigma}} = a(\widehat{\psi}) \left[\frac{1}{n} \sum_{i=1}^{n} \left\{ \frac{G'(\eta_{i,last})^2}{V(\mu_{i,last})} \right\} \mathbf{X}_i \mathbf{X}_i^\top \right]^{-1} = a(\widehat{\psi}) \cdot n \cdot \left(\mathbf{X}^\top \mathbf{W} \mathbf{X} \right)^{-1},$$

with the subscript *last* denoting the values from the last iteration step. Using this estimated covariance we can make inference about the components of β such as tests of significance. For selection between two nested models, typically a likelihood ratio test (LR test) is used.

Example 5.6.
Let us illustrate the GLM using the data on East-West German migration from Table 5.1. This is a sample of East Germans who have been surveyed in 1991 in the German Socio-Economic Panel, see GSOEP (1991). Among other questions the participants have been asked if they can imagine moving to the Western part of Germany or West Berlin. We give the value 1 for those who responded positively and 0 if not.

Recall that the economic model is based on the idea that a person will migrate if the utility (wage differential) exceeds the costs of migration. Of course neither one of the variables, wage differential and costs, are directly available. It is obvious that age has an important influence on migration intention. Younger people will have a higher wage differential. A currently low household income and unemployment will also increase a possible gain in wage after migration. On the other hand, the presence of friends or family members in the Western part of Germany will reduce the costs of migration. We also consider a city size indicator and gender as interesting variables (Table 5.1).

Table 5.3. Logit coefficients for migration data

	Coefficients	t-value
constant	0.512	2.39
FAMILY/FRIENDS	0.599	5.20
UNEMPLOYED	0.221	2.31
CITY SIZE	0.311	3.77
FEMALE	-0.240	-3.15
AGE	$-4.69 \cdot 10^{-2}$	-14.56
INCOME	$1.42 \cdot 10^{-4}$	2.73

Now, we are interested in estimating the probability of migration in dependence of the explanatory variables x. Recall, that

$$P(Y = 1|\boldsymbol{X}) = E(Y|\boldsymbol{X}).$$

A useful model is a GLM with a binary (Bernoulli) Y and the logit link for example:

$$P(Y = 1|\boldsymbol{X} = \boldsymbol{x}) = G(\boldsymbol{x}^\top \boldsymbol{\beta}) = \frac{\exp(\boldsymbol{x}^\top \boldsymbol{\beta})}{1 + \exp(\boldsymbol{x}^\top \boldsymbol{\beta})}.$$

Table 5.3 shows in the middle column the results of this logit fit. The migration intention is definitely determined by age. However, also the unemployment, city size and household income variables are highly significant, which is indicated by their high t-values $(\widehat{\beta}_j / \sqrt{\widehat{\Sigma}_{jj}})$. □

Bibliographic Notes

For general aspects on semiparametric regression we refer to the textbooks of Pagan & Ullah (1999), Yatchew (2003), Ruppert, Wand & Carroll (1990). Comprehensive presentations of the generalized linear model can be found in Dobson (2001), McCullagh & Nelder (1989) and Hardin & Hilbe (2001). For a more compact introduction see Müller (2004), Venables & Ripley (2002, Chapter 7) and Gill (2000).

In the following notes, we give some references for topics we consider related to the considered models. References for specific models are listed in the relevant chapters later on.

The transformation model in (5.4) was first introduced in an econometric context by Box & Cox (1964). The discussion was revised many years later by Bickel & Doksum (1981). In a more recent paper, Horowitz (1996) estimates this model by considering a nonparametric transformation.

For a further reference of dimension reduction in nonparametric estimation we mention projection pursuit and sliced inverse regression. The projection pursuit algorithm is introduced and investigated in detail in Friedman & Stuetzle (1981) and Friedman (1987). Sliced inverse regression means the estimation of $Y = m\left(\boldsymbol{X}^\top \beta_1, \boldsymbol{X}^\top \beta_2, \ldots, \boldsymbol{X}^\top \beta_k, \varepsilon\right)$, where ε is the disturbance term and k the unknown dimension of the model. Introduction and theory can be found e.g. in Duan & Li (1991), Li (1991) or Hsing & Carroll (1992).

More sophisticated models like censored or truncated dependent variables, models with endogenous variables or simultaneous equation systems (Maddala, 1983) will not be dealt with in this book. There are two reasons: On one hand the non- or semiparametric estimation of those models is much more complicated and technical than most of what we aim to introduce in this book. Here we only prepare the basics enabling the reader to consider more special problems. On the other hand, most of these estimation problems are rather particular and the treatment of them presupposes good knowledge of the considered problem and its solution in the parametric world. Instead of extending the book considerably by setting out this topic, we limit ourselves here to some more detailed bibliographic notes.

The non- and semiparametric literature on this is mainly separated into two directions, parametric modeling with unknown error distribution or modeling non-/semiparametrically the functional forms. In the second case a principal question is the identifiability of the model.

For an introduction to the problem of truncation, sample selection and limited dependent data, see Heckman (1976) and Heckman (1979). See also the survey of Amemiya (1984). An interesting approach was presented by Ahn & Powell (1993) for parametric censored selection models with nonparametric selection mechanism. This idea has been extended to general pairwise difference estimators for censored and truncated models in Honoré & Powell (1994). A mostly comprehensive survey about parametric and semiparametric methods for parametric models with non- or semiparametric selection bias can be found in Vella (1998). Even though implementation of and theory on these methods is often quite complicated, some of them turned out to perform reasonably well.

The second approach, i.e. relaxing the functional forms of the functions of interest, turned out to be much more complicated. To our knowledge, the first articles on the estimation of triangular simultaneous equation systems have been Newey, Powell & Vella (1999) and Rodríguez-Póo, Sperlich & Fernández (1999), from which the former is purely nonparametric, whereas the latter considers nested simultaneous equation systems and needs to specify the error distribution for identifiability reasons. Finally, Lewbel & Linton (2002) found a smart way to identify nonparametric censored and truncated regression functions; however, their estimation procedure is quite technical. Note that so far neither their estimator nor the one of Newey, Powell & Vella (1999) have been proved to perform well in practice.

Exercises

Exercise 5.1. Assume model (5.6) and consider $\boldsymbol{X}$ and ε to be independent. Show that

$$P(Y = 1|\boldsymbol{X}) = E(Y|\boldsymbol{X}) = G_\varepsilon\{v(\boldsymbol{X})\}$$

where G_ε denotes the cdf of ε. Explain that (5.7) holds if we do not assume independence of $\boldsymbol{X}$ and ε.

Exercise 5.2. Recall the paragraph about partial linear models. Why may it be sufficient to include $\beta_1 X_1$ in the model when X_1 is binary? What would you do if X_1 were categorical?

Exercise 5.3. Compute $\mathcal{H}(\boldsymbol{\beta})$ and $E\mathcal{H}(\boldsymbol{\beta})$ for the logit and probit models.

Exercise 5.4. Verify the canonical link functions for the logit and Poisson model.

Exercise 5.5. Recall that in Example 5.6 we have fitted the model

$$E(Y|\boldsymbol{X}) = P(Y = 1|\boldsymbol{X}) = G(\boldsymbol{X}^\top \boldsymbol{\beta}),$$

where G is the standard logistic cdf. We motivated this model through the latent-variable model $Y^* = \boldsymbol{X}^\top \boldsymbol{\beta} - \varepsilon$ with ε having cdf G. How does the logit model change if the latent-variable model is multiplied by a factor c? What does this imply for the identification of the coefficient vector $\boldsymbol{\beta}$?

Summary

---

* ⋆ The basis for many semiparametric regression models is the generalized linear model (GLM), which is given by

$$E(Y|\boldsymbol{X}) = G\{\boldsymbol{X}^\top \boldsymbol{\beta}\}.$$

Here, $\boldsymbol{\beta}$ denotes the parameter vector to be estimated and G denotes a known link function. Prominent examples of this type of regression are binary choice models (logit or probit) or count data models (Poisson regression).

---

* ⋆ The GLM can be generalized in several ways: Considering an unknown smooth link function (instead of G) leads to the single index model (SIM). Assuming a nonparametric additive argument of G leads to the generalized additive model (GAM), whereas a combination of additive linear and nonparametric components in the argument of G give a generalized partial linear model (GPLM) or generalized partial linear partial additive model (GAPLM). If there is no link function (or G is the identity function) then we speak of additive models (AM) or partial linear models (PLM) or additive partial linear models (APLM).

---

* ⋆ The estimation of the GLM is performed through an interactive algorithm. This algorithm, the iteratively reweighted least squares (IRLS) algorithm, applies weighted least squares to the adjusted dependent variable Z in each iteration step:

$$\boldsymbol{\beta}^{new} = (\mathbf{X}^\top \mathbf{W}\mathbf{X})^{-1}\mathbf{X}^\top \mathbf{W}\mathbf{Z}$$

This numerical approach needs to be appropriately modified for estimating the semiparametric modifications of the GLM.

6

Single Index Models

A single index model (SIM) summarizes the effects of the explanatory variables $X_1, \ldots, X_d$ within a single variable called the index. As stated at the beginning of Part II, the SIM is one possibility for generalizing the GLM or for restricting the multidimensional regression $E(Y|\mathbf{X})$ to overcome the curse of dimensionality and the lack of interpretability. For more examples of motivating the SIM see Ichimura (1993). Among others, this reference mentions duration, truncated regression (Tobit) and errors-in-variables modeling.

As already indicated, the estimation of a single index model

$$E(Y|\mathbf{X}) = m(\mathbf{X}) = g\left\{v_{\boldsymbol{\beta}}(\mathbf{X})\right\} \tag{6.1}$$

is carried out in two steps. First we estimate the coefficients vector $\boldsymbol{\beta}$, then using the index values for our observations we estimate g by ordinary univariate nonparametric regression of Y on $v_{\widehat{\boldsymbol{\beta}}}(\mathbf{X})$.

Before we proceed to the estimation problem we first have to clarify identification of the model. Next we will turn your attention to estimation methods, introducing iterative approaches as *semiparametric least squares* and *pseudo maximum likelihood* and a the non-iterative technique based on *(weighted) average derivative* estimation. Whereas the former methods work for both discrete and continuous explanatory variables, the latter needs to be modified in presence of discrete variables.

The possibility of estimating the link function g nonparametrically suggests using the SIM for a model check. Thus, at the end of this chapter we will also present a test to compare parametric models against semiparametric alternatives based on the verification of the link specification.

6.1 Identification

Before discussing how single index models can be estimated, we first have to mention an important caveat you should always keep in mind when interpreting estimation results. Consider the following binary response model with logit link G and linear index function v:

$$G(\bullet) = \frac{1}{1 + \exp\left(-\frac{\bullet - \gamma}{\tau}\right)}, \quad v(x) = \beta_0 + \beta_1 x.$$

(For the sake of simplicity we restrict ourselves to a one-dimensional explanatory variable X.) The previous equations yield

$$E(Y|X = x) = \frac{1}{1 + \exp\left(-\frac{\beta_0 + \beta_1 x - \gamma}{\tau}\right)}. \tag{6.2}$$

Note that in this model the parameters γ and τ control location and scale of the link function, respectively, whereas the parameters β_0 and β_1 represent intercept and scale of the index v. We will show now that without further specification γ and τ cannot be identified.

First, we see that γ and β_0 cannot be estimated separately but the difference $\beta_0 - \gamma$ can. To make this clear, add an arbitrary real constant c to both the intercept of the index and the location parameter of the link function:

$$G^+(\bullet) = \frac{1}{1 + \exp(-\frac{\bullet - (\gamma + c)}{\tau})}, \quad v^+(x) = \beta_0 + c + \beta_1 x.$$

It is easily verified that this specification leads to the identical model as in (6.2). As a result, we will clearly not be able to empirically distinguish between the two specifications, i.e. between the intercept parameters γ and β_0. We conclude that neither for the intercept of the index nor for the location of the link function can a unique estimate be found. For further identification we set $c = -\gamma$, respectively the mean of this link to zero, i.e. $\gamma = 0$.

Next, we demonstrate that for arbitrary τ the slope coefficient β_1 cannot be identified either. Multiplying all coefficients by a non-zero constant c yields:

$$G^*(\bullet) = \frac{1}{1 + \exp\left(-\frac{\bullet - \gamma c}{\tau c}\right)}, \quad v^*(x) = \beta_0 c + (\beta_1 c)x.$$

Again, the resulting regression $E(Y|X = x)$ will empirically not be distinguishable from the original specification (6.2). $c = 1/\tau$ would normalize the scale of the link function to 1. But since c can take any non-zero value, we are again faced with the inability to find a unique estimate of the coefficients τ, β_0 and β_1. In this case, these parameters are said to be *identified up to scale*.

The model defined by the normalization $\gamma = 0$ and $\tau = 1$ is labeled *standard logit model*. This is the typical normalization used for logit analysis. Other normalizations are conceivable. Note that in treatments of the SIM (unknown link) many authors assume β_0 to be part of the nonparametric function $g(\bullet)$. Moreover, in cases with several explanatory variables one of the following scale normalizations can be applied:

- set one of the slope coefficients equal to 1, or
- set the length of the coefficient vector to 1.

Hence, normalization in the standard logit model usually differs from that made for the SIM. Consequently, estimated coefficients for logit/probit and SIM can be only compared if the same type of normalization is applied to both coefficient vectors.

An additional identification problem arises when the distribution of the error term in the latent-variable model (5.6) depends on the value of the index. We have already seen some implications of this type of heteroscedasticity in Example 1.6. To make this point clear, consider the latent variable

$$Y^* = v_\beta(\mathbf{X}) - \varepsilon, \quad \varepsilon = \varpi\{v_\beta(\mathbf{X})\}\cdot\zeta$$

with $\varpi(\bullet)$ an unknown function and ζ a standard logistic error term independent of $\mathbf{X}$. (In Example 1.6 we studied a linear index function v combined with $\varpi(u) = \frac{1}{2}\{1+u^2\}$.) The same calculation as for (5.7) (cf. Exercise 5.1) shows now

$$P(Y=1 \mid \mathbf{X}=\boldsymbol{x}) = E(Y \mid \mathbf{X}=\boldsymbol{x}) = G_\zeta\left[\frac{v_\beta(\boldsymbol{x})}{\varpi\{v_\beta(\mathbf{X})\}}\right]. \tag{6.3}$$

This means, even if we know the functional form of the link (G_ζ is the standard logistic cdf) the regression function is unknown because of its unknown component $\varpi(\bullet)$. As a consequence, the resulting link function

$$g(\bullet) = G_\zeta\left[\frac{\bullet}{\varpi\{\bullet\}}\right]$$

is not necessarily monotone increasing any more. For instance, in Figure 1.9 we plotted a graph of the link function g as a result of the heteroscedastic error term.

For another very simple example, consider Figure 6.1. Here, we have drawn two link functions, $G(\bullet)$ (upward sloping) and $G^-(\bullet)$ (downward sloping). Note that both functions are symmetric (the symmetry axis is the dashed vertical line through the origin). Thus, $G(\bullet)$ and $G^-(\bullet)$ are related to each other by

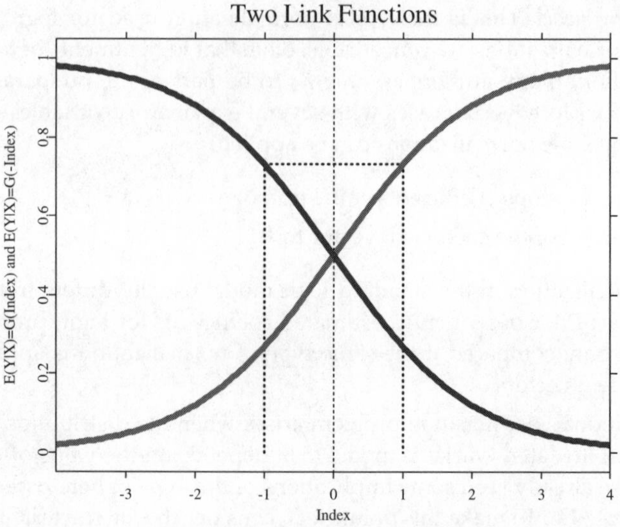

Figure 6.1. Two link functions

$$G^{-}(u) = G(-u).$$

Obviously now, two distinct index values, $u = x^{\top}\beta$ and $u^{-} = -x^{\top}\beta$ will yield the same values of $E(Y|X = x)$ if the link functions are chosen accordingly. Since the two indices differ only in that the coefficients have different sign, we conclude that the coefficients are identified only up to scale and sign in the most general case. In summary, to compare the effects of two particular explanatory variables, we can only use the ratio of their coefficients.

6.2 Estimation

When estimating a SIM, we have to take into account that the functional form of the link function is unknown. Moreover, since the shape of $g(\bullet)$ will determine the value of

$$E(Y|X) = g\{v_{\beta}(X)\}$$

for a given index $v_{\beta}(X)$, estimation of the index weights β will have to adjust to a specific estimate of the link function to yield a "correct" regression value. Thus, in SIM both the index and the link function have to be estimated, even though only the link function is of nonparametric character.

Recall that $\boldsymbol{X}$ is the vector of explanatory variables, $\boldsymbol{\beta}$ is a d-dimensional vector of unknown coefficients (or weights), and $g(\bullet)$ is an arbitrary smooth function. Let

$$\varepsilon = m(\boldsymbol{X}) - Y = E(Y|\boldsymbol{X}) - Y$$

be the deviation of Y from its conditional expectation w.r.t. $\boldsymbol{X}$. Using ε we can write down the single index model as

$$Y = g\left\{v_{\boldsymbol{\beta}}(\boldsymbol{X})\right\} + \varepsilon. \tag{6.4}$$

Our goal is to find efficient estimators for $\boldsymbol{\beta}$ and $g(\bullet)$. As $\boldsymbol{\beta}$ is inside the nonparametric link, the challenge is to find an appropriate estimator for $\boldsymbol{\beta}$, in particular one that reaches the $\sqrt{n}$-rate of convergence. (Recall that $\sqrt{n}$-convergence is typically achieved by parametric estimators.)

Two essentially different approaches exist for this purpose:

- An iterative approximation of $\boldsymbol{\beta}$ by semiparametric least squares (SLS) or pseudo maximum likelihood estimation (PMLE),
- a direct (non-iterative) estimation of $\boldsymbol{\beta}$ through the average derivative of the regression function (ADE, WADE).

In both cases the estimation procedure can be summarized as:

SIM Algorithm
(a) estimate $\boldsymbol{\beta}$ by $\widehat{\boldsymbol{\beta}}$
(b) compute index values $\widehat{\eta} = \boldsymbol{X}^{\top}\widehat{\boldsymbol{\beta}}$
(c) estimate the link function $g(\bullet)$ by using a (univariate) nonparametric method for the regression of Y on $\widehat{\eta}$

The final step is relevant for all SIM estimators considered in the following sections. More precisely: Suppose that $\boldsymbol{\beta}$ has already been estimated by $\widehat{\boldsymbol{\beta}}$. Set

$$\widehat{\eta}_i = \boldsymbol{X}_i^{\top}\widehat{\boldsymbol{\beta}}, \quad i = 1, \ldots, n. \tag{6.5}$$

Once we have created these "observations" of the new dependent variable $\widehat{\eta} = \boldsymbol{X}^{\top}\widehat{\boldsymbol{\beta}}$ we have a standard univariate regression problem. For example, a possible and convenient choice for estimating $g(\bullet)$ would be the Nadaraya-Watson estimator introduced in Chapter 4.1:

$$\widehat{g}_h(z) = \frac{\sum_{i=1}^{n} K_h(z - \widehat{\eta}_i)Y_i}{\sum_{i=1}^{n} K_h(z - \widehat{\eta}_i)}. \tag{6.6}$$

It can be shown that if β is $\sqrt{n}$-consistent, $\widehat{g}_h(z)$ converges for $h \sim n^{-1/5}$ at rate $n^{2/5} = \sqrt{nh}$, i.e. like a univariate kernel regression estimator:

$$n^{2/5}\left\{\widehat{g}_h(z) - g(z)\right\} \stackrel{L}{\longrightarrow} N(b_z, v_z)$$

with bias b_z and variance v_z, see e.g. Powell, Stock & Stoker (1989).

As in the first part of this book we concentrate on kernel based methods here. The choice of h and K in (6.6) is independent of the choice of these parameters in the following sections. For other than kernel based SIM estimators we refer to the bibliographic notes.

6.2.1 Semiparametric Least Squares

As indicated in the introduction we are going to concentrate on the estimation of β now. The methods that we consider here under *semiparametric least squares* (SLS) and *pseudo maximum likelihood estimation* (PMLE) have the following idea in common: establish an appropriate objective function to estimate β with parametric $\sqrt{n}$-rate. Certainly, inside the objective function we use the conditional distribution of Y, or the link function g, or both of them. As these are unknown they need to be replaced by nonparametric estimates. The objective function then will be maximized (or minimized) with respect to the parameter β.

Why is this extension of least squares or maximum likelihood not a trivial one? The reason is that when β changes, the link g (respectively its nonparametric substitute) may change simultaneously. Thus, it is not clear if the necessary iteration will converge and even if it does, that it will converge to a unique optimum.

SLS and its weighted version (WSLS) have been introduced by Ichimura (1993). As SLS is just a special case of WSLS with a weighting equal to the identity matrix, we concentrate here on WSLS. An objective function of least squares type can be motivated by minimizing the variation in the data that can not be explained by the fitted regression. This "left over" variation can be written as

$$\text{Var}\{Y|v_\beta(\boldsymbol{X})\} = E\left[\left\{Y - E[Y|v_\beta(\boldsymbol{X})]\right\}^2 |v_\beta(\boldsymbol{X})\right]$$

with $v_\beta(\bullet)$ being the index function specified up to β. The previous equation leads us to a variant of the well known LS criterion

$$\min_\beta E\left[Y - E\{Y|v_\beta(\boldsymbol{X})\}\right]^2 \tag{6.7}$$

in which $v_\beta(\boldsymbol{X})$ on the right hand side has to be replaced by a (consistent) estimate. The outer expectation in (6.7) can be replaced by an average.

We can account for possible heteroscedasticity by using proper weights. This motivates

$$\min_{\beta} \frac{1}{n} \sum_{i=1}^{n} \left[Y_i - E\{Y_i | v_{\beta}(\mathbf{X}_i)\} \right]^2 w(\mathbf{X}_i) \tag{6.8}$$

with $w(\bullet)$ an appropriate weight function. Next, employ the nonparametric technique to substitute the (inner) unknown conditional expectation. As the index function v_{β} is univariate, we could take any univariate consistent smoother. For simplicity of the presentation, we will use a Nadaraya-Watson type estimator here.

Define the WSLS estimator as

$$\widehat{\beta} = \min_{\beta} \frac{1}{n} \sum_{i=1}^{n} \{ Y_i - \widehat{m}_{\beta}(\mathbf{X}_i) \}^2 w(\mathbf{X}_i) \, \mathrm{I}(\mathbf{X}_i \in \mathcal{X}) \tag{6.9}$$

where $\mathrm{I}(\mathbf{X}_i \in \mathcal{X})$ is a trimming factor and $\widehat{m}_{\beta}$ a leave-one-out estimator of m assuming the parameter β would be known. In more detail, $\widehat{m}_{\beta}$ is a (weighted) Nadaraya-Watson estimator

$$\widehat{m}_{\beta}(\mathbf{X}_i) = \frac{\sum_{j \neq i} Y_j K_h \{ v_{\beta}(\mathbf{X}_i) - v_{\beta}(X_j) \} \, w(X_j) \, \mathrm{I}(X_j \in \mathcal{X}_n)}{\sum_{j \neq i} K_h \{ v_{\beta}(\mathbf{X}_i) - v_{\beta}(X_j) \} \, w(X_j) \, \mathrm{I}(X_j \in \mathcal{X}_n)} \tag{6.10}$$

with h denoting a bandwidth and K_h a scaled (compact support) kernel. The trimming factor $\mathrm{I}(\mathbf{X}_i \in \mathcal{X})$ has been introduced to guarantee that the density of the index v is bounded away from zero. $\mathcal{X}$ has to be chosen accordingly. The set $\mathcal{X}_n$ in the trimming factor of (6.10) is constructed in such a way that all boundary points of $\mathcal{X}$ are interior to $\mathcal{X}_n$:

$$\mathcal{X}_n = \{ x : \|x - z\| \le 2h \text{ for some } z \in \mathcal{X} \}.$$

In practice trimming can often be skipped, but it is helpful for establishing asymptotic theory. Also the choice of taking the leave-one-out version of the Nadaraya-Watson estimator in (6.10) is for that reason. For the estimator given (6.9) the following asymptotic results can be proved, for the details see Ichimura (1993).

Theorem 6.1.
Assume that Y has a κth absolute moment ($\kappa \ge 3$) and follows model (6.4). Suppose further that the bandwidth h converges to 0 at a certain rate. Then, under regularity conditions on the link g, the error term ε, the regressors $\mathbf{X}$ and the index function v, the WSLS estimator (6.9) fulfills

$$\sqrt{n}(\widehat{\beta} - \beta) \xrightarrow{L} N(0, \mathbf{V}^{-1} \mathbf{\Sigma} \mathbf{V}^{-1}),$$

where, using the notation

$$\nabla_v = \left.\frac{\partial v_\beta(\mathbf{X})}{\partial \beta}\right|_\beta, \quad \nabla_m = g'\{v_\beta(\mathbf{X})\}\left(\nabla_v - \frac{E\{\nabla_v w(\mathbf{X})|v_\beta(\mathbf{X}), \mathbf{X} \in \mathcal{X}\}}{E\{w(\mathbf{X})|v_\beta(\mathbf{X}), \mathbf{X} \in \mathcal{X}\}}\right)$$

and $\sigma^2(\mathbf{X}) = \mathrm{Var}(Y|\mathbf{X})$, *the matrices* $\mathbf{V}$ *and* $\mathbf{\Sigma}$ *are defined by*

$$\mathbf{V} = E\left\{w(\mathbf{X})\,\nabla_v\,\nabla_v{}^\top|\mathbf{X} \in \mathcal{X}\right\},$$
$$\mathbf{\Sigma} = E\left\{w^2(\mathbf{X})\sigma^2(\mathbf{X})\,\nabla_v\,\nabla_v{}^\top|\mathbf{X} \in \mathcal{X}\right\}.$$

As we can see, the estimator $\widehat{\beta}$ is unbiased and converges at parametric $\sqrt{n}$-rate. Additionally, by choosing the weight

$$w(\mathbf{X}) = \frac{1}{\sigma^2(\mathbf{X})},$$

the WSLS reaches the efficiency bound for the parameter estimates in SIM (Newey, 1990). This means, $\widehat{\beta}$ is an unbiased parameter estimator with optimal rate and asymptotically efficient covariance matrix. Unfortunately, $\sigma^2(\bullet)$ is an unknown function as well. In practice, one therefore uses an appropriate pre-estimate for the variance function.

How do we estimate the (asymptotic) variance of the estimator? Not surprisingly, the expressions $\mathbf{V}$ and $\mathbf{\Sigma}$ are estimated consistently by their empirical analogs. Denote

$$\widehat{\nabla}_m = \left.\frac{\partial \widehat{m}_\beta(\mathbf{X}_i)}{\partial \beta}\right|_{\widehat{\beta}}.$$

Ichimura (1993) proves that

$$\widehat{\mathbf{V}} = \frac{1}{n}\sum_{i=1}^n w(\mathbf{X}_i)\,\mathrm{I}(\mathbf{X}_i \in \mathcal{X}_n)\widehat{\nabla}_m\widehat{\nabla}_m^\top,$$
$$\widehat{\mathbf{\Sigma}} = \frac{1}{n}\sum_{i=1}^n w^2(\mathbf{X}_i)\,\mathrm{I}(\mathbf{X}_i \in \mathcal{X}_n)\,\{Y_i - \widehat{m}_\beta(\mathbf{X}_i)\}^2\,\widehat{\nabla}_m\widehat{\nabla}_m^\top$$

are consistent estimators for $\mathbf{V}$ and $\mathbf{\Sigma}$ under appropriate bandwidth conditions.

6.2.2 Pseudo Likelihood Estimation

For motivation of *pseudo maximum likelihood estimation* (PMLE) we rely on the ideas developed in the previous section. Let us first discuss why pseudo maximum likelihood always leads to an unbiased $\sqrt{n}$-consistent estimators with minimum achievable variance: In fact, the computation of the PMLE reduces to a formal parametric MLE problem with as many parameters as observations. In this case (as we have also seen above), the inverse Fisher information turns out to be a consistent estimator for the covariance matrix of the PMLE. Gill (1989) and Gill & van der Vaart (1993) explain this as follows: a sensibly defined nonparametric MLE can be seen as a MLE in any parametric submodel which happens to include or pass through the point given by the PMLE. For smooth parametric submodels, the MLE solves the likelihood equations. Consequently, also in nonparametric problems the PMLE can be interpreted as the solution of the likelihood equations for every parametric submodel passing through it.

You may have realized that the mentioned properties coincide with the results made for the WSLS above. Indeed, looking closer at the objective function in (6.7) we could re-interpret it as the result of a maximum likelihood consideration. We only have to set weight w equal to the inverse of the (known) variance function (compare the discussion of optimal weighting). We refer to the bibliographic notes for more details.

To finally introduce the PMLE, let us come back to the issue of a binary response model, i.e., observe Y only in $\{0,1\}$. Recall that this means

$$Y = \begin{cases} 1 & \text{if } v_{\boldsymbol{\beta}}(\boldsymbol{X}) > \varepsilon, \\ 0 & \text{otherwise,} \end{cases} \tag{6.11}$$

where the index function $v_{\boldsymbol{\beta}}$ is known up to $\boldsymbol{\beta}$. We further assume that

$$E(Y|\boldsymbol{X}) = E\{Y|v_{\boldsymbol{\beta}}(\boldsymbol{X})\}.$$

This is indeed an additional restriction but still allowing for multiplicative heteroscedasticity as discussed around equation (6.3). Unfortunately, under heteroscedasticity — when the variance function of the error term depends on the index to estimate — a change of $\boldsymbol{\beta}$ changes the variance function and thus implicitly affects equation (6.11) through ε. This has consequences for PMLE as the likelihood function is determined by (6.11) and the error distribution given the index $v_{\boldsymbol{\beta}}(\bullet)$. For this reason we recall the notation

$$\varepsilon = \varpi\{v(\boldsymbol{X}, \boldsymbol{b})\} \cdot \zeta$$

where ϖ is the variance function and ζ an error term independent of $\boldsymbol{X}$ and v. In the case of homoscedasticity we would have indeed $\varepsilon = \zeta$. From (5.7) we know that

$$E\{Y|v_{\boldsymbol{\beta}}(\boldsymbol{X})\} = G_{\varepsilon|\boldsymbol{X}}\{v_{\boldsymbol{\beta}}(\boldsymbol{X})\},$$

where $G_{\varepsilon|\boldsymbol{X}}$ is the conditional distribution of the error term. Since Y is binary, the log-likelihood function for this model (cf. (5.17)) is given by

$$\frac{1}{n}\sum_{i=1}^{n}\left(Y_i \log\left[G_{\varepsilon|X_i}\{v_\beta(X_i)\}\right] + (1-Y_i)\log\left[1 - G_{\varepsilon|X_i}\{v_\beta(X_i)\}\right]\right). \quad (6.12)$$

The problem is now to obtain a substitute for the unknown function $G_{\varepsilon|X_i}$. We see that

$$G_{\varepsilon|x}(v) = P(\varepsilon < v|v) = P(\varepsilon < v)\frac{\psi_{\varepsilon<v}(v)}{\psi(v)} \quad (6.13)$$

with ψ being the pdf of the index $v_\beta(X)$ and $\psi_{\varepsilon<v}$ the conditional pdf of $v_\beta(X)$ given $\varepsilon < v_\beta(X)$. Since $\varepsilon < v_\beta(X)$ if and only if $Y = 1$, this is equivalent to

$$G_{\varepsilon|x}(v) = P(Y=1)\frac{\psi_{Y=1}(v)}{\psi(v)}.$$

In the last expression, we can estimate all expressions nonparametrically. Instead of estimating $P(Y=1)$ by $\overline{Y}$, Klein & Spady (1993) propose to consider $\psi_y = P(Y=y)\psi_{Y=1}$ and $\psi = \psi_0 + \psi_1$. One can estimate

$$\widehat{\psi}_y(v) = \frac{1}{n-1}\sum_{j\neq i} \mathrm{I}(Y_j = y) K_h\{v_\beta(X_j) - v\}, \quad (6.14)$$

where $K_h(\bullet)$ denotes the scaled kernel as before. Hence, an estimate for $G_{\varepsilon|X_i}$ in (6.12) is given by

$$\widehat{G}_{\varepsilon|X_i}\{v_\beta(X_i)\} = \frac{\sum_{j\neq i}\mathrm{I}(Y_j=1)K_h\{v_\beta(X_j) - v_\beta(X_i)\}}{\sum_{j\neq i}K_h\{v_\beta(X_j) - v_\beta(X_i)\}}.$$

To obtain the $\sqrt{n}$-rate for $\widehat{\beta}$, one uses either bias reduction via higher order kernels or an adaptive undersmoothing bandwidth h. Problems in the denominator when the density estimate becomes small, can be avoided by adding small terms in both the denominator and the numerator. These terms have to vanish at a faster rate than that for the convergence of the densities.

We can now define the pseudo log-likelihood version of (6.12) by

$$\frac{1}{n}\sum_{i=1}^{n} w(X_i)\left\{Y_i \log\left[\widehat{G}_{\varepsilon|X_i}\{v_\beta(X_i)\}\right]^2 + (1-Y_i)\log\left[1 - \widehat{G}_{\varepsilon|X_i}\{v_\beta(X)\}\right]^2\right\} \quad (6.15)$$

The weight function $w(\bullet)$ can be introduced for numerical or technical reasons. Taking the squares inside the logarithms avoids numerical problems when using higher order kernels. (Otherwise these terms could be negative.) The estimator $\widehat{\beta}$ is found by maximizing (6.15).

Klein & Spady (1993) prove the following result on the asymptotic behavior of the PMLE $\widehat{\beta}$. More details about conditions for consistency, the choice of the weight function w and the appropriate adaptive bandwidth h can be found in their article. We summarize the asymptotic distribution in the following theorem.

Theorem 6.2.
Under some regularity conditions it holds

$$\sqrt{n}\left(\widehat{\beta}-\beta\right) \stackrel{L}{\longrightarrow} N(0,\Sigma)$$

where

$$\Sigma^{-1} = E\left\{\frac{\partial\Gamma}{\partial\beta}\frac{\partial\Gamma}{\partial\beta^\top}\frac{1}{\Gamma(1-\Gamma)}\right\} \quad \textit{and} \quad \Gamma = G_{\varepsilon|X_i}\{v_\beta(X_i)\}.$$

It turns out that the Klein & Spady estimator $\widehat{\beta}$ reaches the efficiency bound for SIM estimators. An estimator for Σ can be obtained by its empirical analog.

As the derivation shows, the Klein & Spady PMLE $\widehat{\beta}$ does only work for binary response models. In contrast, the WSLS was given for an arbitrary distribution of Y. For that reason we consider now an estimator that generalizes the idea of Klein & Spady to arbitrary distributions for Y.

Typically, the pseudo log-likelihood is based on the density f (if Y is continuous) or on the probability function f (if Y is discrete). The main idea of the following estimator first proposed by Delecroix, Härdle & Hristache (2003) is that the function f which defines the distribution of Y given X only depends on the index function v, i.e.,

$$f(y|x) = f\{y|v_\beta(x)\}.$$

In other words, the index function v contains all relevant information about X. The objective function to maximize is

$$E\left[\log f\{Y|v_\beta(X)\}\right].$$

As for SLS we proxy this expectation by averaging and introduce a trimming function:

$$\frac{1}{n}\sum_{i=1}^{n} \mathrm{I}\{(Y_i, X_i) \in \mathcal{S}\} \log f\{Y_i|v_\beta(X_i)\} \tag{6.16}$$

where $\mathcal{S}$ denotes a suitable subset of the support of (Y, X). Here, all we have to do is to estimate the conditional density or probability mass function $f\{y|v_\beta(x)\}$. An estimator is given by:

$$\widehat{f}\{Y_i|v_\beta(X_i)\} = \frac{\sum_{j\neq i} \mathrm{I}\{(Y_j, X_j) \in \mathcal{S}\} K_h\left\{Y_i - Y_j\right\} K_h\left\{v_\beta(X_i) - v_\beta(X_j)\right\}}{\sum_{j\neq i} \mathrm{I}\{(Y_j, X_j) \in \mathcal{S}\} K_h\left\{v_\beta(X_i) - v_\beta(X_j)\right\}}. \tag{6.17}$$

To reach the $\sqrt{n}$-rate, fourth order kernels and a bandwidth of rate $n^{-\delta}, \delta \in (1/8, 1/6)$, needs to be used. Delecroix, Härdle & Hristache (2003) present their result for the linear index function

$$v_{\beta}(X) = X^{\top}\beta.$$

Therefore the following theorem only considers the asymptotic results for that special, but most common case.

Theorem 6.3.
Let $\widehat{\beta}$ be the vector that maximizes (6.16) under the use of (6.17). Then under the above mentioned and further regularity conditions it holds

$$\sqrt{n}\left(\widehat{\beta} - \beta\right) \rightarrow N(0, \mathbf{V}^{-1}\mathbf{\Sigma}\mathbf{V}^{-1})$$

where

$$\mathbf{\Sigma} = E\left\{\frac{\partial \log f(Y|X^{\top}\beta)}{\partial \beta}\frac{\partial \log f(Y|X^{\top}\beta)}{\partial b^{\top}} \mathrm{I}\{(Y, X) \in \mathcal{S}\}\right\}$$

and

$$\mathbf{V} = E\left\{-\frac{\partial^2 \log f(Y|X^{\top}\beta)}{\partial \beta \partial \beta^{\top}} \mathrm{I}\{(Y, X) \in \mathcal{S}\}\right\}.$$

Again, it can be shown that the variance of this estimator reaches the lower efficiency bound for SIM. That means this procedure provides efficient estimators for β, too. Estimates for the matrices $\mathbf{V}$ and $\mathbf{\Sigma}$ can be found by their empirical analogs.

6.2.3 Weighted Average Derivative Estimation

We will now turn to a different type of estimator with two advantages: (a) we do not need any distributional assumption on Y and (b) the resulting estimator is direct, i.e. non-iterative. The basic idea is to identify β as the average derivative and thus the studied estimator is called *average derivative estimator* (ADE) or *weighted average derivative estimator* (WADE). The advantages of ADE or WADE estimators come at a cost, though, as they are inefficient. Furthermore, the average derivative method is only directly applicable to models with continuous explanatory variables.

At the end of this section we will discuss how to estimate the coefficients of discrete explanatory variables in SIM, a method that can be combined with the ADE/WADE method. For this reason we introduce the notation

$$X = (T, U)$$

for the regressors. Here, T (q-dimensional) refers explicitly to continuous variables and U (p-dimensional) to discrete (or categorical) variables.

Let us first consider a model with a d-dimensional vector of continuous variables only, i.e.,

$$E(Y|\boldsymbol{T}) = m(\boldsymbol{T}) = g(\boldsymbol{T}^\top \gamma). \tag{6.18}$$

Then, the vector of weighted average derivatives δ is given by

$$\delta = E\left\{\nabla_m(\boldsymbol{T})\, w(\boldsymbol{T})\right\} = E\left\{g'(\boldsymbol{T}^\top\gamma) w(\boldsymbol{T})\right\} \gamma, \tag{6.19}$$

where $\nabla_m(\boldsymbol{T}) = \left(\partial_1 m(\boldsymbol{T}), \ldots, \partial_q m(\boldsymbol{T})\right)^\top$ is the vector of partial derivatives of $m(\bullet)$, g' the derivative of $g(\bullet)$ and $w(\bullet)$ a weight function. (By ∂_k we denote the partial derivative w.r.t. the kth argument of the function.)

Looking at (6.19) shows that δ equals γ up to scale. Hence, if we find a way to estimate δ then we can also estimate γ up to scale. The approach studied in Powell, Stock & Stoker (1989) uses the density $f(\bullet)$ of $\boldsymbol{T}$ as the weight function:

$$w(t) = f(t).$$

This estimator is sometimes referred to as *density weighted* ADE or DWADE. We will concentrate on this particular weight function. Generalizations to other weight functions are possible.

For deriving the estimator, it is instructive to write (6.19) in more detail:

$$\int_{\mathbb{R}^q} \nabla_m(\boldsymbol{t}) f^2(\boldsymbol{t})\, d\boldsymbol{t} = \underbrace{\int_{\mathbb{R}} \cdots \int_{\mathbb{R}}}_{q} \left(\partial_1 m(\boldsymbol{t}), \ldots, \partial_q m(\boldsymbol{t})\right)^\top f^2(\boldsymbol{t})\, dt_1 \cdots dt_q.$$

Partial integration yields

$$\delta = \begin{pmatrix} -2\int\cdots\int \partial_1 f(\boldsymbol{t})\, f(\boldsymbol{t})\, m(\boldsymbol{t})\, dt_1 \cdots dt_q \\ \vdots \\ -2\int\cdots\int \partial_q f(\boldsymbol{t})\, f(\boldsymbol{t})\, m(\boldsymbol{t})\, dt_1 \cdots dt_q \end{pmatrix} = -2E\{\nabla_f(\boldsymbol{t}) m(\boldsymbol{t})\}, \tag{6.20}$$

if we assume that $f(\boldsymbol{t})\, m(\boldsymbol{t}) \to 0$ for $\|\boldsymbol{t}\| \to \infty$. Noting that $m(\boldsymbol{t}) = E(Y|\boldsymbol{t})$ and using the law of iterated expectations we finally arrive at

$$E\{\nabla_f(\boldsymbol{T}) m(\boldsymbol{T})\} = E[E\{\nabla_f(\boldsymbol{T}) Y|\boldsymbol{T}\}] = E\{\nabla_f(\boldsymbol{T}) Y\}. \tag{6.21}$$

We can now estimate δ by using the sample analog of the right hand side of (6.21):

$$\widehat{\delta} = -\frac{2}{n}\sum_{i=1}^{n} \widehat{\nabla}_{f_h}(\boldsymbol{T}_i) Y_i, \tag{6.22}$$

where we estimate ∇_f by

$$\widehat{\nabla}_{f_h}(\boldsymbol{t}) = \left(-\partial_1 \widehat{f}_h(\boldsymbol{t}), \ldots, -\partial_q \widehat{f}_h(\boldsymbol{t})\right)^\top. \tag{6.23}$$

Here, the $\partial_k \widehat{f}_h(\boldsymbol{T})$, are the partial derivatives of the multivariate kernel density estimator from Section 3.6, i.e.,

$$\partial_k \widehat{f}_h(\boldsymbol{t}) = \frac{1}{n\, h_1 \cdots h_q} \sum_{j=1}^{n} \partial_k \mathcal{K}\left(\frac{t_1 - T_{j1}}{h_1}, \ldots, \frac{t_q - T_{jd}}{h_q}\right).$$

Regarding the sampling distribution of the estimator $\widehat{\delta}$ Powell, Stock & Stoker (1989) have shown the following theorem.

Theorem 6.4.
Under regularity conditions we have

$$\sqrt{n}(\widehat{\delta} - \delta) \xrightarrow{L} N(0, \Sigma).$$

The covariance matrix is given by $\Sigma = 4E(\boldsymbol{r} - \delta)(\boldsymbol{r} - \delta)^\top$ *where* $\boldsymbol{r}$ *is given by* $\boldsymbol{r} = f(\boldsymbol{T})\, \nabla_m(\boldsymbol{T}) - \{Y - m(\boldsymbol{T})\}\, \nabla_f(\boldsymbol{T})$.

Note that although $\widehat{\delta}$ is based on a multidimensional kernel density estimator, it achieves $\sqrt{n}$-convergence as the SIM estimators considered previously which were all based on univariate kernel smoothing.

Example 6.1.
We cite here an example on unemployment after completion of an apprenticeship in West Germany which was first analyzed in Proença & Werwatz (1995). The data comprise individuals from the first nine waves of the German Socio-Economic Panel (GSOEP), see GSOEP (1991). The dependent variable Y takes on the value 1 if the individual is unemployed one year after completing the apprenticeship. Explanatory variables are gross monthly earnings as an apprentice (EARNINGS), city size (CITY SIZE), percentage of people apprenticed in a certain occupation divided by the percentage of people employed in this occupation in the entire company (DEGREE) and unemployment rate in the state where the individual lived (URATE).

Table 6.1 shows the results of a GLM fit (using the logistic link function) and the WADE coefficients (for different bandwidths). For easier comparison the coefficients are rescaled such that all coefficients of EARNINGS are equal to -1. To eliminate the possible effects of the correlation between the variables and to standardize the data, a Mahalanobis transformation had been applied before computing the WADE. Note that in particular for $h = 1.5$ the coefficients of both the GLM and WADE are very close. One may thus argue that the parametric logit model is not grossly misspecified. □

Let us now turn to the problem of estimating coefficients for discrete or categorical variables. By definition, derivatives can only be calculated if the

Table 6.1. WADE fit of unemployment data

	GLM	WADE				
	(logit)	$h=1$	$h=1.25$	$h=1.5$	$h=1.75$	$h=2$
constant	-5630	—	—	—	—	—
EARNINGS	-1.00	-1.00	-1.00	-1.00	-1.00	-1.00
CITY SIZE	-0.72	-0.23	-0.47	-0.66	-0.81	-0.91
DEGREE	-1.79	-1.06	-1.52	-1.93	-2.25	-2.47
URATE	363.03	169.63	245.75	319.48	384.46	483.31

variable under study is continuous. Thus, the WADE method fails for discrete explanatory variables. Before presenting a more general solution, let us explain how the coefficient of one dichotomous variable is introduced to the model. We extend model (6.18) by an additional term:

$$E(Y|\boldsymbol{T},\boldsymbol{U}) = g(\boldsymbol{T}^\top\gamma + \boldsymbol{U}^\top\beta) \tag{6.24}$$

with $\boldsymbol{T}$ the continuous and $\boldsymbol{U}$ the discrete part of the regressors. In the simplest case we suppose that discrete part is a univariate binary variable U and that Y is binary as well. In this case, the model "splits" into two submodels

$$\begin{aligned} E(Y|\boldsymbol{T},U) &= P(Y=1|\boldsymbol{T},U) = g(\boldsymbol{T}^\top\gamma) \quad &\text{if } U=0, \\ E(Y|\boldsymbol{T},U) &= P(Y=1|\boldsymbol{T},U) = g(\boldsymbol{T}^\top\gamma+\beta) &\text{if } U=1. \end{aligned}$$

There are in fact two models to be estimated, one for $U=0$ and one for $U=1$. Note that γ alone could be estimated from the first equation only.

Theoretically, the same $\boldsymbol{T}_i$ can be associated with either $U_i=0$ yielding an index value of $\boldsymbol{T}_i^\top\gamma$ or with $U_i=1$ leading to an index value of $\boldsymbol{T}_i^\top\gamma+\beta$. Thus the difference between the two indices is exactly β, see the left panel of Figure 6.2.

In practice finding these *horizontal* differences will be rather difficult. A common approach is based on the observation that the *integral* difference between the two link functions also equals β, see the right panel of Figure 6.2.

A very simple estimator is proposed in Korostelev & Müller (1995). Essentially, the coefficient of the binary explanatory variable can be estimated by

$$\widehat{\beta} = \widehat{J}^{(1)} - \widehat{J}^{(0)}$$

with

$$\widehat{J}^{(0)} = \sum_{i=0}^{n_0} Y_i^{(0)}(\boldsymbol{T}_{i+1}^{(0)} - \boldsymbol{T}_i^{(0)})^\top\gamma, \quad \widehat{J}^{(1)} = \sum_{i=0}^{n_1} Y_i^{(1)}(\boldsymbol{T}_{i+1}^{(1)} - \boldsymbol{T}_i^{(1)})^\top\gamma,$$

where the superscripts $^{(0)}$ and $^{(1)}$ denote the observations from the subsamples according to $U=0$ and $U=1$. The estimator is in the simplest case of

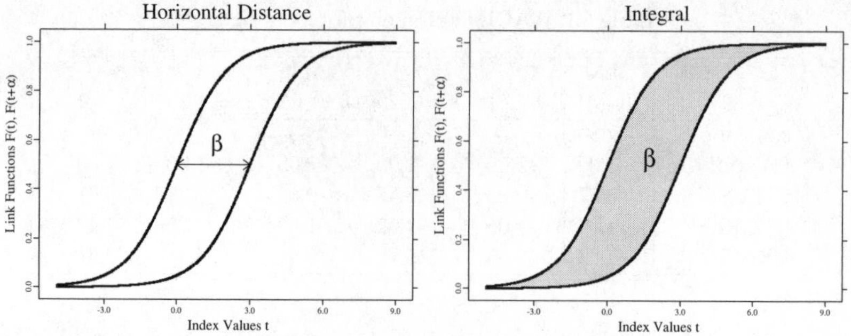

Figure 6.2. The horizontal and the integral approach

a binary Y variable $\sqrt{n}$-consistent and can be improved for efficiency by a one-step estimator (Korostelev & Müller, 1995).

Horowitz & Härdle (1996) extend this approach to multivariate multicategorical $\boldsymbol{U}$ and an arbitrary distribution of Y. Recall the model (6.24)

$$E(Y|\boldsymbol{T},\boldsymbol{U}) = g(\boldsymbol{T}^\top \gamma + \boldsymbol{U}^\top \boldsymbol{\beta}).$$

Again, the approach for this model is based on a split of the whole sample into subsamples according to the categories of $\boldsymbol{U}$. However, this subsampling makes the estimator infeasible for more than one or two discrete variables. To compute integral differences of the link functions according to the realizations of $\boldsymbol{U}$, we consider the truncated link function

$$\widetilde{g} = c_o \, \mathrm{I}(g < c_o) + g \, \mathrm{I}(c_o \le g \le c_1) + c_1 \, \mathrm{I}(g > c_1).$$

Denote now $\boldsymbol{u}^{(k)}$ a possible realization of $\boldsymbol{U}$, then the integrated link function conditional on $\boldsymbol{u}^{(k)}$ is

$$J^{(k)} = \int_{v_o}^{v_1} \widetilde{g}(v + \boldsymbol{u}^{(k)\top} \boldsymbol{\beta}) \, dv.$$

Now compare the integrated link functions for all $\boldsymbol{U}$-categories $\boldsymbol{u}^{(k)}$ ($k = 1,\ldots,M$) to the first $\boldsymbol{U}$-category $\boldsymbol{u}^{(0)}$. It holds

$$J^{(k)} - J^{(0)} = (c_1 - c_0) \left\{ \boldsymbol{u}^{(k)} - \boldsymbol{u}^{(0)} \right\} \boldsymbol{\beta},$$

hence with

$$\Delta J = \begin{pmatrix} J^{(1)} - J^{(0)} \\ \cdots \\ J^{(M)} - J^{(0)} \end{pmatrix}, \Delta \boldsymbol{u} = \begin{pmatrix} \boldsymbol{u}^{(1)} - \boldsymbol{u}^{(0)} \\ \cdots \\ \boldsymbol{u}^{(M)} - \boldsymbol{u}^{(0)} \end{pmatrix}$$

we obtain $\Delta J = (c_1 - c_0)\, \Delta \boldsymbol{u}\, \boldsymbol{\beta}$. This yields finally

$$\boldsymbol{\beta} = (c_1 - c_o)^{-1} (\Delta \boldsymbol{u}^\top \Delta \boldsymbol{u})^{-1} \Delta \boldsymbol{u}^\top \Delta J \tag{6.25}$$

to determine the coefficients $\boldsymbol{\beta}$. The estimation of $\boldsymbol{\beta}$ is based on replacing $J^{(k)}$ in 6.25 by

$$\widehat{J}^{(k)} = \int_{v_0}^{v_1} \widehat{\widetilde{g}}(v + \boldsymbol{\beta}^\top \boldsymbol{u}^{(k)})\, dv$$

with $\widehat{\widetilde{g}}$ a nonparametric estimate of the truncated link function $\widetilde{g}$. This estimator is obtained by a univariate regression of the estimated "continuous" index values $\widehat{\gamma}^\top \boldsymbol{T}_i^{(k)}$ on $Y_i^{(k)}$. Horowitz & Härdle (1996) show that using a $\sqrt{n}$-consistent estimate $\widehat{\gamma}$ and a Nadaraya-Watson estimator $\widehat{\widetilde{g}}$ the estimated coefficient $\widehat{\boldsymbol{\beta}}$ is itself $\sqrt{n}$-consistent and has an asymptotic normal distribution.

6.3 Testing the SIM

Given the computational expense of estimating single index models, it is desirable to know whether the distributional flexibility of these models justifies the extra computational cost. This implies that the performance of a specified parametric model must be compared with that of an estimated single index model as given in (5.9). To this end, Horowitz & Härdle (1994) designed a test that considers the following hypotheses:

$$\begin{aligned} H_0 &: E(Y|\boldsymbol{X} = \boldsymbol{x}) = G(\boldsymbol{x}^\top \boldsymbol{\beta}) \\ H_1 &: E(Y|\boldsymbol{X} = \boldsymbol{x}) = g(\boldsymbol{x}^\top \boldsymbol{\beta}) \end{aligned} \tag{6.26}$$

Here G (the link under H_0) is a known function and H (the link under H_1) an unspecified function. For example, the null hypothesis could be a logit model and the alternative a semiparametric model of SIM type.

The main idea that inspires the test relies on the fact that if the model under the null is true then a nonparametric estimation of $E(Y|\boldsymbol{X}^\top \widehat{\boldsymbol{\beta}} = v)$ gives a correct estimate of $F(v)$. Thus, the specification of the parametric model can be tested by comparing the nonparametric estimate of $E(Y|\boldsymbol{X}^\top \widehat{\boldsymbol{\beta}} = v)$ with the parametric fit using the known link G.

The test statistic is defined as

$$T = \sqrt{h} \sum_{i=1}^{n} w(\boldsymbol{X}_i \top \widehat{\boldsymbol{\beta}}) \left\{ Y_i - G(\boldsymbol{X}_i^\top \widehat{\boldsymbol{\beta}}) \right\} \left\{ \widehat{g}_{-i}(\boldsymbol{X}_i^\top \widehat{\boldsymbol{\beta}}) - G(\boldsymbol{X}_i^\top \widehat{\boldsymbol{\beta}}) \right\} \tag{6.27}$$

where $\widehat{g}_{-i}(\bullet)$ is a leave-one-out Nadaraya-Watson estimate for the regression of Y on the estimated index values, h is the bandwidth used in the kernel regression. $w(\bullet)$ is a weight function that downweights extreme observations.

In practice the weight function is defined as such that it considers only 90% or 95% of the central range of the index values values of $\boldsymbol{X}_i^\top \widehat{\boldsymbol{\beta}}$. Horowitz & Härdle (1994) propose to take $\widehat{\boldsymbol{\beta}}$, the estimate under H_0. That is, the same index values $\boldsymbol{X}_i^\top \widehat{\boldsymbol{\beta}}$ are used to compute both the parametric and the semiparametric regression values.

Let us take a closer look at the intuition behind this test statistic. The first difference term in the sum measures the deviation of the estimated regression from the true realization, that is it measures $Y_i - E(Y|\boldsymbol{X}_i)$. If H_0 holds, then this measure ought to be very small on average. If, however, the parametric model under the null fails to replicate the observed values Y_i well, then T will increase. Obviously, we reject the hypothesis that the data were generated by a parametric model if T becomes unplausibly large.

The second difference term measures the distance between the regression values obtained under the null and under the semiparametric alternative. Suppose the parametric model captures the characteristics of the data well so that $Y_i - G(\boldsymbol{X}_i^\top \widehat{\boldsymbol{\beta}})$ is small. Then even if the semiparametric link deviates considerably from the parametric alternative *on average*, these deviations will be downweighted by the first difference term. Seen differently, the small residuals of the parametric fit are blown up by large differences in the parametric and semiparametric fits, $\widehat{g}_{-i}(\boldsymbol{X}_i^\top \widehat{\boldsymbol{\beta}}) - G(\boldsymbol{X}_i^\top \widehat{\boldsymbol{\beta}})$. Thus, if H_0 is true, the residuals should be small enough to accommodate possible strong differences in the alternative fits. Again, a small statistic will lead to maintaining the null hypothesis.

It can be shown that under H_0 and under some suitable regularity conditions T is asymptotically distributed as a $N(0, \sigma_T^2)$ where σ_T^2 denotes the asymptotic sampling variance of the statistic.

Bibliographic Notes

For related presentations of the topic we refer to Horowitz (1993), Horowitz (1998b) and Pagan & Ullah (1999, Chapter 7).

There is a large amount of literature that investigates the efficiency bound for estimators in semiparametric models. Let us mention Begun, Hall, Huang & Wellner (1983) and Cosslett (1987) as being two of the first references. Newey (1990) and Newey (1994) are more recent articles. The latter treats the variance in a very general and abstract way. A comprehensive resource for efficient estimation in semiparametric models is Bickel, Klaassen, Ritov & Wellner (1993).

The idea of using parametric objective functions and substituting unknown components by nonparametric estimates has first been proposed in Cosslett (1983). The maximum score estimator of Manski (1985) and the maximum rank correlation estimator from Han (1987) are of the same type. Their resulting estimates are still very close to the parametric estimates. For that reason the SLS method of Ichimura (1993) may outperform them when the parametric model is misspecified.

The pseudo maximum likelihood version of the SLS was found independently by Weisberg & Welsh (1994). They present it as a straightforward generalization of the GLM algorithm and discuss numerical details. A different idea for adding nonparametric components to the the maximum likelihood function is given by Gallant & Nychka (1987). They use a Hermite series to expand the densities in the objective function.

The different methods presented in this chapter have been compared in a simulation study by Bonneu, Delecroix & Malin (1993). They also include a study of Bonneu & Delecroix (1992) for a slightly modified pseudo likelihood estimator.

An alternative ADE method (without weight function) was proposed by Härdle & Stoker (1989). This estimator shares the asymptotic properties of the weighted ADE, but requires for practical computation a trimming factor to guarantee that the estimated density is bounded away from zero.

Exercises

Exercise 6.1. Discuss why and how WSLS could also be motivated by considering the log-likelihood.

Exercise 6.2. Recall the PMLE for the binary response model. After equation (6.11) we introduced the restriction $E(Y|\boldsymbol{X}) = E\{Y|v_{\beta}(\boldsymbol{X})\}$. Discuss the issue of heteroscedasticity under this restriction.

Exercise 6.3. The main interest in this chapter has been the estimation of the parametric part. What kind of conditions are common (and typical) for the nonparametric components in estimation, respectively the substitutes used for the unknown terms? What are they good for?

Exercise 6.4. Show that the ADE approach without using a weight function (or equivalently $w(\bullet) \equiv 1$) leads to the estimation of $E\{\nabla_f(\boldsymbol{T})m(\boldsymbol{T})/f(\boldsymbol{T})\}$.

Summary

- ⋆ A single index model (SIM) is of the form

$$E(Y|\mathbf{X}) = m(\mathbf{X}) = g\left\{v_{\boldsymbol{\beta}}(\mathbf{X})\right\},$$

where $v_{\boldsymbol{\beta}}(\bullet)$ is an up to the parameter vector $\boldsymbol{\beta}$ known index function and $g(\bullet)$ is an unknown smooth link function. In most applications the index function is of linear form, i.e., $v_{\boldsymbol{\beta}}(\boldsymbol{x}) = \boldsymbol{x}^\top \boldsymbol{\beta}$).

- ⋆ Due to the nonparametric form of the link function, neither an intercept nor a scale parameter can be identified. For example, if the index is linear, we have to estimate

$$E(Y|\mathbf{X}) = g\left\{\mathbf{X}^\top \boldsymbol{\beta}\right\}.$$

There is no intercept parameter and $\boldsymbol{\beta}$ can only be estimated up to unknown scale factor. To identify the slope parameter of interest, $\boldsymbol{\beta}$ is usually assumed to have one component identical to 1 or to be a vector of length 1.

- ⋆ The estimation of a SIM usually proceeds in the following steps: First the parameter $\boldsymbol{\beta}$ is estimated. Then, using the index values $\eta_i = \mathbf{X}_i^\top \widehat{\boldsymbol{\beta}}$, the nonparametric link function g is estimated by an univariate nonparametric regression method.

- ⋆ For the estimation of $\boldsymbol{\beta}$, two approaches are available: Iterative methods as semiparametric least squares (SLS) or pseudo-maximum likelihood estimation (PMLE) and direct methods as (weighted) average derivative estimation (WADE/ADE). Iterative methods can easily handle mixed discrete-continuous regressors but may need to employ sophisticated routines to deal with possible local optima of the optimization criterion. Direct methods as WADE/ADE avoid the technical difficulties of an optimization but do require continuous explanatory variables. An extension of the direct approach is only possible for a small number of additional discrete variables.

- ⋆ There are specification tests available to test whether we have a GLM or a true SIM.

7

Generalized Partial Linear Models

As indicated in the overview in Chapter 5, a *partial linear model* (PLM) consists of two additive components, a linear and a nonparametric part:

$$E(Y|\boldsymbol{U},\boldsymbol{T}) = \boldsymbol{U}^\top \boldsymbol{\beta} + m(\boldsymbol{T})$$

where $\boldsymbol{\beta} = (\beta_1, \ldots, \beta_p)^\top$ is a finite dimensional parameter and $m(\bullet)$ a smooth function. Here, we assume again a decomposition of the explanatory variables into two vectors, $\boldsymbol{U}$ and $\boldsymbol{T}$. The vector $\boldsymbol{U}$ denotes a p-variate random vector which typically covers categorical explanatory variables or variables that are known to influence the index in a linear way. The vector $\boldsymbol{T}$ is a q-variate random vector of continuous explanatory variables which is to be modeled in a nonparametric way. Economic theory or intuition should guide you as to which regressors should be included in $\boldsymbol{U}$ or $\boldsymbol{T}$, respectively.

Obviously, there is a straightforward generalization of this model to the case with a known link function $G(\bullet)$. We denote this semiparametric extension of the GLM

$$E(Y|\boldsymbol{U},\boldsymbol{T}) = G\{\boldsymbol{U}^\top \boldsymbol{\beta} + m(\boldsymbol{T})\} \tag{7.1}$$

as *generalized partial linear model* (GPLM).

7.1 Partial Linear Models

Let us first consider the PLM in order to discuss the main ideas for the estimation of the components of the model. Our goal is to find the coefficients $\boldsymbol{\beta}$ and the values $m(\bullet)$ in the following structural equation

$$Y = \boldsymbol{U}^\top \boldsymbol{\beta} + m(\boldsymbol{T}) + \varepsilon, \tag{7.2}$$

where ε shall denote an error term with zero mean and finite variance. In the following, we outline how such a model can be estimated and state the properties of the resulting estimators.

Now, let us take expectations conditioned on $\boldsymbol{T}$, i.e.

$$E(Y|\boldsymbol{T}) = E(\boldsymbol{U}^\top \boldsymbol{\beta}|\boldsymbol{T}) + E\{m(\boldsymbol{T})|\boldsymbol{T}\} + E(\varepsilon|\boldsymbol{T}) \tag{7.3}$$

We subtract equation (7.3) from equation (7.2) to obtain

$$Y - E(Y|\boldsymbol{T}) = \{\boldsymbol{U} - E(\boldsymbol{U}|\boldsymbol{T})\}^\top \boldsymbol{\beta} + \varepsilon - E(\varepsilon|\boldsymbol{T}) \tag{7.4}$$

since $E\{m(\boldsymbol{T})|\boldsymbol{T}\} = m(\boldsymbol{T})$. Note that by definition $E(\varepsilon|\boldsymbol{U},\boldsymbol{T}) = 0$. Applying the law of iterated expectations it can be shown that $E\{\varepsilon - E(\varepsilon|\boldsymbol{T})\} = 0$ holds as well.

Once we have calculated one component ($\boldsymbol{\beta}$ or m) of a PLM, the computation of the remaining component is straightforward. There are two alternative approaches to this task:

- Subtract $\boldsymbol{U}^\top \boldsymbol{\beta}$ from Y to get
$$E\{Y - \boldsymbol{U}^\top \boldsymbol{\beta}|\boldsymbol{T}\} = m(\boldsymbol{T}). \tag{7.5}$$
Estimate the parametric coefficient $\boldsymbol{\beta}$ by least squares regression of Y on $\boldsymbol{U}$. Plugging $\boldsymbol{U}^\top \widehat{\boldsymbol{\beta}}$ into (7.5) yields now a classic nonparametric regression problem, so the Nadaraya-Watson or any other nonparametric estimator may be employed to estimate the values of $\widehat{m}(\boldsymbol{T})$. Consider now
$$E\{Y - \widehat{m}(\boldsymbol{T})|\boldsymbol{U}\} = \boldsymbol{U}^\top \boldsymbol{\beta}, \tag{7.6}$$
i.e. use the estimate $\widehat{m}$ to update $\boldsymbol{\beta}$. Update then $\widehat{m}(\boldsymbol{T})$ etc. This is essentially the backfitting approach, which goes back to Buja, Hastie & Tibshirani (1989), see Subsection 7.2.3.
- The alternative approach is based on (7.4). Note that the conditional expectations $E(Y|\boldsymbol{T})$ and $E(\boldsymbol{U}|\boldsymbol{T})$ can be replaced by their empirical counterparts, denoted by $\widehat{Y}$ and $\widehat{\boldsymbol{U}}$, respectively. These quantities can be obtained by using any nonparametric regression technique.

  Note in particular that $E(\boldsymbol{U}|\boldsymbol{T})$ is a q-dimensional column vector just like $\boldsymbol{U}$. Thus, $Y - E(Y|\boldsymbol{T})$ and $\boldsymbol{U} - E(\boldsymbol{U}|\boldsymbol{T})$ in equation (7.4) can be replaced by $Y - \widehat{Y}$ and $\boldsymbol{U} - \widehat{\boldsymbol{U}}$. Obviously, (7.4) is now simply a version of the standard linear model. Applying the familiar standard linear regression, we can easily estimate the vector of coefficients $\boldsymbol{\beta}$. Using this estimated vector to replace $\boldsymbol{U}^\top \boldsymbol{\beta}$ in (7.2) allows us to estimate $m(\bullet)$ by nonparametric regression of $Y - \boldsymbol{U}^\top \widehat{\boldsymbol{\beta}}$ on $\boldsymbol{T}$. This estimation idea has been studied by Speckman (1988) and Robinson (1988a) for univariate T using Nadaraya-Watson kernel regression. See Subsection 7.2.2 for more details.

  Under regularity conditions $\widehat{\beta}$ can be shown to be $\sqrt{n}$-consistent for β and asymptotically normal, and there exists a consistent estimator of its limiting covariance matrix. The nonparametric function m can be estimated (in the univariate case) with the usual univariate rate of convergence.

In the following section, we review both estimation procedures for the PLM and its extension for the GPLM in more detail. We restrict ourselves to Nadaraya-Watson type regression for the nonparametric component. It will become clear, however, that other techniques can be used in the same way. In the later part of this chapter we also discuss tests on the correct specification of the GPLM (vs. a parametric GLM).

7.2 Estimation Algorithms for PLM and GPLM

In order to estimate the GPLM we consider the same distributional assumptions for Y as in the GLM. Thus we have two cases: (a) the distribution of Y belongs to the exponential family (5.13), or (b) the first two (conditional) moments of Y are specified in order to use the quasi-likelihood function (5.18).

To summarize, the estimation of the GPLM will be based on

$$E(Y|\boldsymbol{U},\boldsymbol{T}) = \mu = G(\eta) = G\{\boldsymbol{U}^\top\boldsymbol{\beta} + m(\boldsymbol{T})\},$$
$$\mathrm{Var}(Y|\boldsymbol{U},\boldsymbol{T}) = a(\psi)V(\mu).$$

Recall, that the nuisance parameter ψ is the dispersion parameter (cf. Table 5.2), which is typically given by $a(\psi) = \sigma^2$. In the following we concentrate on the estimation of $\boldsymbol{\beta}$ and $m(\bullet)$ since as in the GLM an estimate for the dispersion parameter ψ is easily obtained from (5.25).

The estimation procedures for $\boldsymbol{\beta}$ and m can be classified into two categories.

- *Profile likelihood* type:
  These comprise the profile likelihood and the (generalized) Speckman method. Both methods coincide for the PLM.
- *Backfitting* type:
  These are based on alternating between parametric and nonparametric estimation. Backfitting was originally developed for AM and GAM. Since the PLM and GPLM can be seen as consisting of two additive components, the backfitting idea applies here as well. We will present backfitting in full detail in Chapters 8 and 9.

7.2.1 Profile Likelihood

The *profile likelihood* method (considered in Severini & Wong (1992) for this type of problems) is based on the fact, that the conditional distribution of Y given $\boldsymbol{U}$ and $\boldsymbol{T}$ is parametric. This method starts from keeping $\boldsymbol{\beta}$ fixed and to estimate a least favorable (in other words a "worst case") nonparametric function $m_{\boldsymbol{\beta}}(\bullet)$ in dependence of the fixed $\boldsymbol{\beta}$. The resulting estimate for $m_{\boldsymbol{\beta}}(\bullet)$ is then used to construct the profile likelihood for $\boldsymbol{\beta}$. As a consequence the resulting $\widehat{\boldsymbol{\beta}}$ is estimated at $\sqrt{n}$-rate, has an asymptotic normal distribution and is asymptotically efficient (i.e. has asymptotically minimum variance). The nonparametric function m can be estimated consistently by $\widehat{m}(\bullet) = \widehat{m}_{\widehat{\boldsymbol{\beta}}}(\bullet)$.

The profile likelihood algorithm can be derived as follows. As explained above we first fix $\boldsymbol{\beta}$ and construct the least favorable curve $m_{\boldsymbol{\beta}}$. The parametric (profile) likelihood function is given by

$$\ell(\boldsymbol{Y}, \boldsymbol{\mu}_{\boldsymbol{\beta}}, \psi) = \sum_{i=1}^{n} \ell\left(Y_i, G\{\boldsymbol{U}_i^\top \boldsymbol{\beta} + m_{\boldsymbol{\beta}}(\boldsymbol{T}_i)\}, \psi\right), \tag{7.7}$$

where $\boldsymbol{Y}$ and $\boldsymbol{\mu}_{\boldsymbol{\beta}}$ denote (in a similar way as in Chapter 5) the corresponding n-dimensional vectors for all observations

$$\boldsymbol{Y} = \begin{pmatrix} Y_1 \\ \vdots \\ Y_n \end{pmatrix}, \quad \boldsymbol{\mu}_{\boldsymbol{\beta}} = \begin{pmatrix} G\{\boldsymbol{U}_1^\top \boldsymbol{\beta} + m_{\boldsymbol{\beta}}(\boldsymbol{T}_1)\} \\ \vdots \\ G\{\boldsymbol{U}_n^\top \boldsymbol{\beta} + m_{\boldsymbol{\beta}}(\boldsymbol{T}_n)\} \end{pmatrix}.$$

We maximize (7.7) to obtain an estimate for $\boldsymbol{\beta}$. A slightly different objective function is used for the nonparametric part of the model. The *smoothed* or *local* likelihood

$$\ell_{\mathbf{H}}(\boldsymbol{Y}, \boldsymbol{\mu}_{m_{\boldsymbol{\beta}}(\boldsymbol{T})}, \psi) = \sum_{i=1}^{n} \mathcal{K}_{\mathbf{H}}(\boldsymbol{t} - \boldsymbol{T}_i)\, \ell\left(Y_i, G\{\boldsymbol{U}_i^\top \boldsymbol{\beta} + m_{\boldsymbol{\beta}}(\boldsymbol{t})\}, \psi\right) \tag{7.8}$$

is maximized to estimate $m_{\boldsymbol{\beta}}(\boldsymbol{T})$ at point $\boldsymbol{T}$. Here we denote $\boldsymbol{\mu}_{m_{\boldsymbol{\beta}}(\boldsymbol{T})}$ the vector with components $G\{\boldsymbol{U}_i^\top \boldsymbol{\beta} + m_{\boldsymbol{\beta}}(\boldsymbol{T})\}$. The local weights $\mathcal{K}_{\mathbf{H}}(\boldsymbol{T} - \boldsymbol{T}_i)$ are kernel weights with $\mathcal{K}$ denoting a (multidimensional) kernel function and $\mathbf{H}$ a bandwidth matrix, see Section 3.6.

Denote by ℓ', ℓ'' the first and second derivatives of the $\ell(Y_i, \bullet, \psi)$ w.r.t. its second argument. The maximization of the local likelihood (7.8) at all observations $\boldsymbol{T}_j$ requires hence to solve

$$0 = \sum_{i=1}^{n} \mathcal{K}_{\mathbf{H}}(\boldsymbol{t} - \boldsymbol{T}_i)\, \ell'\left(Y_i, G\{\boldsymbol{U}_i^\top \boldsymbol{\beta} + m_{\boldsymbol{\beta}}(\boldsymbol{t})\}, \psi\right). \tag{7.9}$$

with respect to $m_{\boldsymbol{\beta}}(\boldsymbol{t})$. In contrast, the maximization of the profile likelihood (7.7) requires the solution of

$$0 = \sum_{i=1}^{n} \ell'\left(Y_i, G\{\boldsymbol{U}_i^\top \boldsymbol{\beta} + m_{\boldsymbol{\beta}}(\boldsymbol{T}_i)\}, \psi\right) \{\boldsymbol{U}_i + \nabla_{m_{\boldsymbol{\beta}}}(\boldsymbol{T}_i)\} \tag{7.10}$$

with respect to the coefficient vector β. The vector $\nabla_{m_{\boldsymbol{\beta}}}$ denotes here the vector of all partial derivatives of $m_{\boldsymbol{\beta}}$ with respect to $\boldsymbol{\beta}$. A further differentiation of (7.9), this time w.r.t. $\boldsymbol{\beta}$, leads to an explicit expression for $\nabla_{m_{\boldsymbol{\beta}}}$:

$$\nabla_{m_\beta}(t) = -\frac{\sum_{i=1}^{n} \mathcal{K}_{\mathbf{H}}(\boldsymbol{t} - \boldsymbol{T}_i)\, \ell''\left(Y_i, G\{\boldsymbol{U}_i^\top \boldsymbol{\beta} + m_{\boldsymbol{\beta}}(\boldsymbol{t})\}, \psi\right) \boldsymbol{U}_i}{\sum_{i=1}^{n} \mathcal{K}_{\mathbf{H}}(\boldsymbol{t} - \boldsymbol{T}_i)\, \ell''\left(Y_i, G\{\boldsymbol{U}_i^\top \boldsymbol{\beta} + m_{\boldsymbol{\beta}}(\boldsymbol{t})\}, \psi\right)}. \tag{7.11}$$

This equation can be used to estimate ∇_{m_β} in the following derivation of estimators for $\boldsymbol{\beta}$ and m.

In general, equations (7.9) and (7.10) can only be solved iteratively. Severini & Staniswalis (1994) present a Newton-Raphson type algorithm for this problem. To write the estimation algorithm in a compact form, abbreviate

$$m_j = m_{\boldsymbol{\beta}}(\boldsymbol{T}_j),$$

$$\ell_i = \ell\left(Y_i, G(\boldsymbol{U}_i^\top \boldsymbol{\beta} + m_i), \psi\right), \quad \ell_{ij} = \ell\left(Y_i, G(\boldsymbol{U}_i^\top \boldsymbol{\beta} + m_j), \psi\right).$$

Furthermore denote ℓ_i', ℓ_i'', ℓ_{ij}', and ℓ_{ij}'' the first and second derivatives of ℓ_i and ℓ_{ij} w.r.t. their second argument. Instead of the free parameter $\boldsymbol{t}$ we will calculate all values at the observations $\boldsymbol{T}_i$. Then, (7.9) and (7.10) transform to

$$0 = \sum_{i=1}^{n} \ell_{ij}' \, \mathcal{K}_{\mathbf{H}}(\boldsymbol{T}_i - \boldsymbol{T}_j) \tag{7.12}$$

$$0 = \sum_{i=1}^{n} \ell_i' \, \{\boldsymbol{U}_i + \nabla_{m_i}\}. \tag{7.13}$$

It is the dependence of $m_i = m_{\boldsymbol{\beta}}(\boldsymbol{T}_i)$ on $\boldsymbol{\beta}$ that adds the additional term ∇_{m_i} (the derivative of m_i w.r.t. $\boldsymbol{\beta}$) to the last equation. The estimate of ∇_{m_i} based on (7.11) is therefore necessary to estimate $\boldsymbol{\beta}$:

$$\nabla_{m_j} = -\frac{\sum_{i=1}^{n} \ell_i'' \, \mathcal{K}_{\mathbf{H}}(\boldsymbol{T}_i - \boldsymbol{T}_j) \boldsymbol{U}_i}{\sum_{i=1}^{n} \ell_i'' \, \mathcal{K}_{\mathbf{H}}(\boldsymbol{T}_i - \boldsymbol{T}_j)}. \tag{7.14}$$

From (7.12) the update step for m (at all observations $\boldsymbol{T}_j$) follows directly:

$$m_j^{new} = m_j - \frac{\sum_{i=1}^{n} \ell_{ij}' \, \mathcal{K}_{\mathbf{H}}(\boldsymbol{T}_i - \boldsymbol{T}_j)}{\sum_{i=1}^{n} \ell_{ij}'' \, \mathcal{K}_{\mathbf{H}}(\boldsymbol{T}_i - \boldsymbol{T}_j)}.$$

The Newton-Raphson solution of (7.13) requires a second derivative of $m_{\boldsymbol{\beta}}$. As an approximation consider ∇_{m_j} to be constant with respect to $\boldsymbol{\beta}$. This leads to

$$\boldsymbol{\beta}^{new} = \boldsymbol{\beta} - \mathbf{B}^{-1} \sum_{i=1}^{n} \ell_i' \, \widetilde{\boldsymbol{U}}_i$$

with a Hessian type matrix

$$\mathbf{B} = \sum_{i=1}^{n} \ell_i'' \, \widetilde{\boldsymbol{U}}_i \widetilde{\boldsymbol{U}}_i^\top$$

and

$$\widetilde{\boldsymbol{U}}_j = \boldsymbol{U}_j + \nabla_{\boldsymbol{m}_j} = \boldsymbol{U}_j - \frac{\sum_{i=1}^{n} \ell_{ij}'' \, \mathcal{K}_{\mathbf{H}}(\boldsymbol{T}_i - \boldsymbol{T}_j) \, \boldsymbol{U}_i}{\sum_{i=1}^{n} \ell_{ij}'' \, \mathcal{K}_{\mathbf{H}}(\boldsymbol{T}_i - \boldsymbol{T}_j)} .$$

The resulting estimation algorithm consists of iterating the following update steps for $\boldsymbol{\beta}$ and m up to convergence. The updating step for m (the nonparametric part) is in general of quite complex structure and cannot be simplified. (Only in some models, in particular for identity and exponential link functions G, equation (7.12) can be solved explicitly for m_j). It is possible however, to rewrite the updating step for $\boldsymbol{\beta}$ in a closed matrix form. Define $\mathbf{S}^P$ a smoother matrix with elements

$$(\mathbf{S}^P)_{ij} = \frac{\ell_{ij}'' \mathcal{K}_{\mathbf{H}}(\boldsymbol{T}_i - \boldsymbol{T}_j)}{\sum_{i=1}^{n} \ell_{ij}'' \mathcal{K}_{\mathbf{H}}(\boldsymbol{T}_i - \boldsymbol{T}_j)} \tag{7.15}$$

and let $\mathbf{U}$ be the design matrix with rows $\boldsymbol{U}_i^\top$. Further denote by $\mathbf{I}$ the $n \times n$ identity matrix and

$$\boldsymbol{v} = (\ell_1', \ldots, \ell_n')^\top , \quad \mathbf{W} = \text{diag}(\ell_1'', \ldots, \ell_n'').$$

Setting $\widetilde{\mathbf{U}} = (\mathbf{I} - \mathbf{S}^P)\mathbf{U}$ and $\widetilde{\mathbf{Z}} = \widetilde{\mathbf{U}}\boldsymbol{\beta} - \mathbf{W}^{-1}\boldsymbol{v}$ we can summarize the iteration in the following form:

Profile Likelihood Iteration Step for GPLM

updating step for $\boldsymbol{\beta}$

$$\boldsymbol{\beta}^{new} = (\widetilde{\mathbf{U}}^\top \mathbf{W} \widetilde{\mathbf{U}})^{-1} \widetilde{\mathbf{U}}^\top \mathbf{W} \widetilde{\mathbf{Z}}$$

updating step for m_j

$$m_j^{new} = m_j - \frac{\sum_{i=1}^{n} \ell_{ij}' \, \mathcal{K}_{\mathbf{H}}(\boldsymbol{T}_i - \boldsymbol{T}_j)}{\sum_{i=1}^{n} \ell_{ij}'' \, \mathcal{K}_{\mathbf{H}}(\boldsymbol{T}_i - \boldsymbol{T}_j)} .$$

The variable $\widetilde{\mathbf{Z}}$ is a sort of adjusted dependent variable. From the formula for $\boldsymbol{\beta}^{new}$ it becomes clear, that the parameter $\boldsymbol{\beta}$ is updated by a parametric method. The only difference to (5.23) is the nonparametric modification of $\mathbf{U}$ and $\mathbf{Z}$ resulting in $\widetilde{\mathbf{U}}$ and $\widetilde{\mathbf{Z}}$. We remark that as in the GLM case, the functions ℓ_i'' can be replaced by their expectations (w.r.t. Y_i) to obtain a Fisher scoring type procedure.

7.2.2 Generalized Speckman Estimator

The profile likelihood estimator is particularly easy to derive in the case of a PLM, i.e. in particular for the identity link G and normally distributed Y_i. Here we have

$$\ell_i' = -(Y_i - \boldsymbol{U}_i^\top \boldsymbol{\beta} - m_j)/\sigma^2, \quad \text{and} \quad \ell_i'' \equiv -1/\sigma^2.$$

The latter yields then a simpler smoother matrix $\mathbf{S}$ defined by its elements

$$(\mathbf{S})_{ij} = \frac{\mathcal{K}_{\mathbf{H}}(\boldsymbol{T}_i - \boldsymbol{T}_j)}{\sum\limits_{i=1}^{n} \mathcal{K}_{\mathbf{H}}(\boldsymbol{T}_i - \boldsymbol{T}_j)}. \tag{7.16}$$

Moreover, the updating step for m_j simplifies to

$$m_j^{new} = \frac{\sum\limits_{i=1}^{n} (Y_i - \boldsymbol{U}_i^\top \boldsymbol{\beta})\, \mathcal{K}_{\mathbf{H}}(\boldsymbol{T}_i - \boldsymbol{T}_j)}{\sum\limits_{i=1}^{n} \mathcal{K}_{\mathbf{H}}(\boldsymbol{T}_i - \boldsymbol{T}_j)}$$

which we can denote in matrix form as

$$\boldsymbol{m}^{new} = \mathbf{S}(\boldsymbol{Y} - \mathbf{U}\boldsymbol{\beta})$$

using again the vector notation for $\boldsymbol{Y}$ and similarly for the nonparametric function estimate: $\boldsymbol{m}^{new} = \left(m_1^{new}, \ldots, m_n^{new}\right)^\top$. For the parameter $\boldsymbol{\beta}$ we obtain

$$\boldsymbol{\beta}^{new} = (\widetilde{\mathbf{U}}^\top \widetilde{\mathbf{U}})^{-1} \widetilde{\mathbf{U}}^\top \widetilde{\boldsymbol{Y}}$$

using $\widetilde{\mathbf{U}} = (\mathbf{I} - \mathbf{S})\mathbf{U}$ and $\widetilde{\boldsymbol{Y}} = (\mathbf{I} - \mathbf{S})\boldsymbol{Y}$. It turns out that no iteration is necessary. These estimators for the components of a PLM can be summarized as follows:

Profile Likelihood / Speckman Estimator for PLM
estimate $\boldsymbol{\beta}$: $\widehat{\boldsymbol{\beta}} = (\widetilde{\mathbf{U}}^\top \widetilde{\mathbf{U}})^{-1} \widetilde{\mathbf{U}}^\top \widetilde{\boldsymbol{Y}}$,
estimate $\boldsymbol{m}$: $\widehat{\boldsymbol{m}} = \mathbf{S}(\boldsymbol{Y} - \mathbf{U}\widehat{\boldsymbol{\beta}})$

Now consider the GPLM. We are going to combine the Speckman estimators with the IRLS technique for the GLM (cf. (5.23)). Recall that each iteration step of a GLM was obtained by weighted least squares regression on the adjusted dependent variable. In the same spirit, we estimate the GPLM by replacing the IRLS with a weighted partial linear fit on the adjusted dependent variable. This variable is here given as

$$\mathbf{Z} = \mathbf{U}\boldsymbol{\beta} + \boldsymbol{m} - \mathbf{W}^{-1}\boldsymbol{v}. \tag{7.17}$$

As previously defined, $\boldsymbol{v}$ denotes the vector and $\mathbf{W}$ denotes the diagonal matrix of the first respectively second derivatives of ℓ_i.

As for (7.15) we have to introduce weights for the GPLM smoother matrix. The basic simplification in comparison to (7.15) is that we use the matrix $\mathbf{S}$ with elements

$$(\mathbf{S})_{ij} = \frac{\ell_i'' \mathcal{K}_{\mathbf{H}}(\boldsymbol{T}_i - \boldsymbol{T}_j)}{\sum\limits_{i=1}^{n} \ell_i'' \mathcal{K}_{\mathbf{H}}(\boldsymbol{T}_i - \boldsymbol{T}_j)}. \tag{7.18}$$

Note the small but important difference in ℓ_i'' compared to $\mathbf{S}^P$. We will come back to the computational simplification in Subsection 7.2.4.

Using the notation $\widetilde{\mathbf{U}} = (\mathbf{I} - \mathbf{S})\mathbf{U}$ and $\widetilde{\mathbf{Z}} = \widetilde{\mathbf{U}}\boldsymbol{\beta} - \mathbf{W}^{-1}\boldsymbol{v}$ (define $\boldsymbol{v}$ and $\mathbf{W}$ as before), an expression for each iteration step in matrix notation is possible here, too:

Generalized Speckman Iteration Step for GPLM
updating step for $\boldsymbol{\beta}$ $$\boldsymbol{\beta}^{new} = (\widetilde{\mathbf{U}}^\top \mathbf{W} \widetilde{\mathbf{U}})^{-1} \widetilde{\mathbf{U}}^\top \mathbf{W} \widetilde{\mathbf{Z}},$$
updating step for $\boldsymbol{m}$ $$\boldsymbol{m}^{new} = \mathbf{S}(\mathbf{Z} - \mathbf{U}\boldsymbol{\beta})$$

There is an important property that this algorithm shares with the backfitting procedure we consider in the next subsection. The updating step for $\boldsymbol{m}$ implies

$$\mathbf{S}\mathbf{Z} = \mathbf{S}\mathbf{U}\boldsymbol{\beta} + \boldsymbol{m}$$

at convergence of the algorithm. Combining this with the definition of $\mathbf{Z}$ yields

$$\begin{aligned}
\mathbf{Z} - \mathbf{S}\mathbf{Z} &= \mathbf{U}\boldsymbol{\beta} + \boldsymbol{m} - \mathbf{W}^{-1}\boldsymbol{v} - (\mathbf{S}\mathbf{U}\boldsymbol{\beta} + \boldsymbol{m}) \\
&= (\mathbf{I} - \mathbf{S})\mathbf{U}\boldsymbol{\beta} - \mathbf{W}^{-1}\boldsymbol{v} \\
&= \widetilde{\mathbf{Z}}.
\end{aligned}$$

Hence,

$$\widetilde{\mathbf{Z}} = (\mathbf{I} - \mathbf{S})\mathbf{Z}, \tag{7.19}$$

i.e. $\widetilde{\mathbf{Z}}$ is obtained in the same way as $\widetilde{\mathbf{U}}$. (Note that we used a different definition when we introduced $\widetilde{\mathbf{Z}}$.) Equation (7.19) shows that updating the index from $\mathbf{U}\boldsymbol{\beta} + \boldsymbol{m}$ to $\mathbf{U}\boldsymbol{\beta}^{new} + \boldsymbol{m}^{new}$ is equivalent to determine

$$\mathbf{U}\boldsymbol{\beta}^{new} + \boldsymbol{m}^{new} = \mathbf{R}^S\mathbf{Z}. \tag{7.20}$$

Hence, each iteration step corresponds to applying the linear estimation (or hat) matrix $\mathbf{R}^S$ on the adjusted dependent variable $\mathbf{Z}$. In order to derive the formula for $\mathbf{R}^S$, we see from (7.19) that we have at convergence

$$\mathbf{U}\boldsymbol{\beta} + \boldsymbol{m} = \mathbf{U}(\widetilde{\mathbf{U}}^\top\mathbf{W}\widetilde{\mathbf{U}})^{-1}\widetilde{\mathbf{U}}^\top\mathbf{W}(\mathbf{I} - \mathbf{S})\mathbf{Z} + \mathbf{S}(\mathbf{Z} - \mathbf{U}\boldsymbol{\beta}), \tag{7.21}$$

$$\mathbf{S}\mathbf{U}\boldsymbol{\beta} = \mathbf{S}\mathbf{U}(\widetilde{\mathbf{U}}^\top\mathbf{W}\widetilde{\mathbf{U}})^{-1}\widetilde{\mathbf{U}}^\top\mathbf{W}(\mathbf{I} - \mathbf{S})\mathbf{Z}. \tag{7.22}$$

Now plugging (7.22) into the right hand side of (7.21) gives

$$\mathbf{U}\boldsymbol{\beta} + \boldsymbol{m} = \left[\widetilde{\mathbf{U}}\{\widetilde{\mathbf{U}}^\top\mathbf{W}\widetilde{\mathbf{U}}\}^{-1}\widetilde{\mathbf{U}}^\top\mathbf{W}(\mathbf{I} - \mathbf{S}) + \mathbf{S}\right]\mathbf{Z}.$$

This finally shows that the $\mathbf{R}^S$ is of the form

$$\mathbf{R}^S = \widetilde{\mathbf{U}}\{\widetilde{\mathbf{U}}^\top\mathbf{W}\widetilde{\mathbf{U}}\}^{-1}\widetilde{\mathbf{U}}^\top\mathbf{W}(\mathbf{I} - \mathbf{S}) + \mathbf{S}. \tag{7.23}$$

The linear estimation matrix $\mathbf{R}^S$ will be used later on to compute approximate degrees of freedom for the semiparametric GPLM estimator.

7.2.3 Backfitting

The backfitting method was originally suggested as an iterative algorithm for fitting an additive model (Buja, Hastie & Tibshirani, 1989; Hastie & Tibshirani, 1990). Its key idea is to regress the additive components separately on partial residuals.

Again, let us first consider the PLM

$$E(Y|\boldsymbol{U},\boldsymbol{T}) = \boldsymbol{U}^\top\boldsymbol{\beta} + m(\boldsymbol{T}),$$

which consists of only two additive components. Denote now $\mathbf{P}$ the projection matrix

$$\mathbf{P} = \mathbf{U}(\mathbf{U}^\top\mathbf{U})^{-1}\mathbf{U}^\top$$

and $\mathbf{S}$ a smoother matrix (e.g. the Nadaraya-Watson smoother matrix given by its elements (7.16)). Backfitting means to solve

$$\mathbf{U}\boldsymbol{\beta} = \mathbf{P}(\boldsymbol{Y} - \boldsymbol{m})$$

$$\boldsymbol{m} = \mathbf{S}(\boldsymbol{Y} - \mathbf{U}\boldsymbol{\beta}).$$

In this case no iteration is necessary and an explicit solution can be given:

$$\widehat{\beta} = \{\mathbf{U}^\top(\mathbf{I}-\mathbf{S})\mathbf{U}\}^{-1}\mathbf{U}^\top(\mathbf{I}-\mathbf{S})Y, \tag{7.24}$$

$$\widehat{\boldsymbol{m}} = \mathbf{S}(Y - \mathbf{U}\widehat{\beta}). \tag{7.25}$$

We can summarize the estimation procedure in the same way we did for the Speckman estimator. Note the small but subtle difference in the estimate for the parametric part:

Backfitting Estimator for PLM

estimate $\boldsymbol{\beta}$

$$\widehat{\beta}^{new} = (\mathbf{U}^\top\widetilde{\mathbf{U}})^{-1}\mathbf{U}^\top\widetilde{Y},$$

estimate $\boldsymbol{m}$

$$\widehat{\boldsymbol{m}} = \mathbf{S}(Y - \mathbf{U}\widehat{\beta})$$

We will now extend this technique to the GPLM. The motivation for the backfitting iteration coincides with that for the Speckman iteration. Again we recognize the only difference to Speckman in the updating step for the parametric part. The matrices $\mathbf{S}$ and $\mathbf{W}$ as well as the vector v are defined in exactly the same way as for the Speckman iteration.

Backfitting Iteration Step for GPLM

updating step for $\boldsymbol{\beta}$

$$\beta^{new} = (\mathbf{U}^\top\mathbf{W}\widetilde{\mathbf{U}})^{-1}\mathbf{U}^\top\mathbf{W}\widetilde{\mathbf{Z}},$$

updating step for $\boldsymbol{m}$

$$\boldsymbol{m}^{new} = \mathbf{S}(\mathbf{Z} - \mathbf{U}\beta),$$

It is easy to see that the backfitting algorithm implies a linear estimation matrix for updating the index, too. We have

$$\mathbf{U}\beta^{new} + \boldsymbol{m}^{new} = \mathbf{R}^B\mathbf{Z}$$

with

$$\mathbf{R}^B = \widetilde{\mathbf{U}}\{\mathbf{U}^\top\mathbf{W}\widetilde{\mathbf{U}}\}^{-1}\mathbf{U}^\top\mathbf{W}(\mathbf{I}-\mathbf{S}) + \mathbf{S}. \tag{7.26}$$

An important caveat of the so defined backfitting procedure is that it may fail with correlated explanatory variables. Hastie & Tibshirani (1990, p. 124ff.) therefore propose a modification of the algorithm ("modified backfitting", which first searches for a (parametric) solution and only fits the remaining parts nonparametrically. We will introduce this modification later on in Chapter 8.

7.2.4 Computational Issues

Whereas the PLM estimators are directly applicable, all presented algorithms for GPLM are iterative and therefore require an initialization step. Different strategies to initialize the iterative algorithm are possible:

- Start with $\widetilde{\boldsymbol{\beta}}$, $\widetilde{m}(\bullet)$ from a parametric (GLM) fit.
- Alternatively, start with $\boldsymbol{\beta} = 0$ and $m(\boldsymbol{T}_j) = G^{-1}(Y_j)$ (for example with the adjustment $m_j = G^{-1}\{(Y_j + 0.5)/2\}$ for binary responses).
- Backfitting procedures often use $\boldsymbol{\beta} = 0$ and $m(\boldsymbol{T}_j) \equiv G^{-1}(\overline{Y})$.

Next, the smoothing step for the nonparametric function $m(\bullet)$ has to be carried out. Consider the profile likelihood algorithm first. The updating step for $m_j = m_{\boldsymbol{\beta}}(\boldsymbol{T}_j)$ requires a ratio with numerator and denominator of convolution type

$$\sum_{i=1}^{n} \delta_{ij} K_{\mathbf{H}}(\boldsymbol{T}_i - \boldsymbol{T}_j), \tag{7.27}$$

where δ_{ij} is a derivative of the log-likelihood. Note, that this has to be done at least for all $\boldsymbol{T}_j$ $(j = 1, \dots, n)$ since the updated values of $m(\bullet)$ at all observation points are required in the updating step for $\boldsymbol{\beta}$. Thus $O(n^2)$ operations are necessary and in each of these operations the kernel function $K_{\mathbf{H}}(\boldsymbol{T}_i - \boldsymbol{T}_j)$ and both likelihood derivatives need to be evaluated.

The essential difference between the profile likelihood and the Speckman algorithm lies in the fact that the former uses ℓ'_i and ℓ''_i instead of ℓ'_{ij} and ℓ''_{ij}. Thus, both algorithms produce very similar results, when the bandwidth $\mathbf{H}$ is small or when $m(\bullet)$ is relatively constant or small with respect to the parametric part. Müller (2001) points out that both estimators very often resemble each other. The comparison of the estimation algorithms can be summarized as follows:

- Backfitting outperforms profile likelihood estimation for a GPLM under independent design. However, if the explanatory variables are dependent, both variants of profile likelihood improve the fit.
- For small sample sizes, the Speckman estimator seems to perform best. The resulting estimator can hence be considered as a good compromise between accuracy and computational efficiency in estimation and specification testing. For larger sample sizes this algorithm gives estimates that are almost identical to the profile likelihood algorithm.

It should be pointed out that this comparison of algorithms is based on Nadaraya-Watson type kernel smoothing. A local linear or local polynomial approach (Fan & Gijbels, 1996) is possible as well. For the particular case of a one-dimensional nonparametric component (i.e. univariate m), spline smoothing approaches are often discussed in the literature.

Example 7.1.
We consider a subsample of the data used in Examples 5.1 and 5.6. Here we present the estimation results for Mecklenburg-Vorpommern (a state in the very North of Eastern Germany, $n = 402$). The data set contains again six explanatory variables, two of them are continuous (age, household income). Descriptive statistics for this subsample are given in Table 7.1.

Table 7.1. Descriptive statistics for migration data (subsample from Mecklenburg-Vorpommern, $n = 402$)

		Yes	No	(in %)	
Y	MIGRATION INTENTION	39.9	60.1		
X_1	FAMILY/FRIENDS IN WEST	88.8	11.2		
X_2	UNEMPLOYED/JOB LOSS CERTAIN	21.1	78.9		
X_3	CITY SIZE 10,000–100,000	35.8	64.2		
X_4	FEMALE	50.2	49.8		
		Min	Max	Mean	S.D.
X_5	AGE (in years)	18	65	39.93	12.89
T	HOUSEHOLD INCOME (in DM)	400	4000	2262.22	769.82

For illustration purposes we restrict ourselves to considering only one continuous variable (household income) for the nonparametric part. Table 7.2 shows on the left the coefficients of the parametric logit fit. The estimated coefficients for the parametric part of the GPLM are given on the right side of Table 7.2. For easier assessment of the coefficients, both continuous variables (age, household income) have been linearly transformed to $[0, 1]$. We see that the migration intention is definitely determined by age (X_5). However, the coefficients of unemployment (X_2), city size (X_3) and household income (T) variables are also highly significant.

The nonparametric estimate $\widehat{m}(\bullet)$ in this example seems to be an obvious nonlinear function, see Figure 7.1. As already observed in the simulations, both profile likelihood methods give very similar results. Also the nonparametric curves from backfitting and modified backfitting do not differ much from those of the profile likelihood approaches. The reason is the very similar smoothing step in all four methods. We show therefore only a plot of the curves obtained from the generalized Speckman algorithm.

Table 7.2 shows on the right hand side the parameter estimates for all semiparametric estimates. For the generalized Speckman method, the ma-

Table 7.2. Logit and GPLM coefficients for migration data, $h = 0.3$

	Logit (t value)	Profile	Speckman	Backfitting	Modified Backfitting
const.	-0.358 (-0.68)	—	—	—	—
X_1	0.589 (1.54)	0.600 (1.56)	0.599 (1.56)	0.395	0.595
X_2	0.780 (2.81)	0.800 (2.87)	0.794 (2.85)	0.765	0.779
X_3	0.822 (3.39)	0.842 (3.47)	0.836 (3.45)	0.784	0.815
X_4	-0.388 (-1.68)	-0.402 (-1.73)	-0.400 (-1.72)	-0.438	-0.394
X_5	-3.364 (-6.92)	-3.329 (-6.86)	-3.313 (-6.84)	-3.468	-3.334
T	1.084 (1.90)	—	—	—	—
	Linear (GLM)	Part. Linear (GPLM)			

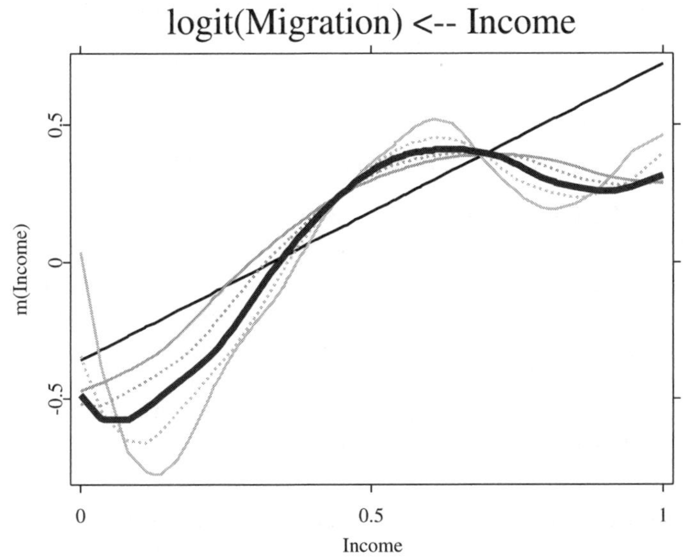

Figure 7.1. GPLM logit fit for migration data, generalized Speckman estimator for m using $h = 0.3$ (thick curve), $h = 0.2$, $h = 0.25$, $h = 0.35$, $h = 0.4$ (thin curves) and parametric logit fit (straight line) `SPMmigmv`

trix $a(\widehat{\psi})(\widetilde{\mathbf{U}}^\top \mathbf{W}\widetilde{\mathbf{U}})^{-1}$ with $\widetilde{\mathbf{U}}$ and $\mathbf{W}$ from the last iteration can be used as an estimate for the covariance matrix of $\widehat{\boldsymbol{\beta}}$. We present the calculated t-values from this matrix. Note that $a(\psi) = 1$ for binary Y, so we omit this factor here. An analogous approach is possible for the profile likelihood method, see Severini & Staniswalis (1994). Here we can estimate the covariance of $\widehat{\boldsymbol{\beta}}$

by $a(\widehat{\psi})(\widetilde{\mathbf{U}}^\top \mathbf{W}\widetilde{\mathbf{U}})^{-1}$, now using the appropriate smoother matrix $\mathbf{S}^P$ instead of $\mathbf{S}$.

The obvious difference between backfitting and the other procedures in the estimated effect of X_1 is an interesting observation. It is most likely due to the multivariate dependence structure within the explanatory variables, an effect which is not easily reflected in simulations. The profile likelihood methods have by construction a similar correction ability for concurvity as the modified backfitting has. □

7.3 Testing the GPLM

Having estimated the function $m(\bullet)$, it is natural to ask whether the estimate $\widehat{m}(\bullet)$ is significantly different from a parametric function obtained by a parametric GLM fit. In the simplest case, this means to consider

$$H_0 : m(t) = \boldsymbol{T}^\top \gamma + \gamma_0$$
$$H_1 : m(\bullet) \text{ is an arbitrary smooth function.}$$

A test statistic for this test problem is typically based on a semiparametric generalization of the parametric likelihood ratio test.

We will discuss two approaches here: Hastie & Tibshirani (1990) propose to use the difference of the deviances of the linear and the semiparametric model, respectively, and to approximate the degrees of freedom in the semiparametric case. The asymptotic behavior of this method is unknown, though. Härdle, Mammen & Müller (1998) derive an asymptotic normal distribution for a slightly modified test statistic.

7.3.1 Likelihood Ratio Test with Approximate Degrees of Freedom

In the following we denote the semiparametric estimates by $\widehat{\mu}_i = G\{\boldsymbol{U}_i^\top \widehat{\beta} + \widehat{m}(\boldsymbol{T}_i)\}$ and the parametric estimates by $\widetilde{\mu}_i = G\{\boldsymbol{U}_i^\top \widetilde{\beta} + \boldsymbol{T}^\top \widetilde{\gamma} + \widetilde{\gamma}_0\}$. A natural approach is to compare both estimates by a likelihood ratio test statistic

$$LR = 2\sum_{i=1}^{n} \ell(Y_i, \widehat{\mu}_i, \widehat{\psi}) - \ell(Y_i, \widetilde{\mu}_i, \widehat{\psi}). \tag{7.28}$$

which would have an asymptotic χ^2 distribution if the estimates $\widehat{\mu}_i$ were from a nesting parametric fit.

This test statistic can be used in the semiparametric case, too. However, an approximate number of degrees of freedom needs to be defined for the GPLM. The basic idea is as follows. Recall that $D(\boldsymbol{Y}, \widehat{\mu}, \psi)$ is the *deviance* in the observations Y_i and fitted values $\widehat{\mu}_i$, see (5.19). Abbreviate the estimated index $\widehat{\eta} = \mathbf{U}\widehat{\beta} + \widehat{m}$ and consider the adjusted dependent variable $\mathbf{Z} = \widehat{\eta} - \mathbf{W}^{-1}v$. If at convergence of the iterative estimation $\widehat{\eta} = \mathbf{R}\mathbf{Z} = \mathbf{R}(\widehat{\eta} - \mathbf{W}^{-1}v)$ with a linear operator $\mathbf{R}$, then

$$a(\psi)\, D(\mathbf{Y}, \widehat{\mu}, \psi) \approx -(\mathbf{Z} - \widehat{\eta})^\top \mathbf{W}(\mathbf{Z} - \widehat{\eta}) \tag{7.29}$$

which has approximately

$$df^{err}(\widehat{\mu}) = n - \operatorname{tr}\left(2\mathbf{R} - \mathbf{R}^\top \mathbf{W}\mathbf{R}\mathbf{W}^{-1}\right) \tag{7.30}$$

degrees of freedom. In practice, the computation of the trace $\operatorname{tr}\left(\mathbf{R}^\top \mathbf{W}\mathbf{R}\mathbf{W}^{-1}\right)$ can be rather difficult. It is also possible to use the simpler approximation

$$df^{err}(\widehat{\mu}) = n - \operatorname{tr}(\mathbf{R}) \tag{7.31}$$

which would be correct if $\mathbf{R}$ were a projection operator and $\mathbf{W}$ were the identity matrix. Now, for the comparison of the semiparametric $\widehat{\mu}$ and the parametric $\widetilde{\mu}$, the test statistic (7.28) can be expressed by

$$LR = D(\mathbf{Y}, \widetilde{\mu}, \psi) - D(\mathbf{Y}, \widehat{\mu}, \psi)$$

and should follow approximately a χ^2 distribution with $df^{err}(\widetilde{\mu}) - df^{err}(\widehat{\mu})$ degrees of freedom.

Property (7.29) holds for backfitting and the generalized Speckman estimator with matrices $\mathbf{R}^B$ and $\mathbf{R}^S$, respectively. A direct application to the profile likelihood algorithm is not possible because of the more involved estimation of the nonparametric function $m(\bullet)$. However a workable approximation can be obtained by using

$$\mathbf{R}^P = \widetilde{\mathbf{U}}\{\widetilde{\mathbf{U}}^\top \mathbf{W}\widetilde{\mathbf{U}}\}^{-1}\widetilde{\mathbf{U}}^\top \mathbf{W}(\mathbf{I} - \mathbf{S}^P) + \mathbf{S}^P, \tag{7.32}$$

where $\widetilde{\mathbf{U}}$ denotes $(\mathbf{I} - \mathbf{S}^P)\mathbf{U}$.

7.3.2 Modified Likelihood Ratio Test

The direct comparison of the semiparametric estimates $\widehat{\mu}_i$ and the parametric estimates $\widetilde{\mu}_i$ can be misleading because $\widehat{m}(\bullet)$ has a non-negligible smoothing bias, even under the linearity hypothesis. Hence, the key idea is to use the estimate $\overline{m}(\boldsymbol{T}_i)$ which introduces a smoothing bias to $\boldsymbol{T}_i^\top \widetilde{\gamma} + \widetilde{\gamma}_0$. This estimate can be obtained from the updating procedure for m_i on the parametric estimate. Note that here the second argument of $L(\bullet, \bullet)$ should be the parametric estimate of $E(Y_i|\boldsymbol{U}_i, \boldsymbol{T}_i)$ instead of Y_i which means to apply the smoothing step according to (7.8) to the artificial data set consisting of $\{G(\boldsymbol{U}_i^\top \widetilde{\beta} + \boldsymbol{T}_i^\top \widetilde{\gamma} + \widetilde{\gamma}_0), \boldsymbol{U}_i, \boldsymbol{T}_i\}$.

Using this "bias-adjusted" parametric estimate $\overline{m}(\bullet)$, one can form the test statistic

$$\widetilde{LR} = 2\sum_{i=1}^{n} \ell(\widehat{\mu}_i, \widehat{\mu}_i, \widehat{\psi}) - \ell(\overline{\mu}_i, \widetilde{\mu}_i, \widehat{\psi}) \tag{7.33}$$

where $\overline{\mu}_i = G\{\boldsymbol{U}_i^\top \widetilde{\boldsymbol{\beta}} + \overline{m}(\boldsymbol{T}_i)\}$ is the bias-adjusted parametric GLM fit and $\widehat{\mu}_i$ is the semiparametric GPLM fit to the observations. Asymptotically, this test statistic is equivalent to

$$\widetilde{\widetilde{LR}} = \frac{1}{a(\widehat{\psi})}\sum_{i=1}^{n} \frac{[G'\{\boldsymbol{U}_i^\top \boldsymbol{\beta} + \widehat{m}(\boldsymbol{T}_i)\}]^2}{V\{G(\widehat{\mu}_i)\}} \left\{\boldsymbol{U}_i^\top(\widehat{\boldsymbol{\beta}} - \widetilde{\boldsymbol{\beta}}) + \widehat{m}(\boldsymbol{T}_i) - \overline{m}(\boldsymbol{T}_i)\right\}^2 . \tag{7.34}$$

Hence, the resulting test statistic can be interpreted as a weighted quadratic difference of the (bias-adjusted) parametric predictor $\boldsymbol{U}_i^\top \widetilde{\boldsymbol{\beta}} + \overline{m}(\boldsymbol{T}_i)$ and the semiparametric predictor $\boldsymbol{U}_i^\top \widehat{\boldsymbol{\beta}} + \widehat{m}(\boldsymbol{T}_i)$.

Both test statistics $\widetilde{LR}$ and $\widetilde{\widetilde{LR}}$ have the same asymptotic normal distribution if the profile likelihood algorithm is used. (A χ^2 approximation does not hold in this case since kernel smoother matrices are not projection operators.) It turns out that the normal approximation does not work well. Therefore, for the calculation of quantiles, it is recommended to use a bootstrap approximation of the quantiles of the test statistic:

(a) Generate samples $Y_1^*, \ldots, Y_n^*$ with

$$E^*(Y_i^*) = G(\boldsymbol{U}_i^\top \widetilde{\boldsymbol{\beta}} + \boldsymbol{T}_i^\top \widetilde{\gamma} + \gamma_0)$$
$$\mathrm{Var}^*(Y_i^*) = a(\widehat{\psi})\, V\{G(\boldsymbol{U}_i^\top \widetilde{\boldsymbol{\beta}} + \boldsymbol{T}_i^\top \widetilde{\gamma} + \gamma_0)\}.$$

(b) Calculate estimates based on the bootstrap samples and finally the test statistics $\widetilde{LR}^*$. The quantiles of the distribution of $\widetilde{LR}$ are estimated by the quantiles of the conditional distributions of $\widetilde{LR}^*$.

There are several possibilities for the choice of the conditional distribution of the Y_i^*s. In a binary response model, the distribution of Y_i is completely specified by $\mu_i = G(\boldsymbol{U}_i^\top \beta + \boldsymbol{T}_i^\top \gamma + \gamma_0)$ and a parametric bootstrap procedure can be used. If the distribution of Y_i cannot be specified (apart from the first two moments) one may use the wild bootstrap procedure of Härdle & Mammen (1993).

Example 7.2.
Finally, consider the testing problem for Example 7.1. From Figure 7.1 it is difficult to judge significance of the nonlinearity. For this real data example, it cannot be excluded that the difference between the nonparametric and the

Table 7.3. Observed significance levels for testing GLM versus GPLM, migration data, 400 bootstrap replications

h	0.20	0.30	0.40
LR (profile likelihood)	0.066	0.048	0.035
LR (generalized Speckman)	0.068	0.047	0.033
LR (backfitting)	0.073	0.062	0.068
LR (modified backfitting)	0.068	0.048	0.035
$\widetilde{LR}$ (profile likelihood, bootstrap)	0.074	0.060	0.052

linear fit may be caused by boundary and bias problems of $\widehat{m}(\bullet)$. Additionally, the variable age (included in a linear way) has dominant influence in the model.

Table 7.3 shows the results of the application of the different test statistics for different choices of the bandwidth h. As we have seen in the simulations, the likelihood ratio test statistic LR and the modified test statistic $\widetilde{LR}$ in combination with bootstrap give very similar results. The number of bootstrap simulations has been chosen as $n_{boot} = 400$. Linearity is clearly rejected (at 10% level) for all bandwidths from 0.2 to 0.4.

The different behavior of the tests for different h gives some indication of possible deviance of $m(\bullet)$ from linear functions. The appearance of small wiggles of small length seems not to be significant for the bootstrap ($h = 0.2$). Also, the bootstrapped $\widetilde{LR}$ still rejects large values of h. This is due to the comparison of the semiparametric estimator with a bias corrected parametric one, yielding more independence of the bandwidth. □

Bibliographic Notes

Partial linear models were first considered by Green & Yandell (1985), Denby (1986), Speckman (1988) and Robinson (1988b). For a combination with spline smoothing see also Schimek (2000a), Eubank, Kambour, Kim, Klipple, Reese & Schimek (1998) and the monograph of Green & Silverman (1994).

The extension of the partial linear and additive models to generalized regression models with link function is mainly considered in Hastie & Tibshirani (1986) and their monograph Hastie & Tibshirani (1990). They employed the observation of Nelder & Wedderburn (1972) and McCullagh & Nelder (1989) that the parametric GLM can be estimated by applying a weighted least squares estimator to the adjusted dependent variable and modified the LS estimator in a semi-/nonparametric way. Formal asymptotic results for the GPLM using Nadaraya-Watson type smoothing were first obtained by Severini & Wong (1992) and applied to this specific model by Severini & Staniswalis (1994). An illustration for the use of the profile likelihood and its efficiency is given by Staniswalis & Thall (2001).

The theoretical ideas for testing the GPLM using a likelihood ratio test and approximate degrees of freedom go back to Buja, Hastie & Tibshirani (1989) and Hastie & Tibshirani (1990). The bootstrap procedure for comparing parametric versus nonparametric functions was formally discussed in Härdle & Mammen (1993). The theoretical results of Härdle, Mammen & Müller (1998) have been empirically analyzed by Müller (2001).

Exercises

Exercise 7.1. Explain why (7.21) and (7.22) hold.

Exercise 7.2. Derive the backfitting estimators (7.24) and (7.25).

Exercise 7.3. Prove that the linear estimation matrix for backfitting in the GPLM case has form (7.26).

Summary

⋆ A partial linear model (PLM) is given by

$$E(Y|\boldsymbol{X}) = \boldsymbol{X}^\top \boldsymbol{\beta} + m(\boldsymbol{T}),$$

where $\boldsymbol{\beta}$ is an unknown parameter vector and $m(\bullet)$ is an unknown smooth function of a multidimensional argument $\boldsymbol{T}$.

⋆ A generalized partial linear model (GPLM) is of the form

$$E(Y|\boldsymbol{X}) = G\{\boldsymbol{X}^\top \boldsymbol{\beta} + m(\boldsymbol{T})\},$$

where G is a know link function, $\boldsymbol{\beta}$ is an unknown parameter vector and $m(\bullet)$ is an unknown smooth function of a multidimensional argument $\boldsymbol{T}$.

⋆ Partial linear models are usually estimated by Speckman's estimator. This estimator determines first the parametric component by applying an OLS estimator to a nonparametrically modified design matrix and response vector. In a second step the nonparametric component is estimated by smoothing the residuals w.r.t. the parametric part.

⋆ The profile likelihood approach is based on the fact that the conditional distribution of Y given $\boldsymbol{U}$ and $\boldsymbol{T}$ is parametric. Its idea is to estimate the least favorable nonparametric function $m_{\boldsymbol{\beta}}(\bullet)$ in dependence of $\boldsymbol{\beta}$. The resulting estimate for $m_{\boldsymbol{\beta}}(\bullet)$ is then used to construct the profile likelihood for $\boldsymbol{\beta}$.

⋆ The generalized Speckman estimator can be seen as a simplification of the profile likelihood method. It is based on a combination of a parametric IRLS estimator (applied to a nonparametrically modified design matrix and response vector) and a nonparametric smoothing method (applied to the adjusted dependent variable reduced by its parametric component).

⋆ Generalized partial linear models should be estimated by the profile likelihood method or by a generalized Speckman estimator.

⋆ To check whether the underlying true model is a parametric GLM or a semiparametric GPLM, one can use specification tests that are modifications or the classical likelihood ratio test. In the semiparametric setting, either an approximate number of degrees of freedom is used or the test statistic itself is modified such that bootstrapping its distribution leads to appropriate critical values.

8

Additive Models and Marginal Effects

Additive models have been proven to be very useful as they naturally generalize the linear regression model and allow for an interpretation of marginal changes, i.e. for the effect of one variable on the mean function $m(\bullet)$ when holding all others constant. This kind of model structure is widely used in both theoretical economics and in econometric data analysis. The standard text of Deaton & Muellbauer (1980) provides many examples in microeconomics for which the additive structure provides interpretability. In econometrics, additive structures have a desirable statistical form and yield many well known economic results. For instance, an additive structure allows us to aggregate inputs into indices; elasticities or rates of substitutions can be derived directly. The separability of the input parameters is consistent with decentralization in decision making or optimization by stages. In summary, additive models can easily be interpreted.

Additive models are also interesting from a statistical point of view. They allow for a componentwise analysis and combine flexible nonparametric modeling of multidimensional inputs with a statistical precision that is typical of a one-dimensional explanatory variable. Let Y be the dependent variable and $\mathbf{X}$ the d-dimensional vector of explanatory variables. Consider the estimation of the general nonparametric regression function $m(\mathbf{X}) = E(Y|\mathbf{X})$. Stone (1985) showed that the optimal convergence rate for estimating $m(\bullet)$ is $n^{-\kappa/(2\kappa+d)}$ with κ an index of smoothness of $m(\bullet)$. Note how high values of d lead to a slow rate of convergence. An additive structure for $m(\bullet)$ is a regression function of the form

$$m(\mathbf{X}) = c + \sum_{\alpha=1}^{d} g_\alpha(X_\alpha), \tag{8.1}$$

where $g_\alpha(\bullet)$ are one-dimensional nonparametric functions operating on each element of the predictor variables. Stone (1985) also showed that for the additive regression function the optimal rate for estimating $m(\bullet)$ is the one-dimensional $n^{-\kappa/(2\kappa+1)}$-rate. One speaks thus of *dimension reduction* through additive modeling.

Additive models of the form (8.1) were first considered in the context of input-output analysis by Leontief (1947a) who called them *additive separable* models. In the statistical literature, additive regression models were introduced in the early eighties, and promoted largely by the work of Buja, Hastie & Tibshirani (1989) and Hastie & Tibshirani (1990). They proposed the iterative *backfitting* procedure to estimate the additive components. This method is presented in detail in Section 8.1.

More recently, Tjøstheim & Auestad (1994b) and Linton & Nielsen (1995) introduced a non-iterative method for estimating marginal effects. Note that marginal effects coincide with the additive components $g_\alpha(\bullet)$ if the true regression function $m(\bullet)$ is indeed additive. The idea of this method is to first estimate a multidimensional functional of $m(\bullet)$ and then use *marginal integration* to obtain the marginal effects. Under additive structure the marginal integration estimator yields the functions $g_\alpha(\bullet)$ up to a constant. This estimator will be introduced starting from Section 8.2. A comparison of backfitting and marginal integration is given in Section 8.3.

8.1 Backfitting

The *backfitting* procedures proposed in Breiman & Friedman (1985), Buja, Hastie & Tibshirani (1989) are widely used to approximate the additive components $g_\alpha(\bullet)$ of (8.1). The latter paper considers the problem of finding the projection of $m(\bullet)$ onto the space of additive models represented by the right hand side of (8.1). Using the observed sample, this leads to a system of normal equations with $nd \times nd$ dimensions. To solve this system in practice, a backfitting or a Gauss-Seidel algorithm is used.

The backfitting technique is iterative which leads to additional difficulties for developing asymptotic theory. Moreover, the final estimates may depend on the starting values or the convergence criterion. Therefore, since its first introduction, the method has been refined considerably and extended to more complicated models. We refer to the bibliographic notes at the end of this and the following chapter for more information.

8.1.1 Classical Backfitting

We will now consider the problem of estimating the additive conditional expectation of Y given $\mathbf{X}$ based on the random sample $\{\mathbf{X}_i, Y_i\}, i = 1, \ldots, n$. The component functions $g_\alpha(\bullet)$ in (8.1) explain the specific impact of the particular component X_α on the response Y.

We first introduce an identification assumption. Let

$$E_{X_\alpha}\{g_\alpha(X_\alpha)\} = 0 \tag{8.2}$$

for all α which consequently yields

$$c = E(Y).$$

Note that condition (8.2) on the marginal effects is no restriction on the model since all we can estimate is the relative impact of each direction X_α. This means the g_α are otherwise only identified up to a shift in the vertical direction. As before we use the model

$$Y = m(\boldsymbol{X}) + \varepsilon,$$

where (by definition) $E(\varepsilon|\boldsymbol{X}) = 0$ and and $(\varepsilon|\boldsymbol{X}) = \sigma^2(\boldsymbol{X})$, i.e. the model is allowed to be heteroscedastic. The constant c can easily be estimated with $\sqrt{n}$-rate (i.e. a rate faster than the nonparametric) by

$$\widehat{c} = \overline{Y} = \frac{1}{n}\sum_{i=1}^{n} Y_i\,.$$

Thus, we assume

$$c = 0$$

for the rest of this section if not indicated differently. This is without loss of generality, since you may replace Y_i by $Y_i = Y_i - \overline{Y}$ otherwise.

Let us now turn to the estimation of the additive components $g_\alpha(\bullet)$. Hastie & Tibshirani (1990) motivate the backfitting method as follows. Consider the optimization problem

$$\min_m E\left\{Y - m(\boldsymbol{X})\right\}^2 \quad \text{such that} \quad m(\boldsymbol{X}) = \sum_{\alpha=1}^{d} g_\alpha(X_\alpha). \tag{8.3}$$

This means, we are searching for the best additive predictor for $E(Y|\boldsymbol{X})$ where "best" is meant with respect to the expected squared distance. By the theory of projections, there exists a unique solution which is characterized by the first order conditions

$$\begin{aligned} &E\left[\left\{Y - m(\boldsymbol{X})\right\}|X_\alpha\right] = 0 \\ &\text{or equivalently} \quad g_\alpha(X_\alpha) = E\left[\left\{Y - \sum_{k\neq\alpha} g_k(X_k)\right\}\Big|X_\alpha\right], \end{aligned} \tag{8.4}$$

for all $\alpha = 1, \ldots, d$. This leads to the matrix representation

$$\begin{pmatrix} \mathcal{I} & \mathcal{P}_1 & \cdots & \mathcal{P}_1 \\ \mathcal{P}_2 & \mathcal{I} & \cdots & \mathcal{P}_2 \\ \vdots & & \ddots & \vdots \\ \mathcal{P}_d & \cdots & \mathcal{P}_d & \mathcal{I} \end{pmatrix} \begin{pmatrix} g_1(X_1) \\ g_2(X_2) \\ \vdots \\ g_d(X_d) \end{pmatrix} = \begin{pmatrix} \mathcal{P}_1 Y \\ \mathcal{P}_2 Y \\ \vdots \\ \mathcal{P}_d Y \end{pmatrix}, \tag{8.5}$$

where $\mathcal{P}_\alpha(\bullet) = E(\bullet|X_\alpha)$. By analogy, let $\mathbf{S}_\alpha$ be a $(n \times n)$ smoother matrix which, when applied to the vector $\boldsymbol{Y} = (Y_1, \ldots, Y_n)^\top$, yields a $n \times 1$ estimate $\mathbf{S}_\alpha \boldsymbol{Y}$ of $\{E(Y_1|X_{1\alpha}), \ldots, E(Y_n|X_{n\alpha})\}^\top$. Substituting the operator $\mathcal{P}_\alpha$ by the smoother matrix $\mathbf{S}_\alpha$ we obtain a system of equations

$$\underbrace{\begin{pmatrix} \mathbf{I} & \mathbf{S}_1 & \cdots & \mathbf{S}_1 \\ \mathbf{S}_2 & \mathbf{I} & \cdots & \mathbf{S}_2 \\ \vdots & & \ddots & \vdots \\ \mathbf{S}_d & \cdots & \mathbf{S}_d & \mathbf{I} \end{pmatrix}}_{nd \times nd} \begin{pmatrix} g_1 \\ g_2 \\ \vdots \\ g_d \end{pmatrix} = \begin{pmatrix} \mathbf{S}_1 \boldsymbol{Y} \\ \mathbf{S}_2 \boldsymbol{Y} \\ \vdots \\ \mathbf{S}_d \boldsymbol{Y} \end{pmatrix}, \tag{8.6}$$

or abbreviated

$$\widehat{\mathbf{P}} g = \widehat{\mathbf{Q}} \boldsymbol{Y}. \tag{8.7}$$

Although $E(\bullet|X_\alpha)$ means in fact the expectation over all directions $X_1, \ldots, X_d$, Buja, Hastie & Tibshirani (1989) suggest using only one-dimensional smoothers as a sufficient approximation to $\mathcal{P}_\alpha$.

The system (8.7) could principally be solved exactly for $\widehat{g}_\alpha(X_{1\alpha}), \ldots, \widehat{g}_\alpha(X_{n\alpha})$. Of course, when nd is large the exact solution of (8.7) is hardly feasible. Furthermore, the matrix $\widehat{\mathbf{P}}$ on the left is often not regular and thus the equation cannot be solved directly. Therefore, in practice the backfitting (Gauss-Seidel) algorithm is used to solve these equations: Given starting values $\widehat{g}_\alpha^{(0)}$, the vectors g_α are updated through

$$\widehat{g}_\alpha^{(l)} = \mathbf{S}_\alpha \left\{ \boldsymbol{Y} - \sum_{k \neq \alpha} \widehat{g}_k^{(l-1)} \right\}, \quad l = 1, 2, \ldots \tag{8.8}$$

until the distance between successive estimates is below a prespecified tolerance. So what one actually does is a successive one-dimensional nonparametric regression on X_α, using the partial residuals $\{\boldsymbol{Y} - \sum_{k \neq \alpha} \widehat{g}_k^{(l-1)}\}$ (instead of simply $\boldsymbol{Y}$). To summarize, we present the whole procedure including the estimation of the (possibly non-zero) constant:

Backfitting Algorithm	
initialization	$\widehat{c} = \overline{Y},\ \widehat{g}_\alpha^{(0)} \equiv 0$ for $\alpha = 1, \ldots, d$,
repeat	for $\alpha = 1, \ldots, d$ the cycles:
	$\boldsymbol{r}_\alpha = \boldsymbol{Y} - \widehat{c} - \sum_{k=1}^{\alpha-1} \widehat{\boldsymbol{g}}_\alpha^{(l+1)} - \sum_{k=\alpha+1}^{d} \widehat{\boldsymbol{g}}_k^{(l)}$,
	$\widehat{g}_\alpha^{(l+1)}(\bullet) = \mathbf{S}_\alpha(\boldsymbol{r}_\alpha)$
until	convergence is reached

Of course, the choice of the convergence criterion is a crucial problem and in practice different software implementations of the algorithm may stop after a different number of iterations. Also, the result of the iteration can obviously depend on the order of the variables $X_1, \ldots, X_d$.

It is important to note that to estimate $\mathcal{P}_\alpha = E(\bullet|X_\alpha)$, the backfitting algorithm uses smoother matrices $\mathbf{S}_\alpha$ that depend only on the components X_α. These smoother matrices are easy and fast to calculate since they employ only a one-dimensional smoothing problem. This is a strong simplification (also from a theoretical point of view) and may lead to problems in theory and in interpretation of what the procedure is fitting in practice. We will come back to this point in Subsection 8.1.3.

For the two-dimensional case we can write down the backfitting estimator explicitly. From equation (8.6) we have

$$\boldsymbol{g}_1 = \mathbf{S}_1(\boldsymbol{Y} - \boldsymbol{g}_2), \quad \boldsymbol{g}_2 = \mathbf{S}_2(\boldsymbol{Y} - \boldsymbol{g}_1).$$

Now it is easy to derive the smoother (or hat) matrices for $\widehat{\boldsymbol{g}}_1$ and $\widehat{\boldsymbol{g}}_2$. Let $\widehat{\boldsymbol{g}}_1^{(l)}$ and $\widehat{\boldsymbol{g}}_2^{(l)}$ denote the estimates in the lth updating step. The backfitting procedure steps are

$$\widehat{\boldsymbol{g}}_1^{(l)} = \mathbf{S}_1\left\{\boldsymbol{Y} - \widehat{\boldsymbol{g}}_2^{(l-1)}\right\}, \quad \widehat{\boldsymbol{g}}_2^{(l)} = \mathbf{S}_2\left\{\boldsymbol{Y} - \widehat{\boldsymbol{g}}_1^{(l)}\right\}.$$

Since the backfitting consists of alternating these steps, we obtain by induction

$$\widehat{\boldsymbol{g}}_1^{(l)} = \boldsymbol{Y} - \sum_{\alpha=0}^{l-1}(\mathbf{S}_1\mathbf{S}_2)^\alpha(\mathbf{I} - \mathbf{S}_1)\boldsymbol{Y} - (\mathbf{S}_1\mathbf{S}_2)^{l-1}\mathbf{S}_1\widehat{\boldsymbol{g}}_2^{(0)},$$

$$\widehat{\boldsymbol{g}}_2^{(l)} = \mathbf{S}_2\sum_{\alpha=0}^{l-1}(\mathbf{S}_1\mathbf{S}_2)^\alpha(\mathbf{I} - \mathbf{S}_1)\boldsymbol{Y} + \mathbf{S}_2(\mathbf{S}_1\mathbf{S}_2)^{l-1}\mathbf{S}_1\widehat{\boldsymbol{g}}_2^{(0)}.$$

Due to the assumptions $E_{X_\alpha}\{g_\alpha(X_\alpha)\} = 0$ and $c = 0$, the initialization $\widehat{\boldsymbol{g}}_2^{(0)} = 0$ is reasonable and we have

$$\widehat{g}_1^{(l)} = \{\mathbf{I} - \sum_{\alpha=0}^{l-1} (\mathbf{S}_1\mathbf{S}_2)^{\alpha}(\mathbf{I} - \mathbf{S}_1)\}\boldsymbol{Y},$$

$$\widehat{g}_2^{(l)} = \{\mathbf{S}_2 \sum_{\alpha=0}^{l-1} (\mathbf{S}_1\mathbf{S}_2)^{\alpha}(\mathbf{I} - \mathbf{S}_1)\}\boldsymbol{Y}\,.$$

This algorithm converges if the operator $\mathbf{S}_1\mathbf{S}_2$ is shrinking (i.e. for its matrix norm holds $\|\mathbf{S}_1\mathbf{S}_2\| < 1$) and hence gives

$$\begin{aligned}\widehat{g}_1 &= \{\mathbf{I} - (\mathbf{I} - \mathbf{S}_1\mathbf{S}_2)^{-1}(\mathbf{I} - \mathbf{S}_1)\}\boldsymbol{Y},\\ \widehat{g}_2 &= \{\mathbf{I} - (\mathbf{I} - \mathbf{S}_2\mathbf{S}_1)^{-1}(\mathbf{I} - \mathbf{S}_2)\}\boldsymbol{Y}.\end{aligned} \tag{8.9}$$

when $l \to \infty$.

In the general case, because of its iterative nature, theoretical results for backfitting have not been known until recently. Opsomer & Ruppert (1997) provide conditional mean squared error expressions albeit under rather strong conditions on the design and smoother matrices. Mammen, Linton & Nielsen (1999) prove consistency and calculate the asymptotic properties under weaker conditions. We will present their and a different backfitting algorithm in Subsection 8.1.3.

Before proceeding we will consider two examples illustrating what precision can be achieved by additive regression and how the backfitting algorithm performs in practice.

Example 8.1.
Consider a regression problem with four input variables X_1 to X_4. When n is small, it is difficult to obtain a precise nonparametric kernel estimate due to the curse of dimensionality. Let us take a sample of $n = 75$ regression values. We use explanatory variables that are independent and uniformly distributed on $[-2.5, 2.5]$ and responses generated from the additive model

$$Y = \sum_{\alpha=1}^{4} g_\alpha(X_\alpha) + \varepsilon, \quad \varepsilon \sim N(0,1).$$

The component functions are chosen as

$$\begin{aligned} g_1(X_1) &= -\sin(2X_1), & g_2(X_2) &= X_2^2 - E(X_2^2),\\ g_3(X_3) &= X_3, & g_4(X_4) &= \exp(-X_4) - E\{\exp(-X_4)\}.\end{aligned}$$

In Figure 8.1 we have plotted the true functions (at the corresponding observations $X_{i\alpha}$) and the estimated curves. We used backfitting with univariate local linear smoothers and set the bandwidth to $h = 1$ for each dimension (using the Quartic kernel). We see that even for this small sample size the estimator gives rather precise results. □

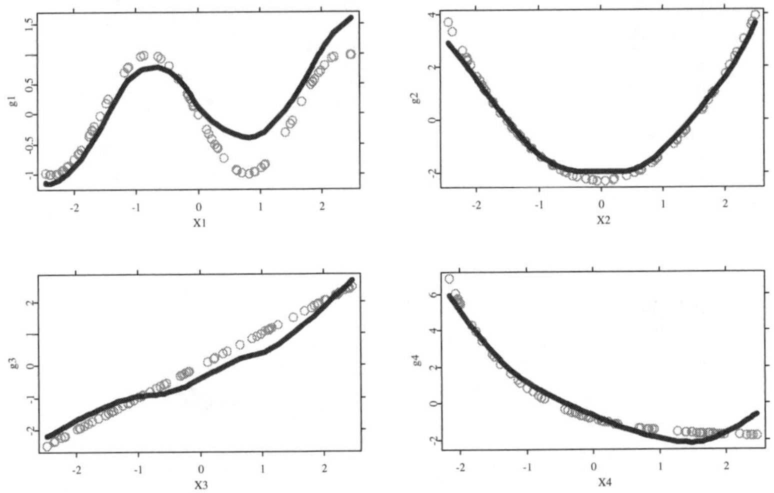

Figure 8.1. Estimated (solid lines) versus true additive component functions (circles at the input values)

Next, we turn to a real data example demonstrating the use of additive regression estimators in practice and manifesting that even for a high dimensional data set the backfitting estimator works reasonably well.

Example 8.2.
We consider the Boston house price data of Harrison & Rubinfeld (1978). with $n = 506$ observations for the census districts of the Boston metropolitan area. We selected ten explanatory variables, given below in the same order as the pictures of the resulting estimates:

- X_1 per capita crime rate by town,
- X_2 proportion of non-retail business acres per town,
- X_3 nitric oxides concentration (parts per 10 million),
- X_4 average number of rooms per dwelling,
- X_5 proportion of owner-occupied units built prior to 1940,
- X_6 weighted distances to five Boston employment centers,
- X_7 full-value property tax rate per \$10,000,
- X_8 pupil-teacher ratio by town,
- X_9 $1{,}000(Bk - 0.63)^2$ where Bk is the proportion of people of Afro-American descent by town,
- X_{10} percent lower status of the population.

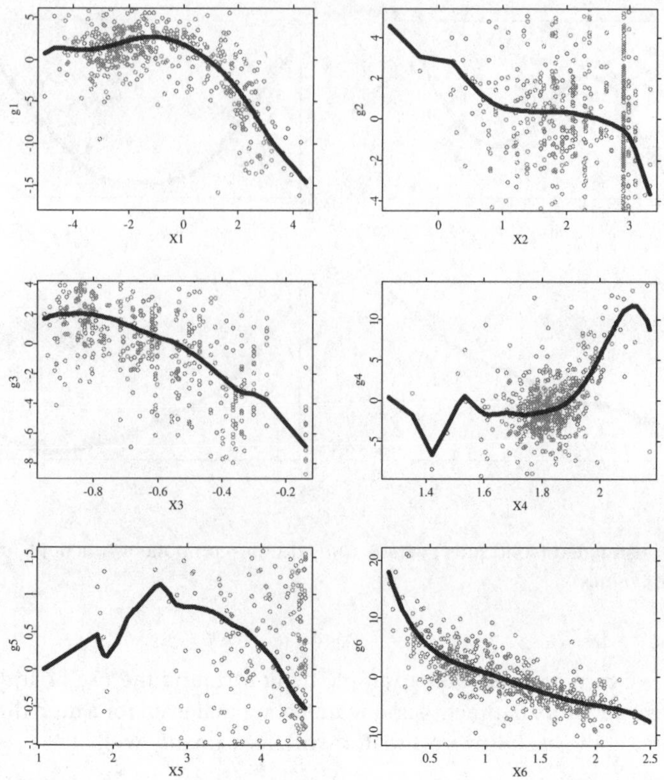

Figure 8.2. Function estimates of additive components g_1 to g_6 and partial residuals (from the left to the right, top to bottom)

The response variable Y is the median value of owner-occupied homes in 1,000 USD. Proceeding on the assumption that the model is

$$E(Y|\mathbf{X}) = c + \sum_{\alpha=1}^{10} g_\alpha\{\log(X_\alpha)\},$$

we get the results for $\widehat{g}_\alpha$ as plotted in Figure 8.2. We used again local linear smoothers (with Quartic kernel) and chose the bandwidths for each dimension α proportional to its standard deviation ($h_\alpha = 0.5\widehat{\sigma}_\alpha$). Our choice of bandwidths is perhaps suboptimal in a theoretical sense, but is reasonable to get an impression of the curves.

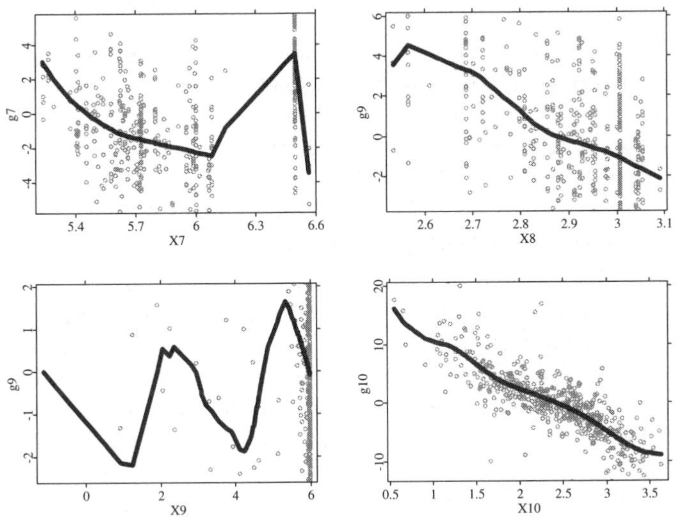

Figure 8.3. Function estimates of additive components g_7 to g_{10} and partial residuals (from the left to the right, top to bottom)

Looking at Figures 8.2, 8.3 we realize that a log-linear model would fit these data for many directions. This was the model actually applied by Harrison & Rubinfeld (1978). In contrast, the explanatory variables X_1 and X_5 show a clear nonlinear impact. □

Note, that in this example smoothing for the variables X_7 and X_9 seems to be problematic due to the distribution of these regressors. In fact, by applying backfitting we have up to now assumed continuous regressors only. To overcome this deficiency, we will later on (in Section 9.3) extend the generalized additive model to a semiparametric model. Among other advantages, this allows us to fit partly in a parametric fashion.

8.1.2 Modified Backfitting

Two of the disadvantages of the above described backfitting procedure are the dependence of the final solution on the starting functions and the possible non-uniqueness of the solution. Recall equation (8.6) which can be written as $\widehat{\mathbf{P}}\boldsymbol{g} = \widehat{\mathbf{Q}}\boldsymbol{Y}$. Non-uniqueness can happen if in (8.6) there exists a vector $\boldsymbol{b}$ such that $\widehat{\mathbf{P}}\boldsymbol{b} = 0$. Then equation (8.6) has an infinite number of solutions because if $\widehat{g}$ is a solution, then so is $\widehat{g} + \gamma \boldsymbol{b}$ for any $\gamma \in \mathbb{R}$. This phenomenon is called *concurvity*.

Concurvity occurs for special interrelations among the $\mathbf{S}_\alpha$s. In mathematical terms this can be explained as follows: Let $\mathbf{S}_\alpha$ be smoother matrices with eigenvalues in $[0,1]$, which is true for all cases presented here. Let further $\mathcal{V}_1(\mathbf{S}_\alpha)$ be the space spanned by all eigenvectors of $\mathbf{S}_\alpha$ that correspond to eigenvalue 1. Then $\widehat{\mathbf{P}}\boldsymbol{b} = 0$ is possible if there exist $\boldsymbol{b}_\alpha \in \mathcal{V}_1(\mathbf{S}_\alpha)$ for all $\alpha = 1,\ldots,d$, and $\sum_{\alpha=1}^{d} \boldsymbol{b}_\alpha = 0$. Note that the elements of $\mathcal{V}_1(\mathbf{S}_\alpha)$ are those which pass unchanged through the smoother $\mathbf{S}_\alpha$.

To reduce the problem of concurvity, a modification of the backfitting algorithm is useful. Let $\mathbf{A}_\alpha$ be the matrix that projects on $\mathcal{V}_1(\mathbf{S}_\alpha)$ and consequently $\mathbf{I} - \mathbf{A}_\alpha$ the matrix that projects on $\mathcal{V}_1(\mathbf{S}_\alpha)^\perp$ (the orthogonal complement of $\mathcal{V}_1(\mathbf{S}_\alpha)$). Let further denote $\mathbf{A}$ is the orthogonal projection onto $\mathcal{V}_1(\mathbf{S}_1) + \ldots + \mathcal{V}_1(\mathbf{S}_d)$. The idea is now to first project the variable to regress onto $\mathcal{V}_1(\mathbf{S}_1) + \ldots + \mathcal{V}_1(\mathbf{S}_d)$, then to smooth by using $\mathbf{S}_\alpha$ and finally to project onto $\mathcal{V}_1(\mathbf{S}_\alpha)^\perp$. In other words, we first use $\mathbf{A}$ to the full and then $\widetilde{\mathbf{S}}_\alpha = (\mathbf{I} - \mathbf{A}_\alpha)\mathbf{S}_\alpha$ to the partial residuals.

We can summarize this modified estimation algorithm as follows:

Modified Backfitting Algorithm

initialization	$\widehat{c} = \overline{Y}$, $\widehat{g}_\alpha^{(0)} \equiv 0$ for $\alpha = 1,\ldots,d$, $\widehat{g}_+ = \widehat{c} + \widehat{g}_1^{(0)} + \cdots \widehat{g}_d^{(0)}$
repeat	regress $\boldsymbol{Y} - \widehat{g}_+^{(l)}$ onto the space $\mathcal{V}_1(\mathbf{S}_1) + \ldots + \mathcal{V}_1(\mathbf{S}_d)$, i.e. set $\boldsymbol{a} = \mathbf{A}(\boldsymbol{Y} - \widehat{g}_+^{(l)})$; apply one cycle of backfitting to $(\boldsymbol{Y} - \boldsymbol{a})$ using $\widetilde{\mathbf{S}}_\alpha$, this yields updated additive function estimates $\widehat{g}^{(l+1)}$
until	convergence is reached

In practice, often $\mathbf{A} = \mathbf{X}(\mathbf{X}^\top\mathbf{X})^{-1}\mathbf{X}^\top$ is used, i.e. each iteration step starts with a parametric linear least squares regression on the explanatory variables $\boldsymbol{X}$.

Hastie & Tibshirani (1990) claim that this algorithm makes it easier to find a unique solution and also eliminates the dependence of the final results on the starting functions.

8.1.3 Smoothed Backfitting

As already mentioned above, Mammen, Linton & Nielsen (1999) give theoretical results for a backfitting procedure that is different from the algorithms presented so far. Their main idea is to estimate the conditional expectations "correctly" instead of approximating them with the one-dimensional smoother matrices $\mathbf{S}_\alpha$. The proof uses rather weak conditions and their modification as well as their proof is going back to the interpretation of the backfitting as a projection of the original data into the space of additive models.

Let us consider the algorithm of Mammen, Linton & Nielsen (1999) in more detail here. As before, set for the ease of notation $c = 0$. Recall (8.4) to (8.8), i.e.

$$g_\alpha(X_\alpha) = E\left[\left\{Y - \sum_{k\neq\alpha} g_k(X_k)\right\} \middle| X_\alpha\right]$$

leading to

$$g_\alpha(X_\alpha) = E(Y|X_\alpha) - \sum_{k\neq\alpha} E\{g_k(X_k)|X_\alpha\}\,. \tag{8.10}$$

Then, replacing the expressions on the right hand side by nonparametric estimates, we obtain for each iteration step

$$\widehat{g}_\alpha^{(l)}(X_{i\alpha}) = \mathbf{S}_{i\alpha}^\top \mathbf{Y} - \sum_{k\neq\alpha} \int \widehat{g}_k^{(l-1)}(t)\, \frac{\widehat{f}_{\alpha k}(X_{i\alpha}, t)}{\widehat{f}_\alpha(X_{i\alpha})}\, dt\,. \tag{8.11}$$

Here, $\widehat{f}_\alpha$ is a univariate density estimator for the marginal pdf of X_α and $\widehat{f}_{\alpha k}$ is a bivariate estimator for the joint pdf of X_α and X_k. $\mathbf{S}_{i\alpha}$ denotes the ith row of the smoother matrix $\mathbf{S}_\alpha$ (introduced above for estimating $\mathcal{P}_\alpha = E(\bullet|X_\alpha)$).

The difference to the earlier backfitting procedures is that here $E(\bullet|X_\alpha)$ in (8.10) is estimated as the expectation over all dimensions and not only over direction α. For more details see Mammen, Linton & Nielsen (1999). They give the following theorem when using kernel regression smoothers inside their backfitting step:

Theorem 8.1.
Under appropriate regularity conditions on the regressor densities, the additive component functions, the error term and the smoother it holds:

(a) There exists a unique solution $\widehat{g}_\alpha$ for the estimation problem of equation (8.11) and the iteration converges.

(b) For the backfitting estimator we have

$$n^{2/5}\left\{\widehat{g}_\alpha(x_\alpha) - g_\alpha(x_\alpha)\right\} \stackrel{L}{\longrightarrow} N\left\{b_\alpha(x_\alpha), v_\alpha(x_\alpha)\right\}.$$

The variance function is of the same rate as if we had estimated the univariate model $Y = g_\alpha(X_\alpha) + \varepsilon$. The same result holds for the bias term b_α if the regression function is truly additive. Note in particular, that for estimating the univariate functions $g_\alpha(\bullet)$, we obtain the univariate rate of convergence as we know from Chapter 4.

Further, for the overall regression estimator $\widehat{m}$ constructed from the estimators $\widehat{g}_\alpha$ the following result holds:

Theorem 8.2.
Assume the same regularity conditions as in Theorem 8.1. For

$$\widehat{m}(\boldsymbol{x}) = \sum_{\alpha=1}^{d} \widehat{g}_\alpha(x_\alpha)$$

we have

$$n^{2/5}\{\widehat{m}(\boldsymbol{x}) - m(\boldsymbol{x})\} \xrightarrow{L} N\left\{\sum_\alpha b_\alpha(x_\alpha), \sum_\alpha v_\alpha(x_\alpha)\right\}.$$

This means that even for the estimate of $m(\bullet)$ we obtain the univariate rate of convergence. Moreover, the correlations between two estimators $\widehat{g}_\alpha$ and $\widehat{g}_k$ are of higher order and hence asymptotically negligible. For details on the implementation of the smoothed backfitting estimator and its computational performance we refer to Nielsen & Sperlich (2002).

8.2 Marginal Integration Estimator

We now turn to the problem of estimating the marginal effects of the regressors X_α. The marginal effect of an explanatory variable tells how Y changes on average if this variable is varying. In other words, the marginal effect represents the conditional expectation $E_{\varepsilon,\boldsymbol{X}_{\underline{\alpha}}}(Y|X_\alpha)$ where the expectation is not only taken on the error distribution but also on all other regressors. (Note, that we usually suppressed the ε in all expectations up to now. This is the only case where we need to explicitely mention on which distribution the expectation is calculated.)

As already indicated, in case of true additivity the marginal effects correspond exactly to the additive component functions g_α. The estimator here is based on an integration idea, coming from the following observation. Denote by f_α the marginal density of X_α. We have from (8.2)

$$E_{X_\alpha}\{g_\alpha(X_\alpha)\} = \int g_\alpha(t) f_\alpha(t)\, dt = 0, \quad \text{for all } \alpha = 1, \ldots, d.$$

Denote further $\boldsymbol{X}_{\underline{\alpha}}$ the vector of all explanatory variables but X_α, i.e.

$$\mathbf{X}_{\underline{\alpha}} = \left(X_{i1}, \ldots, X_{i(\alpha-1)}, X_{i(\alpha+1)}, \ldots, X_{id}\right)$$

and $f_{\underline{\alpha}}$ their joint pdf. If now $m(\mathbf{X}) = m(X_\alpha, \mathbf{X}_{\underline{\alpha}})$ is of additive form (8.1), then

$$\begin{aligned}\int m(\mathbf{x}) f_{\underline{\alpha}}(\mathbf{x}_{\underline{\alpha}}) \prod_{k \neq \alpha} dX_k &= E_{\mathbf{X}_{\underline{\alpha}}}[m(X_\alpha, \mathbf{X}_{\underline{\alpha}})] \\ &= E_{\mathbf{X}_{\underline{\alpha}}}\{c + g_\alpha(X_\alpha) + \sum_{k \neq \alpha} g_k(X_k)\} \\ &= c + g_\alpha(X_\alpha) \, . \end{aligned} \tag{8.12}$$

You see that indeed we calculate $E_{\varepsilon, \mathbf{X}_{\underline{\alpha}}}(Y|X_\alpha)$ instead of $E_\varepsilon(Y|X_\alpha)$. We give a simple example to illustrate marginal integration:

Example 8.3.
Suppose we have a data generating process of the form

$$Y = 4 + X_1^2 + 2 \cdot \sin(X_2) + \varepsilon \, ,$$

where $X_1 \sim U[-2, 2]$ and $X_2 \sim U[-3, 3]$ uniformly distributed and ε a regular, possibly heteroscedastic noise term. The regression function obviously is

$$m(x_1, x_2) = E(Y|\mathbf{X} = \mathbf{x}) = 4 + x_1^2 + 2 \cdot \sin(x_2).$$

We have consequently marginal expectations

$$E_{X_2}\{m(X_1, X_2)\} = \int_{-3}^{3} \frac{1}{6} \left\{4 + X_1^2 + 2 \cdot \sin(u)\right\} du = 4 + X_1^2 \, ,$$

$$E_{X_1}\{m(X_1, X_2)\} = \int_{-2}^{2} \frac{1}{4} \left\{4 + u^2 + 2 \cdot \sin(X_2)\right\} du = \frac{16}{3} + 2 \cdot \sin(X_2) \, .$$

This yields the component functions

$$g_1(x_1) = x_1^2 - \frac{4}{3}, \quad g_2(x_2) = 2 \sin(x_2) \, , \quad \text{and} \quad c = \frac{16}{3}$$

which are normalized to $E_{X_\alpha} g_\alpha(X_\alpha) = 0$ □.

Many extensions and modifications of the integration approach have been developed recently. We consider now the simultaneous estimation of both the functions and their derivatives by combining the procedure with a local polynomial approach (Subsections 8.2.1, 8.2.2) and the estimation of interaction terms (Subsection 8.2.3).

8.2.1 Estimation of Marginal Effects

In order to estimate the marginal effect $g_\alpha(X_\alpha)$, equation (8.12) suggests the following idea: First estimate the function $m(\bullet)$ with a multidimensional pre-smoother $\widetilde{m}$, then integrate out the variables different from X_α. In the estimation procedure integration can be replaced by averaging (over the directions not of interest, i.e. $\boldsymbol{X}_{\underline{\alpha}}$) resulting in

$$\widehat{\{g_\alpha(\bullet)+c\}} = \frac{1}{n}\sum_{i=1}^{n} \widetilde{m}(\bullet, \boldsymbol{X}_{i\underline{\alpha}}) . \tag{8.13}$$

Note that to get the marginal effects, we just integrate $\widetilde{m}$ over all other (the nuisance) directions $\underline{\alpha}$. In case of additivity these marginal effects are the additive component functions g_α plus the constant c. As for backfitting, the constant c can be estimated consistently by $\widehat{c} = \overline{Y}$ at $\sqrt{n}$-rate. Hence, a possible estimate for g_α is

$$\widehat{g}_\alpha(\bullet) = \frac{1}{n}\sum_{i=1}^{n} \widetilde{m}(\bullet, \boldsymbol{X}_{i\underline{\alpha}}) - \overline{Y} . \tag{8.14}$$

Centering the marginals yields the same asymptotic result, i.e.

$$\begin{aligned}\widehat{g}_\alpha(\bullet) &= \widehat{\{g_\alpha(\bullet)+c\}} - \frac{1}{n}\sum_{i=1}^{n} \widehat{\{g_\alpha(X_{i\alpha})+c\}} \\ &= \frac{1}{n}\sum_{i=1}^{n} \widetilde{m}(\bullet, \boldsymbol{X}_{i\underline{\alpha}}) - \frac{1}{n}\sum_{i=1}^{n}\frac{1}{n}\sum_{l=1}^{n} \widetilde{m}(X_{i\alpha}, \boldsymbol{X}_{l\underline{\alpha}}) . \end{aligned} \tag{8.15}$$

It remains to discuss how to obtain a reasonable pre-estimator $\widetilde{m}(x_\alpha, \boldsymbol{x}_{l\underline{\alpha}})$. Principally this could be any multivariate nonparametric estimator. We make use here of a special type of multidimensional local linear kernel estimators, cf. Ruppert & Wand (1994) and Severance-Lossin & Sperlich (1999). This estimator is given by minimizing

$$\begin{aligned}\sum_{i=1}^{n}\{Y_i - \beta_0 - \beta_1(X_{i\alpha}-x_\alpha) - \ldots - \beta_p(X_{i\alpha}-x_\alpha)^p\}^2 \\ \cdot K_h(X_{i\alpha}-x_\alpha)\mathcal{K}_{\mathbf{H}}(\boldsymbol{X}_{i\underline{\alpha}}-\boldsymbol{x}_{l\underline{\alpha}})\end{aligned} \tag{8.16}$$

w.r.t. to $\beta_0, \ldots, \beta_p$. Here, K_h denotes a (scaled) univariate and $\mathcal{K}_{\mathbf{H}}$ a (scaled) $(d-1)$-dimensional kernel function. h and $\mathbf{H} = \widetilde{h}\mathbf{I}_{d-1}$ are the bandwidth parameters. To obtain the estimated marginal function, we need to extract the estimated β_0. This means we use

$$\widetilde{m}(x_\alpha, \boldsymbol{x}_{l\underline{\alpha}}) = e_0^T(\mathbf{X}_\alpha^\top \mathbf{W}_{l\alpha} Z_\alpha)^{-1}\mathbf{X}_\alpha^\top \mathbf{W}_{l\alpha}\boldsymbol{Y},$$

where

$$\mathbf{W}_{l\alpha} = \operatorname{diag}\left(\left\{\frac{1}{n}K_h(X_{i\alpha} - x_\alpha)\mathcal{K}_{\mathbf{H}}(\mathbf{X}_{i\underline{\alpha}} - \mathbf{x}_{l\underline{\alpha}})\right\}_{i=1,\ldots,n}\right),$$

$$\mathbf{X}_\alpha = \begin{pmatrix} 1 & X_{1\alpha} - x_\alpha & (X_{1\alpha} - x_\alpha)^2 & \ldots & (X_{1\alpha} - x_\alpha)^p \\ \vdots & \vdots & \vdots & \vdots & \vdots \\ 1 & X_{n\alpha} - x_\alpha & (X_{n\alpha} - x_\alpha)^2 & \ldots & (X_{n\alpha} - x_\alpha)^p \end{pmatrix}.$$

This estimator is a local polynomial smoother of degree p for the direction α and a local constant one for all other directions. Note that the resulting estimate is simply a weighted least squares estimate. For a more detailed discussion recall Subsection 4.1.3, where local polynomial estimators have been introduced. Bringing the estimate together, we have

$$\widehat{g}_\alpha(x_\alpha) = \frac{1}{n}\sum_{l=1}^{n} e_0^\top (\mathbf{X}_\alpha^\top W_{l\alpha}\mathbf{X}_\alpha)^{-1}\mathbf{X}_\alpha^\top \mathbf{W}_{l\alpha}\mathbf{Y} - \overline{Y}\,. \tag{8.17}$$

To derive asymptotic properties of these estimators the concept of equivalent kernels is used, see e.g. Ruppert & Wand (1994). The main idea is that the local polynomial smoother of degree p is asymptotically equivalent (i.e. has the same leading term) to a kernel estimator using a higher order kernel given by

$$K_\nu^\star(u) = \sum_{t=0}^{p} s_{\nu t}u^t K(u), \tag{8.18}$$

where $\mathbf{S} = \left(\left(\int u^{t+r}K(u)\,du\right)_{0\le t,r\le p}\right)$ and $\mathbf{S}^{-1} = \left((s_{\nu t})_{0\le \nu,t\le p}\right)$. For the resulting asymptotics and some real data examples we refer to the following subsections.

8.2.2 Derivative Estimation for the Marginal Effects

We now extend the marginal integration method to the estimation of derivatives of the functions $g_\alpha(\bullet)$. For additive linear functions their first derivatives are constants and so all higher order derivatives vanish. However, very often in economics the derivatives of the marginal effects are of essential interest, e.g. to determine the elasticities or returns to scale as in Example 8.5.

To estimate the derivatives of the additive components, we do not need any further extension of our method, since using a local polynomial estimator of order p for the pre-estimator $\widetilde{m}$ provides us simultaneously with both component functions and derivative estimates up to degree p. The reason is that the optimal β_ν in equation (8.16) is an estimate for $g_\alpha^{(\nu)}(X_\alpha)/\nu!$, provided that dimension α is separable from the others. In case of additivity this is automatically given.

Thus, we can use

$$\widehat{g}_\alpha^{(\nu)}(x_\alpha) = \frac{\nu!}{n} \sum_{l=1}^{n} e_\nu^\top (\mathbf{X}_\alpha^\top \mathbf{W}_{l\alpha} \mathbf{X}_\alpha)^{-1} \mathbf{X}_\alpha^\top \mathbf{W}_{l\alpha} Y \tag{8.19}$$

for estimating the νth derivative. Compare this with equation (8.17). Here, e_ν is now the $(\nu+1)$th unit vector in order to extract the estimate for β_ν Asymptotic properties are derived in Severance-Lossin & Sperlich (1999). Recall that the integration estimator requires to choose two bandwidths, h for the direction of interest and $\widetilde{h}$ for the nuisance direction.

Theorem 8.3.
Consider kernels K and $\mathcal{K}$, where $\mathcal{K}$ is a product of univariate kernels of order $q > 2$, h, $\widetilde{h}$ bandwidths such that $nh\widetilde{h}^{(d-1)} / \log^2(n) \to \infty$, $\widetilde{h}^q h^{\nu-p-1} \to 0$ and $h = O\{n^{-1/(2p+3)}\}$. We assume $p-\nu$ odd and some regularity conditions. Then,

$$n^{(p+1-\nu)/(2p+3)} \left\{ \widehat{g}_\alpha^{(\nu)}(x_\alpha) - g_\alpha^{(\nu)}(x_\alpha) \right\} \xrightarrow{L} N\left\{ b_\alpha(x_\alpha), v_\alpha(x_\alpha) \right\},$$

where

$$b_\alpha(x_\alpha) = \frac{\nu! h_0^{p+1-\nu}}{(p+1)!} \mu_{p+1}(K_\nu^*) \left\{ g_\alpha^{(p+1)}(x_\alpha) \right\},$$

and

$$v_\alpha(x_\alpha) = \frac{(\nu!)^2}{h_0^{2\nu+1}} \|K_\nu^*\|_2^2 \int \sigma^2(x_\alpha, x_{\underline{\alpha}}) \frac{f_{\underline{\alpha}}^2(x_{\underline{\alpha}})}{f(x_\alpha, x_{\underline{\alpha}})} dx_{\underline{\alpha}}.$$

Additionally, for the regression function estimate constructed by the additive component estimates $\widehat{g}_\alpha$ and $\widehat{c}$, i.e.

$$\widehat{m}(\mathbf{X}) = \widehat{c} + \sum_{\alpha=1}^{d} \widehat{g}_\alpha(X_\alpha)$$

we have:

Theorem 8.4.
Using the same assumptions as in Theorem 8.3 it holds

$$n^{(p+1)/(2p+3)} \left\{ \widehat{m}(\boldsymbol{x}) - m(\boldsymbol{x}) \right\} \xrightarrow{L} N\left\{ b(\boldsymbol{x}), v(\boldsymbol{x}) \right\},$$

where $\boldsymbol{x} = (x_1, \ldots, x_d)^\top$, $b(\boldsymbol{x}) = \sum_{\alpha=1}^d b_\alpha(x_\alpha)$ and $v(\boldsymbol{x}) = \sum_{\alpha=1}^d v_\alpha(x_\alpha)$.

In the following example we illustrate the smoothing properties of this estimator for $\nu = 0$ and $p = 1$.

Example 8.4.
Consider the same setup as in Example 8.1, but now using $n = 150$ observations. In Figure 8.4 we have plotted the true functions (at the corresponding observations $X_{i\alpha}$) and the estimated component functions using marginal integration with the local linear smoothers. The bandwidths are $h = 1$ for the first and $h = 1.5$ for the other dimensions. Further we set $\widetilde{h} = 3$ for all nuisance directions. We used the (product) Quartic kernel for all estimates. As in Example 8.1 the estimated curves reflect almost perfectly the underlying true curves. □

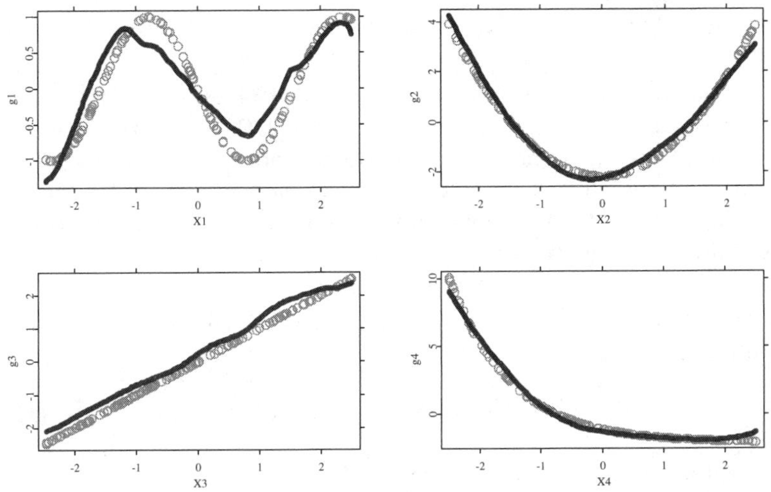

Figure 8.4. Estimated local linear (solid line) versus true additive component functions (circles at the input values)

8.2.3 Interaction Terms

As pointed out before, marginal integration estimates marginal effects. These are identical to the additive components, if the model is truly additive. But what happens if the underlying model is not purely additive? How do the estimators behave when we have some interaction between explanatory variables, for example given by an additional term $g_{\alpha j}(X_\alpha, X_j)$?

An obvious weakness of the truly additive model is that those interactions are completely ignored, and in certain econometric contexts — production function modeling being one of them — the absence of interaction terms has often been criticized. For that reason we will now extend the regression model by pairwise interactions resulting in

$$m(x) = c + \sum_{\alpha=1}^{d} g_\alpha(X_\alpha) + \sum_{1\le\alpha<j\le d} g_{\alpha j}(X_\alpha, X_j) . \tag{8.20}$$

Here we use $1 \le \alpha < j \le d$ to make sure that we include each pairwise interaction only once. In other words, we assume $g_{\alpha j} = g_{j\alpha}$. Principally, we could also consider interaction terms of higher order than two, but this would make visualization and interpretation hardly possible. Furthermore, the advantage of avoiding the curse of dimensionality would get lost step by step. We will therefore restrict ourselves to the case of only bivariate interactions.

For the marginal integration estimator bivariate interaction terms have been studied in Sperlich, Tjøstheim & Yang (2002). They provide asymptotic properties and additionally introduce test procedures to check for significance of the interactions. In the following we will only sketch the construction of the relevant estimation procedure and its application. For the theoretical results we remark that they are higher dimensional extensions of Theorem 8.3 and refer to the above mentioned article.

For the estimation of (8.20) by marginal integration we have to extend our identification condition

$$E g_\alpha(X_\alpha) = \int g_\alpha(x_\alpha) f_\alpha(x_\alpha)\, dx_\alpha = 0 \quad \text{for all } \alpha, \tag{8.21}$$

by further ones for the interaction terms:

$$\int g_{\alpha j}(x_\alpha, x_j) f_\alpha(x_\alpha)\, dx_\alpha = \int g_{\alpha j}(x_\alpha, x_j) f_j(x_j)\, dx_j = 0 , \tag{8.22}$$

with $f_\alpha(\bullet)$, $f_j(\bullet)$ being the marginal densities of the X_α and X_j.

As before, equations (8.21) and (8.22) should not be considered as restrictions. It is always possible to shift the functions g_α and $g_{\alpha j}$ in the vertical direction without changing the functional forms or the overall regression function. Moreover, all models of the form (8.20) are equivalent to exactly one model satisfying (8.21) and (8.22).

According to the definition of $\boldsymbol{X}_{\underline{\alpha}}$, let $\boldsymbol{X}_{\underline{\alpha j}}$ now denote the $(d-2)$-dimensional random variable obtained by removing X_α and X_j from $\boldsymbol{X} =$

$(X_1, \dots, X_d)^\top$. With some abuse of notation we will write $\boldsymbol{X} = (X_\alpha, X_j, \boldsymbol{X}_{\underline{\alpha j}})$ to highlight the directions in d-dimensional space represented by the α and j coordinates. We denote the marginal densities of X_α, $\boldsymbol{X}_{\underline{\alpha j}}$ and $\boldsymbol{X}$ by $f_\alpha(x_\alpha)$, $f_{\underline{\alpha j}}(\boldsymbol{x}_{\underline{\alpha j}})$, and $f(\boldsymbol{x})$, respectively.

Again consider marginal integration as used before

$$\theta_\alpha(x_\alpha) = \int m(x_\alpha, \boldsymbol{x}_{\underline{\alpha}}) f_{\underline{\alpha}}(\boldsymbol{x}_{\underline{\alpha}}) d\boldsymbol{x}_{\underline{\alpha}}, \quad 1 \le \alpha \le d, \tag{8.23}$$

and in addition

$$\theta_{\alpha j}(x_\alpha, x_j) = \int m(x_\alpha, x_j, \boldsymbol{x}_{\underline{\alpha j}}) f_{\underline{\alpha j}}(\boldsymbol{x}_{\underline{\alpha j}}) d\boldsymbol{x}_{\underline{\alpha j}}, \tag{8.24}$$

$$c_{\alpha j} = \int g_{\alpha j}(x_\alpha, x_j) f_{\alpha j}(x_\alpha, x_j)\, dx_\alpha\, dx_j$$

for every pair $1 \le \alpha < j \le d$. It can be shown that

$$\theta_{\alpha j}(x_\alpha, x_j) - \theta_\alpha(x_\alpha) - \theta_j(x_j) + \int m(\boldsymbol{x}) f(\boldsymbol{x})\, d\boldsymbol{x} = g_{\alpha j}(x_\alpha, x_j) + c_{\alpha j}.$$

Centering this function in an appropriate way would hence give us the interaction function of interest.

Using the same estimation procedure as described above, i.e., replacing the expectations by averages and the function m by an appropriate pre-estimator, we get estimates for g_α and for the interaction terms $g_{\alpha j}$. For the ease of notation we give only the formula for $p = 1$, i.e., the local linear estimator, in the pre-estimation step. We obtain

$$\widehat{\{g_{\alpha j} + c_{\alpha j}\}} = \widehat{\theta}_{\alpha j} - \widehat{\theta}_\alpha - \widehat{\theta}_j + \widehat{c}, \tag{8.25}$$

where

$$\widehat{\theta}_{\alpha j} = \frac{1}{n} \sum_{j=1}^{n} e_0^\top (\mathbf{X}_{\alpha j}^\top \mathbf{W}_{l\alpha j} \mathbf{X}_{\alpha j})^{-1} \mathbf{X}_{\alpha j}^\top \mathbf{W}_{l\alpha j} \boldsymbol{Y},$$

and

$$\mathbf{W}_{l\alpha j} = \operatorname{diag}\left(\left\{ \frac{1}{n} \mathcal{K}_{\mathbf{H}}(X_{i\alpha} - x_\alpha, X_{ij} - x_j) \widetilde{\mathcal{K}}_{\widetilde{\mathbf{H}}}(\boldsymbol{X}_{i\underline{\alpha j}} - \boldsymbol{x}_{l\underline{\alpha j}}) \right\}_{i=1,\dots,n} \right),$$

$$\mathbf{X}_{\alpha j} = \begin{pmatrix} 1 & X_{1\alpha} - x_\alpha & X_{1j} - x_j \\ \vdots & \vdots & \vdots \\ 1 & X_{n\alpha} - x_\alpha & X_{nj} - x_j \end{pmatrix}.$$

This is a local linear estimator in the directions α, j and a local constant one for the nuisance directions $\underline{\alpha j}$. $\widehat{\theta}_\alpha = \widehat{\{g_\alpha + c\}}$, $\widehat{\theta}_j = \widehat{\{g_j + c\}}$ and $\widehat{c}$ are exactly as defined above.

Finally, let us turn to an example, which presents the application of marginal integration estimation, derivative (elasticity) estimation and allows us to illustrate the use of interaction terms.

Example 8.5.
Our illustration is based on the example and the data used in Severance-Lossin & Sperlich (1999) who investigated a production function for livestock in Wisconsin. The main interest here is to estimate the impact of various regressors, their return to scale and hence on derivative estimation. Additionally, we validate the additivity assumption by estimating the interaction terms $g_{\alpha j}$.

We use a subset of $n = 250$ observations of an original data set of more than 1,000 Wisconsin farms collected by the Farm Credit Service of St. Paul, Minnesota in 1987. Severance-Lossin & Sperlich (1999) removed outliers and incomplete records and selected farms which only produced animal outputs. The data consist of farm level inputs and outputs measured in dollars. In more detail, output Y is livestock, and the input variables are

X_1 family labor force,

X_2 hired labor force,

X_3 miscellaneous inputs (as e.g. repairs, rent, custom hiring, supplies, insurance, gas),

X_4 animal inputs (as e.g. purchased feed, breeding, or veterinary services),

X_5 intermediate run assets, that is assets with a useful life of one to ten years.

To get an idea of the distribution of the regressors one could plot kernel density estimates for each of them. You would recognize that the regressors are behaving almost normally distributed so that an application of kernel smoother methods should not cause serious numerical problems.

A purely additive model (ignoring any possible interaction) is of the form

$$\log(Y) = c + \sum_{\alpha=1}^{d} g_\alpha \{\log(X_\alpha)\} + \varepsilon . \tag{8.26}$$

This model can be viewed as a generalization of the Cobb-Douglas production, where $g_\alpha \{\log(X_\alpha)\} = \beta_\alpha \log(X_\alpha)$ (see (5.4)). Additionally, we allow for inclusion of interaction terms $g_{\alpha j}$ and obtain

$$\log(Y) = c + \sum_{\alpha=1}^{d} g_\alpha \{\log(X_\alpha)\} + \sum_{1\le\alpha<j\le d} g_{\alpha j} \{\log(X_\alpha), \log(X_j)\} + \varepsilon . \tag{8.27}$$

The important point to understand in marginal integration is that the estimation of the one-dimensional component functions is not affected by the inclusion of interaction terms. This means, whether we estimate in model (8.26) or in (8.27) does not change the results for the estimation of the marginal functions g_α.

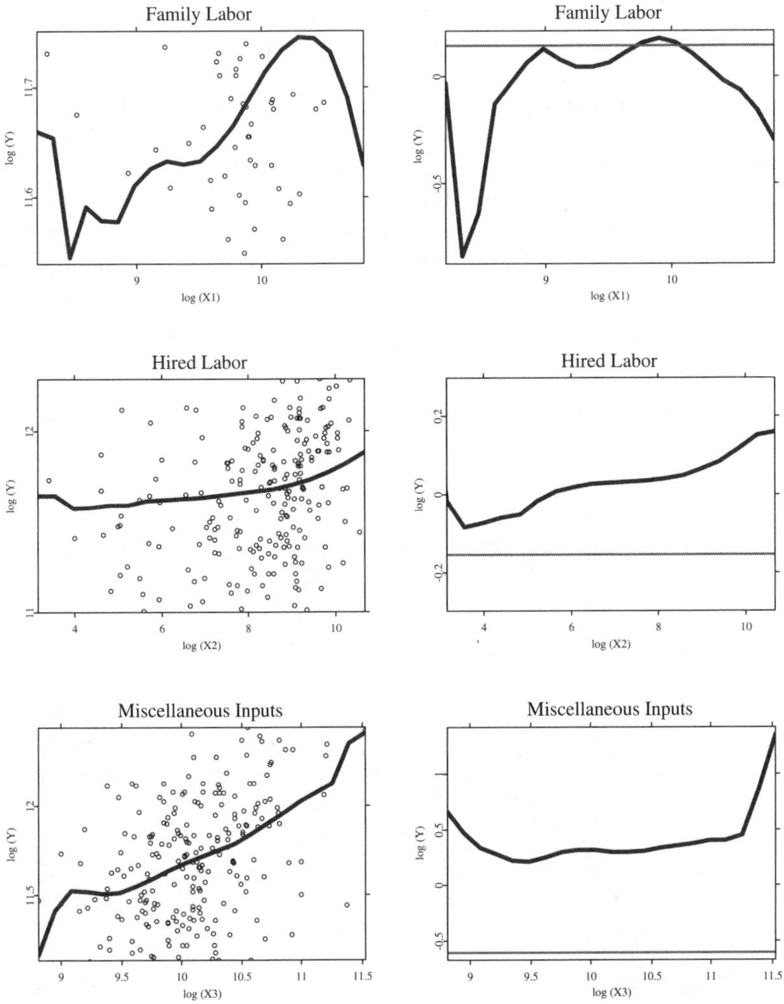

Figure 8.5. Function estimates for the additive components and observations (left), derivative estimates for the parametric (thin lines) and the nonparametric case (right), variables X_1 to X_3

The results are given in Figures 8.5, 8.6. We use (product) Quartic kernels for all dimensions and bandwidths proportional to the standard deviation of X_α ($h_\alpha = 1.5\,\sigma_\alpha$, $\widetilde{h}_\alpha = 4\,h_\alpha$). It is known that the integration estimator is quite robust against different choices of bandwidths, see e.g. Sperlich, Linton & Härdle (1999).

To highlight the shape of the estimates we display the main part of the point clouds including the function estimates. The graphs give some indication of nonlinearity, in particular for X_1, X_2 and X_5. The derivatives seem to indicate that the elasticities for these inputs increase and could finally lead to increasing returns to scale. Note that for all dimensions (especially where the mass of the observations is located) the nonparametric results differ a lot from the parametric case. An obvious conclusion from the economic point of view is that for instance larger farms are more productive (intuitively quite reasonable).

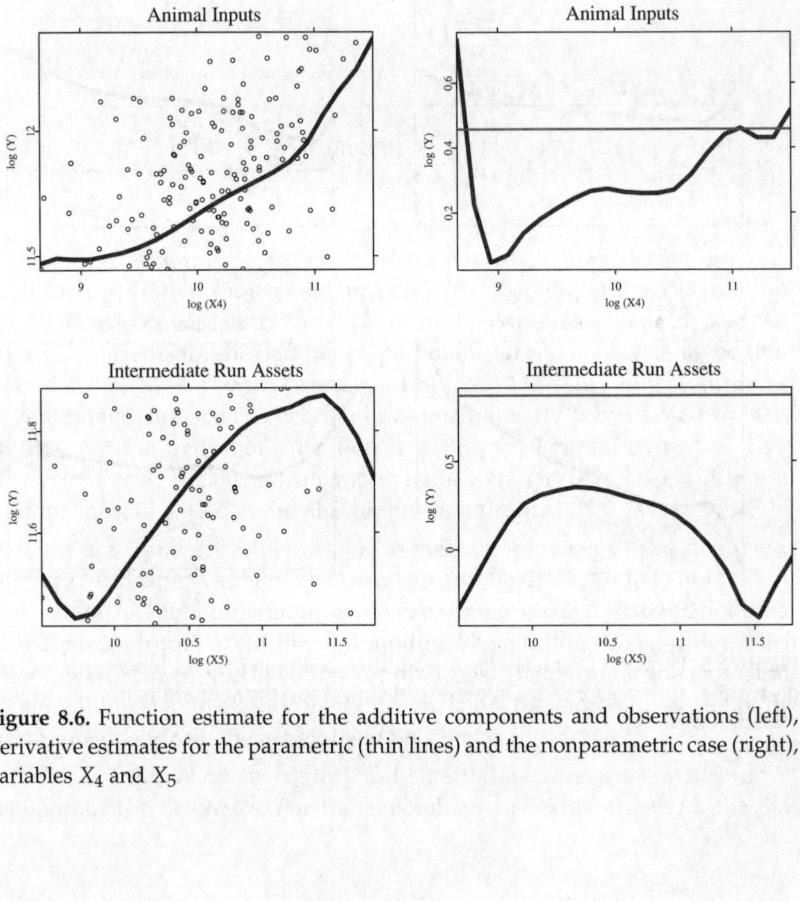

Figure 8.6. Function estimate for the additive components and observations (left), derivative estimates for the parametric (thin lines) and the nonparametric case (right), variables X_4 and X_5

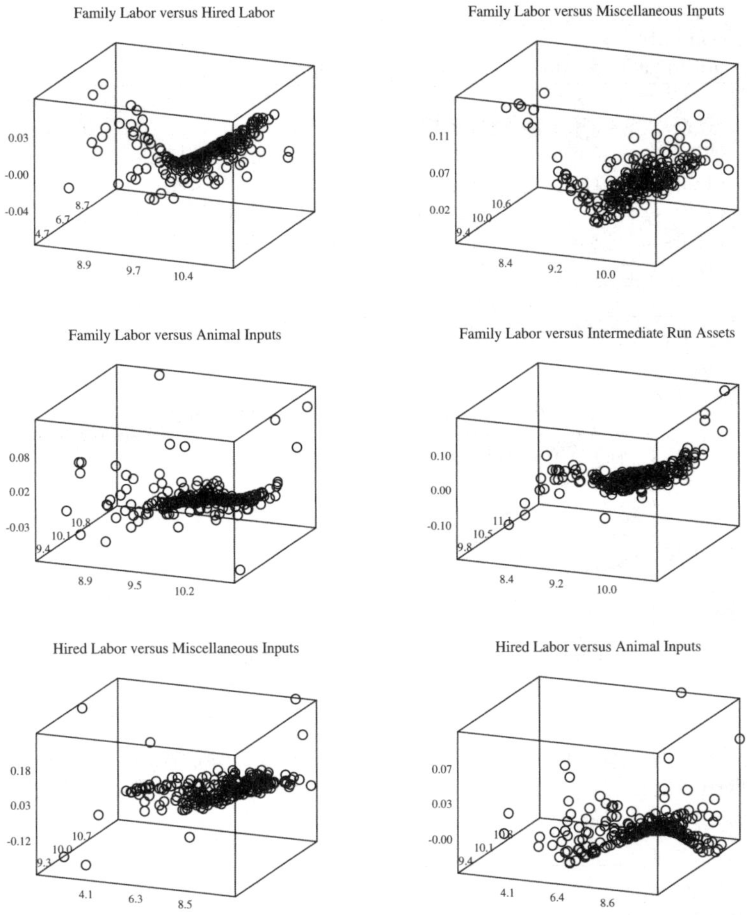

Figure 8.7. Estimates for interaction terms for Wisconsin farm data

In Figures 8.7 to 8.8 we present the estimates of the bivariate interaction terms $g_{\alpha j}$. For their estimation and graphical presentation we trimmed the data by removing 2% of the most extreme observations. Again Quartic kernels were used, here with bandwidths $h_\alpha = 1.7\,\sigma_\alpha$ and $\widetilde{h}_\alpha$ as above.

Obviously, often it is hard to interpret those interaction terms. But as long as we can visualize relationships a careful interpretation can be tried. Sperlich, Tjøstheim & Yang (2002) find that a weak form of interaction is present. The variable X_1 (family labor) plays an important role in the interactions, es-

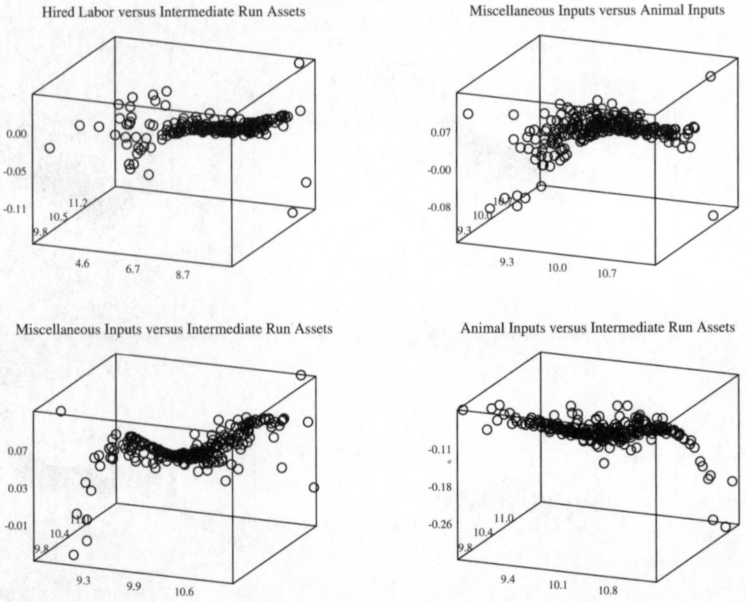

Figure 8.8. Estimates for interaction terms for Wisconsin farm data

pecially $g_{1,3}$ (family labor and miscellaneous inputs), $g_{1,5}$ (family labor and intermediate run assets) and $g_{3,5}$ (miscellaneous inputs and intermediate run assets) should be taken into account. □

8.3 Finite Sample Behavior

Asymptotic statistical properties are only one part of the story. For the applied researcher knowledge of the finite sample behavior of a method and its robustness are essential. This section is reserved for this topic.

We present and examine now results on the finite sample performance of the competing backfitting and integration approaches. To keep things simple we only use local constant and local linear estimates here. For a more detailed discussion see Sperlich, Linton & Härdle (1999) and Nielsen & Linton (1998). Let us point out again, that

- backfitting always projects the regression problem into the space of additive models,

- marginal integration estimates the marginal impact of a regressor independently of the underlying model.

Thus, both techniques are only comparable in truly additive models. We concentrate here only on the backfitting algorithm as introduced by Hastie & Tibshirani (1990) and the marginal integration as presented in Severance-Lossin & Sperlich (1999).

The following two subsections refer to results for the specific models

$$Y_i = g_a(X_{i1}) + g_b(X_{i2}) + \varepsilon_i, \quad i = 1, \ldots, n$$

where $a \neq b$ are chosen from $1, \ldots, 4$ and

$$\begin{aligned} g_1(X_{i\bullet}) &= 2X_{i\bullet}, & g_2(X_{i\bullet}) &= X_{i\bullet}^2 - E(X_{i\bullet}^2), \\ g_3(X_{i\bullet}) &= \exp(X_{i\bullet}) - E\{\exp(X_{i\bullet})\} & g_4(X_{i\bullet}) &= 0.5 \cdot \sin(-1.5X_{i\bullet}). \end{aligned}$$

Here, the $X_{i\bullet}$ stand for i.i.d. regressor variables that are either uniform on $[-3, 3] \times [-3, 3]$ distributed or (correlated) bivariate normals with mean 0, variance 1 and covariance $\rho \in \{0, 0.4, 0.8\}$. The error terms ε_i are independently $N(0, 0.5)$ distributed. If not mentioned differently, we consider a sample size $n = 100$.

Note that all estimators presented in this section (two-dimensional backfitting and marginal integration estimators) are linear in $\boldsymbol{Y}$, i.e. of the form

$$\widehat{g}_\alpha(\bullet) = \sum_{i=1}^{n} w_{\alpha i}(\bullet, \boldsymbol{X}_i) Y_i$$

for some weights $w_{\alpha i}(\bullet, \boldsymbol{X}_{i\underline{\alpha}})$. Consequently the conditional bias and variance can be calculated by

$$\begin{aligned} \text{Bias}\,\{\widehat{g}_\alpha(\bullet)|\boldsymbol{X}\} &= \sum_{i=1}^{n} w_{\alpha i}(\bullet, \boldsymbol{X}_i)\, m(\boldsymbol{X}_i) - g_\alpha(X_\alpha), \\ \text{Var}\,\{\widehat{g}_\alpha(\bullet)|\boldsymbol{X}\} &= \sigma_\varepsilon^2 \sum_{i=1}^{n} w_{\alpha i}^2(\bullet, \boldsymbol{X}_i)\,. \end{aligned}$$

An analog representation holds for the overall regression estimate $\widehat{m}$. We introduce the notation

$$\text{MSE}_t = \text{MSE}\{\widehat{g}_\alpha(t)\} = E\,\{\widehat{g}_\alpha(t) - g_\alpha(t)\}^2$$

for the mean squared error, and

$$\text{MASE} = \text{MASE}(\widehat{g}_\alpha) = \frac{1}{n} \sum_i \{\widehat{g}_\alpha(X_{i\alpha}) - g_\alpha(X_{i\alpha})\}^2$$

for the mean averaged squared error, the density weighted empirical version of the MSE. We will also use the analog definitions for $\widehat{m}$.

8.3.1 Bandwidth Choice

As we have already discussed in the first part of this book, the choice of the smoothing parameter is crucial in practice. The integration estimator requires to choose two bandwidths, h for the direction of interest and $\widetilde{h}$ for the nuisance direction. Possible practical approaches are the rule of thumb of Linton & Nielsen (1995) and the plug-in method suggested in Severance-Lossin & Sperlich (1999). Both methods use the MASE-minimizing bandwidth, the former approximating it by means of parametric pre-estimators, the latter one by using nonparametric pre-estimators.

For example, the formula for the MASE-minimizing (and thus asymptotically optimal) bandwidth in the local linear case is given by

$$h = \left\{ \frac{\|K\|_2^2 \int \sigma^2 f_{\underline{\alpha}}^2(x_{\underline{\alpha}}) f_\alpha(x_\alpha) \{f(\boldsymbol{x})\}^{-1} \, dx_{\underline{\alpha}} \, dx_\alpha}{\mu_2^2(K) \int \{g_\alpha''(x_\alpha)\}^2 f_\alpha(x_\alpha) \, dx_\alpha} \right\}^{1/5} n^{-1/5} . \tag{8.28}$$

However, simulations in the above cited papers show that in small samples the bandwidths are far away from those obtained by numerical MASE minimization. Note also, that formula (8.28) is not valid for the $\widetilde{h}$. For these bandwidths, the literature recommends undersmoothing. It turns out that this is not essential (in practice). The reason is that the multiplicative term corresponding to $\widetilde{h}$ is often already very small compared to the bias term corresponding to h.

In case of backfitting the procedure becomes possible due to the fact that we only consider one-dimensional smoothers. Here, the MASE-minimizing bandwidth is commonly approximated by the MASE-minimizing bandwidth for the corresponding one-dimensional kernel regression case.

Example 8.6.
To demonstrate the sensitivity, respectively the robustness of the estimators w.r.t. the choice of the bandwidth we plot MASE and MSE_0 against h for the two models $m = g_2 + g_3$ and $m = g_2 + g_4$. We show here only the results for the local linear smoother and the independent designs $\boldsymbol{X} \sim U[-3,3] \times U[-3,3]$, $\boldsymbol{X} \sim N(0,1) \times N(0,1)$. The MASE and MSE_0 curves are displayed in Figures 8.9 and 8.10. In all of them, we use thick solid lines for marginal integration and dashed lines for backfitting. □

Obviously, the backfitting estimator is rather sensitive to the choice of bandwidth. To get small MASE values it is important for the backfitting method to choose the smoothing parameter appropriately. For the integration estimator the results differ depending on the model. This method is nowhere near as sensitive to the choice of bandwidth as the backfitting. Focusing on the MSE_0 we have similar results as for the MASE but weakened

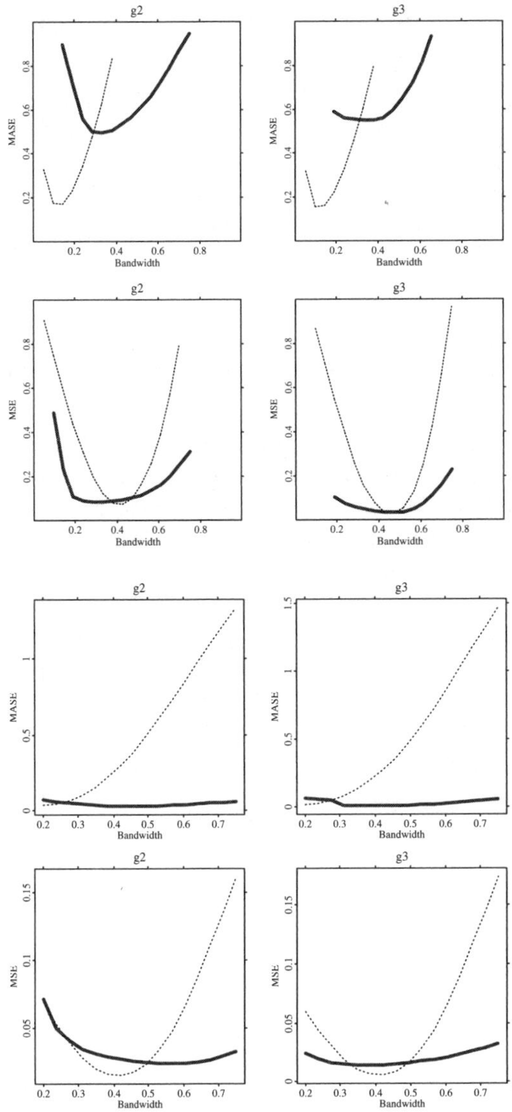

Figure 8.9. Performance of MASE (top and third row) and MSE_0 (second and bottom row) by bandwidth h, overall model is $m = g_2 + g_3$, the columns represent g_2 (left) and g_3 (right) under uniform design (upper two rows) and normal design (lower two rows)

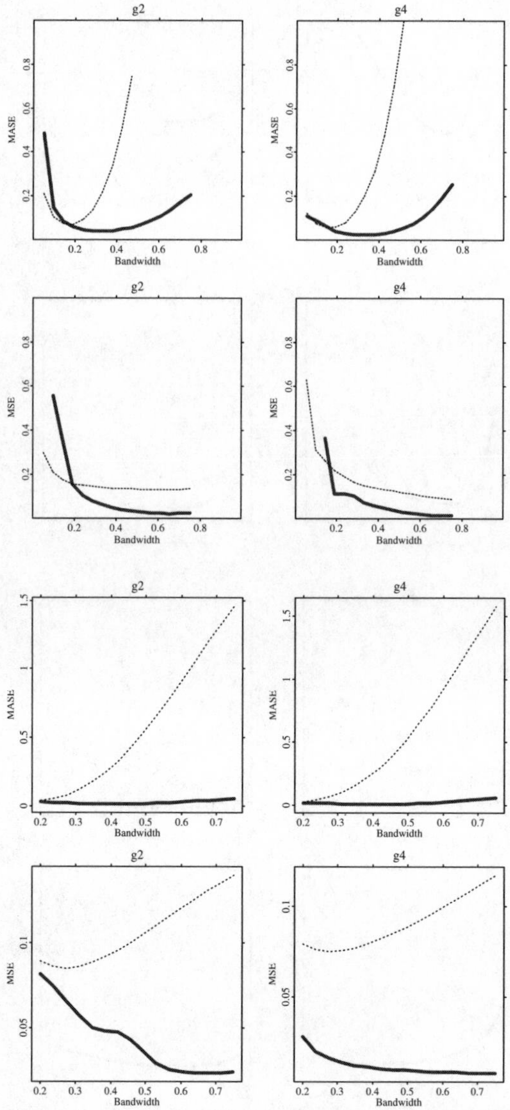

Figure 8.10. Performance of MASE (top and third row) and MSE_0 (second and bottom row) by bandwidth h, overall model is $m = g_2 + g_4$, the columns represent g_2 (left) and g_3 (right) under uniform design (upper two rows) and normal design (lower two rows)

concerning the sensitivity. Here the results differ more depending on the data generating model.

8.3.2 MASE in Finite Samples

Table 8.1 presents the MASE when using local linear smoothers and the asymptotically optimal bandwidths. To exclude boundary effects each entry of the table consists of two rows: evaluation on the complete data set in the upper row, and evaluation on trimmed data in the lower row. The trimming was implemented by cutting off 5% of the data on each side of the support.

Table 8.1. MASE for backfitting (back) and marginal integration (int) for estimating g_α and m, normal designs with different covariances (first row), MASE calculated for the complete (upper row) and the trimmed data (lower row)

covariance		0.0	0.4	0.8	0.0	0.4	0.8	0.0	0.4	0.8	0.0	0.4	0.8
model		$m = g_1 + g_2$			$m = g_1 + g_3$			$m = g_2 + g_4$			$m = g_3 + g_4$		
g_a	back	0.047	0.041	0.020	0.046	0.028	0.053	0.124	0.135	0.128	0.068	0.081	0.111
		0.038	0.031	0.014	0.037	0.018	0.033	0.107	0.116	0.099	0.046	0.055	0.081
	int	0.019	0.030	0.057	0.031	0.075	0.081	0.047	0.048	0.089	0.053	0.049	0.056
		0.013	0.017	0.047	0.024	0.059	0.071	0.022	0.026	0.078	0.026	0.022	0.041
g_b	back	0.083	0.079	0.047	0.073	0.053	0.058	0.112	0.121	0.110	0.051	0.062	0.096
		0.071	0.060	0.024	0.058	0.032	0.028	0.101	0.110	0.091	0.039	0.048	0.075
	int	0.090	0.116	0.530	0.137	0.234	0.528	0.048	0.480	1.32	0.057	0.603	2.41
		0.028	0.029	0.205	0.027	0.031	0.149	0.032	0.061	0.151	0.040	0.265	1.02
m	back	0.052	0.054	0.049	0.051	0.054	0.057	0.061	0.063	0.068	0.065	0.066	0.064
		0.032	0.031	0.028	0.030	0.029	0.035	0.035	0.035	0.037	0.038	0.037	0.038
	int	0.115	0.145	0.619	0.175	0.285	0.608	0.118	0.561	1.37	0.085	0.670	2.24
		0.041	0.041	0.252	0.043	0.053	0.194	0.076	0.083	0.189	0.044	0.257	0.681

We see that no estimator is uniformly superior to the others. All results depend more significantly on the design distribution and the underlying model than on the particular estimation procedure. The main conclusion is that backfitting almost always fits the overall regression better whereas the marginal integration often does better for the additive components. Recalling the construction of the procedures this is not surprising, but exactly what one should have expected.

Also not surprisingly, the integration estimator suffers more heavily from boundary effects. For increasing correlation both estimators perform worse, but this effect is especially present for the integration estimator. This is in line with the theory saying that the integration estimator is inefficient for correlated designs, see Linton (1997). Here a bandwidth matrix with appropriate non-zero arguments in the off diagonals can help in case of high correlated regressors, see a corresponding study in Sperlich, Linton & Härdle (1999). They point out that the fit can be improved significantly by using well defined off-diagonal elements in the bandwidth matrices. A similar analysis would be harder to do for the backfitting method as it depends only on one-dimensional smoothers. We remark that internalized marginal integration estimators (Dette, von Lieres und Wilkau & Sperlich, 2004) and smoothed backfitting estimators (Mammen, Linton & Nielsen, 1999; Nielsen & Sperlich, 2002) are much better suited to deal with correlated regressors.

8.3.3 Equivalent Kernel Weights

How do the additive approaches overcome the curse of dimensionality? We compare now the additive estimation method with the bivariate Nadaraya-Watson kernel smoother. We define *equivalent kernels* as the linear weights w_i used in the estimates for fitting the regression function at a particular point. In the following we take the center point $(0,0)$, which is used in Figures 8.11 to 8.13. All estimators are based on univariate or bivariate Nadaraya-Watson smoothers (in the latter case using a diagonal bandwidth matrix).

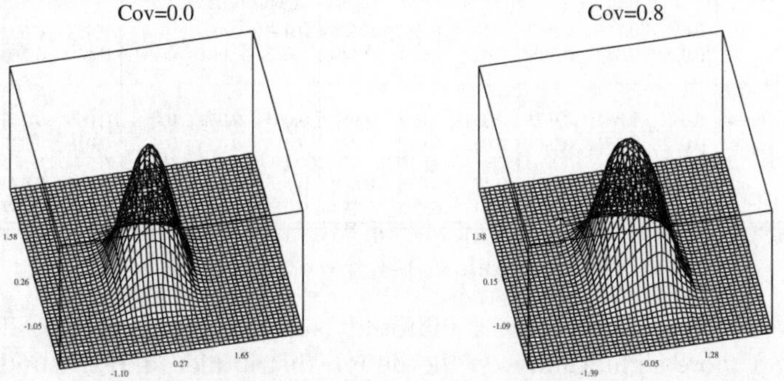

Figure 8.11. Equivalent kernels for the bivariate Nadaraya-Watson estimator, normal design with covariance 0 (left) and 0.8 (right)

Obviously, additive methods (Figures 8.12, 8.13) are characterized by their local panels along the axes instead of being uniformly equal in all directions like the bivariate Nadaraya-Watson (Figure 8.11). Since additive estimators are made up of components that behave like univariate smoothers,

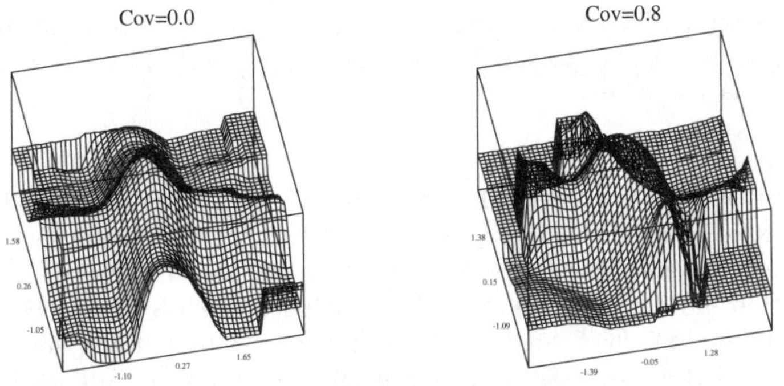

Figure 8.12. Equivalent kernels for backfitting based on univariate Nadaraya-Watson smoothers, normal design with covariance 0 (left) and 0.8 (right)

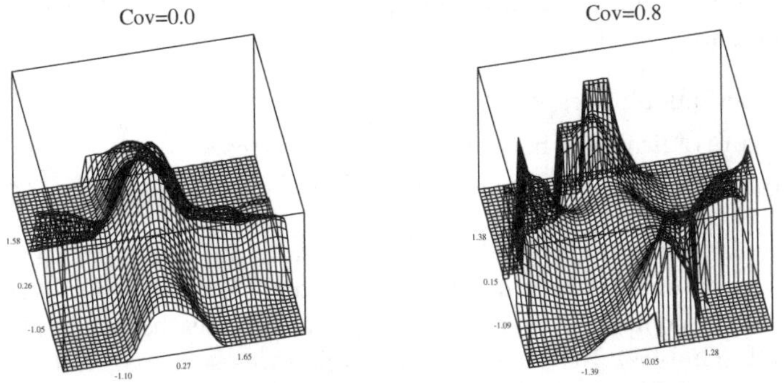

Figure 8.13. Equivalent kernels for marginal integration based on bivariate Nadaraya-Watson smoothers, normal design with covariance 0 (left) and 0.8 (right)

they can overcome the curse of dimensionality. The pictures for the additive smoothers look very similar (apart from some negative weights for the backfitting).

Finally, we see clearly how both additive methods run into problems when the correlation between the regressors is increasing. In particular for the marginal integration estimator recall that before we apply the integration over the nuisance directions, we pre-estimate m on all combinations of realizations of X_1 and X_2. For example, since X_1, X_2 are both uniform on $[-1,1]$ it may happen that we have to pre-estimate the regression function at the point $(0.9, -0.9)$. Now imagine that X_1 and X_2 are positively correlated. In small samples, the pre-estimate for $m(0.9, -0.9)$ is then usually obtained by extrapolation. The insufficient quality of the pre-estimate does then transfer to the final estimate.

Example 8.7.
Let us further illustrate the differences and common features of the discussed estimators by means of an explorative real data analysis. We investigate the relation of managerial compensation to firm size and financial performance based on the data used in Grasshoff, Schwalbach & Sperlich (1999). Empirical studies show a high pay for firm size sensitivity and a low pay for financial performance sensitivity. These studies use linear, log-linear or semi-log-linear relations.

Consider the empirical model

$$\log(C_i) = \beta_0 + \beta_1 \log(S_i) + \beta_2 P_i + \varepsilon_i \tag{8.29}$$

for a sample of n firms at different time points $t = 1, \ldots, T$. The explanatory variables are

C_i compensation per capita,
S_i measure of firm size, here number of employees,
P_i measure of financial performance, here the profit to sales ratio (ROS).

The data base for this analysis is drawn from the Kienbaum Vergütungsstudie, containing data about top management compensation of German AGs and the compensation of managing directors (Geschäftsführer) of incorporateds (GmbHs). To measure compensation we use managerial compensation per capita due to the lack of more detailed information. The analysis is based on the following four industry groups.

- group 1 (basic industries)
  consists of companies for chemicals/ pharmaceuticals/ plastics, iron and steel, rubber, mineral oil, non iron ores and cement,
- group 2 (capital goods)
  consists of companies for metal ware /sheet metal, electrical industry, vehicles, precision mechanics and optics, mechanical engineering and ship building,
- group 3 (consumer goods)
  consists of companies for glass/ ceramics and textiles/ clothing,
- group 4 (food, drink, and tobacco)
  consists of companies for breweries, other food, drink and tobacco.

Table 8.2. Parameter estimates for the log-linear model (asterisks indicate significance at the 1% level)

Group	1	2	3	4
# observations	131	148	41	38
constant	4.128*	4.547*	3.776*	4.120*
ROS	1.641	0.959	15.01*	8.377
log(SIZE)	0.258*	0.201*	0.283*	0.249*

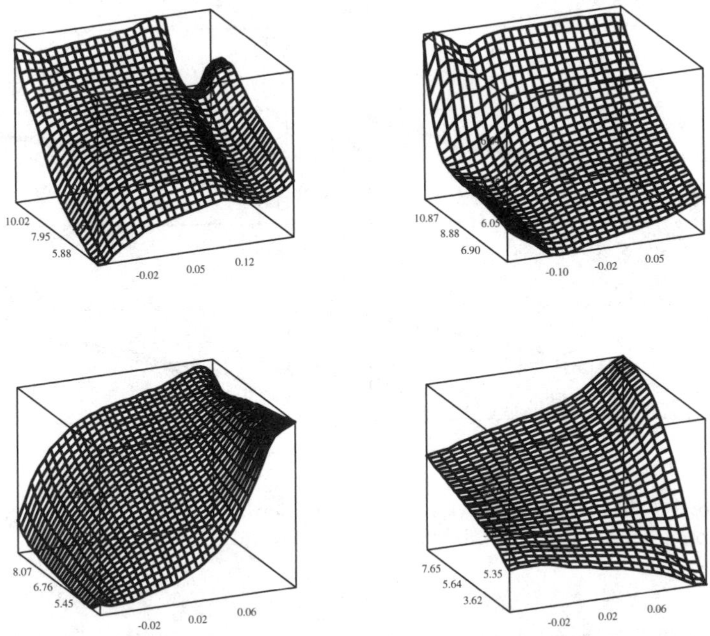

Figure 8.14. 3D surface estimates for branch groups 1 to 4 (upper left to lower right)

We first present the results of the parametric analysis for each group, see Table 8.2. The sensitivity parameter for the size variable can be directly interpreted as the size elasticity in each case.

We now check for a possible heterogeneous behavior over the groups. A two-dimensional Nadaraya-Watson estimate is shown in Figure 8.14. Considering the plots we realize that the estimated surfaces are similar for the industry groups 1 and 2 (upper row) while the surfaces for the two other

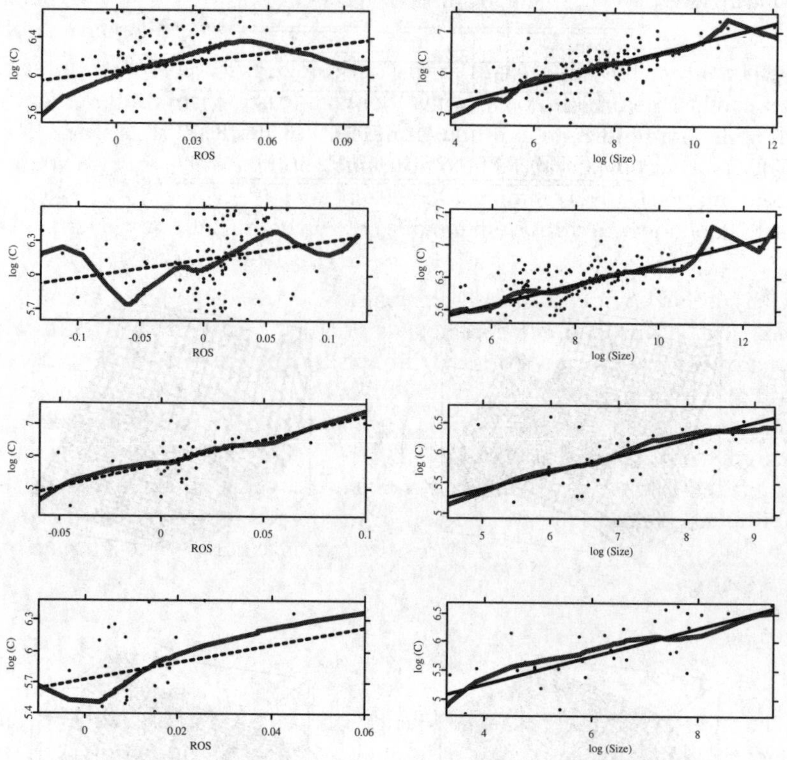

Figure 8.15. Backfitting additive and linear function estimates together with selected data points, branch groups 1 to 4 (from above to below)

groups clearly differ. Further, we see a strong positive relation for compensation to firm size at least in groups 1 and 2, and a weaker one to the performance measure varying over years and groups. Finally, interaction of the regressors — especially in groups 3 and 4 — can be recognized.

The backfitting procedure projects the data into the space of additive models. We used for the backfitting estimators univariate local linear kernel smoother with Quartic kernel and bandwidth inflation factors 0.75, 0.5 for group 1 and 2 and 1.25, 1.0 for groups 3 and 4. In the Figure 8.15 we compare the nonparametric (additive) components with the parametric (linear) functions. Over all groups we observe a clear nonlinear impact of ROS. Note, that the low values for significance in the parametric model describe only the linear impact, which here seem to be caused by functional misspecification (or interactions).

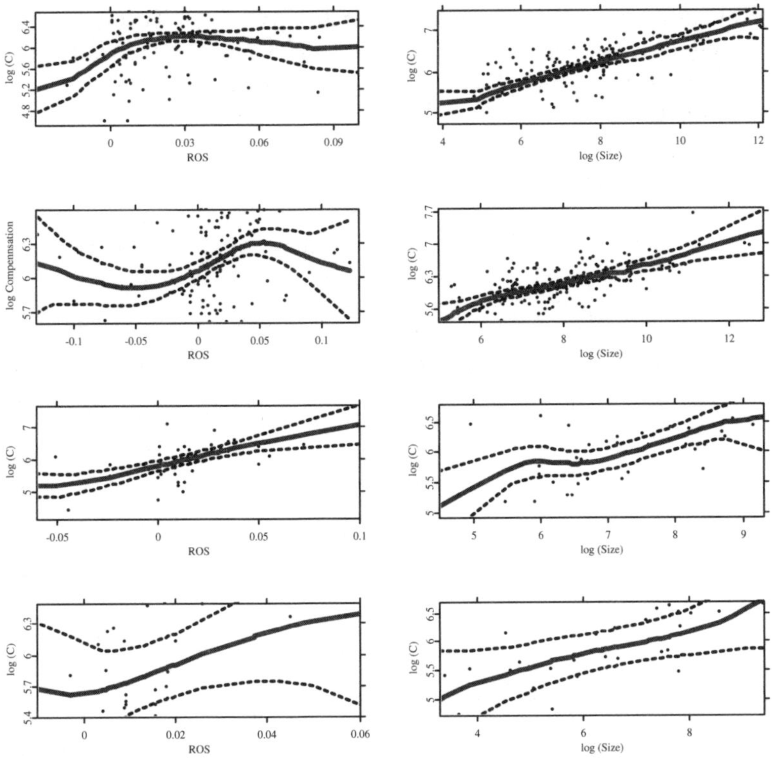

Figure 8.16. Marginal integration estimates with 2σ bands, branch groups 1 to 4 (from above to below)

Finally, in Figure 8.16 we estimate the marginal effects of the regressors using local linear smoothers. The estimated marginal effects are presented together with 2σ-bands, where we use for σ the variance functions of the estimates. Note that for ROS in group 1 the ranges are slightly different as in Figure 8.15.

Generally, the results are consistent with the findings above. The nonlinear effects in the impact of ROS are stronger, especially in groups 1 and 2. Since the abovementioned bumps in the firm size do not exist here, we can conclude that indeed interaction effects are responsible for this. The backfitting results differ substantially from the estimated marginal effects in group 3 and 4 what again underlines the presence of interaction effects.

To summarize, we conclude that the separation into groups is useful, but groups 1 and 2 respectively 3 and 4 seem to behave similarly. The assumption of additivity seems to be violated for groups 3 and 4. Furthermore, the nonparametric estimates yield different results due to nonlinear effects and interaction, so that parametric elasticities underestimate the true elasticity in our example. □

Bibliographic Notes

Additive models were first considered for economics and econometrics by Leontief (1947a,b). Intensive discussion of their application to economics can be found in Deaton & Muellbauer (1980) and Fuss, McFadden & Mundlak (1978). Wecker & Ansley (1983) introduced especially the backfitting method in economics.

The development of the backfitting procedure has a long history. The procedure goes back to algorithms of Friedman & Stuetzle (1982) and Breiman & Friedman (1985). We also refer to Buja, Hastie & Tibshirani (1989) and the references therein. Asymptotic theory for backfitting has been studied first by Opsomer & Ruppert (1997), and later on (under more general conditions) in the above mentioned paper of Mammen, Linton & Nielsen (1999).

The marginal integration estimator was first presented by Tjøstheim & Auestad (1994a) and Linton & Nielsen (1995), the idea can also be found in Newey (1994) or Boularan, Ferré & Vieu (1994) for estimating growth curves. Hengartner, Kim & Linton (1999) introduce modifications leading to computational efficiency. Masry & Tjøstheim (1995, 1997) use marginal integration and prove its consistency in the context of time series analysis. Dalelane (1999) and Achmus (2000) prove consistency of bootstrap methods for marginal integration. Linton (1997) combines marginal integration and a one step backfitting iteration to obtain an estimator that is both efficient and easy to analyze.

Interaction models have been considered in different papers. Stone, Hansen, Kooperberg & Truong (1997) and Andrews & Whang (1990) developed estimators for interaction terms of any order by polynomial splines. Spline estimators have also been used by Wahba (1990). For series estimation we refer in particular to Newey (1995) and the references therein. Härdle, Sperlich & Spokoiny (2001) use wavelets to test for additive models. Testing additivity is a field with a growing amount of literature, such as Chen, Liu & Tsay (1995), Eubank, Hart, Simpson & Stefanski (1995) and Gozalo & Linton (2001).

A comprehensive resource for additive modeling is is the textbook by Hastie & Tibshirani (1990) who focus on the backfitting approach. Further references are Sperlich (1998) and Ruppert, Wand & Carroll (1990).

Exercises

Exercise 8.1. Assume that the regressor variable X_1 is independent from $X_2, \ldots, X_d$. How does this change the marginal integration estimator now? Interpret this change.

Exercise 8.2. When using the local linear smoother for backfitting, what would be the optimal bandwidth for estimating linear functions?

Exercise 8.3. Give at least two methods of constructing confidence intervals for the component function estimates. Consider both the backfitting and the marginal integration method. Discuss also approaches for the construction of confidence bands.

Exercise 8.4. We mentioned that backfitting procedures could be implemented using any one-dimensional smoother. Discuss the analog issue for the marginal integration method.

Exercise 8.5. Assume we have been given a regression problem with five explanatory variables and a response Y. Our aim is to predict Y but we are not really interested in the particular impact of each input. Although we do not know whether the model is additive, the curse of dimensionality forces us to think about dimension reduction and we decide upon an additive approach. Which estimation procedure will you recommend?

Exercise 8.6. Recall Subsection 8.3.3. Does the underlying model matter for construction or interpretation of Figures 8.11 to 8.13? Justify your answer.

Exercise 8.7. Recall Subsection 8.2.3 where we considered models of the form

$$m(\mathbf{X}) = c + \sum_{\alpha=1}^{d} g_\alpha(X_\alpha) + \sum_{1 \le \alpha < j \le d} g_{\alpha j}(X_\alpha, X_j),$$

i.e. additive models with pairwise interaction terms. We introduced identification and estimation in the context of marginal integration. Discuss the problem now for the backfitting method. In what sense does the main difference between backfitting and marginal integration kick in here?

Exercise 8.8. Again, recall Subsection 8.2.3, but now extend the model to

$$m(\mathbf{X}) = c + \sum_{\alpha=1}^{d} g_\alpha(X_\alpha) + \sum_{1 \le \alpha < j \le d} g_{\alpha j}(X_\alpha, X_j) + \sum_{1 \le \alpha < j < k \le d} g_{\alpha jk}(X_\alpha, X_j, X_k).$$

How could this model be identified and estimated?

Exercise 8.9. Discuss the possibility of modeling some of the component functions g_α parametrically. What would be the advantage of doing so?

Exercise 8.10. Assume we have applied backfitting and marginal integration in a high dimensional regression problem. It turns out that we obtain very different estimates for the additive component functions. Discuss the reasons which can cause this effect.

Exercise 8.11. How should the formula for the MASE-minimizing bandwidth (8.28) be modified if we consider the local constant case?

Summary

★ Additive models are of the form

$$E(Y|\boldsymbol{X}) = m(\boldsymbol{X}) = c + \sum_{\alpha=1}^{d} g_\alpha(X_\alpha).$$

In estimation they can combine flexible nonparametric modeling of many variables with statistical precision that is typical of just one explanatory variable, i.e. they circumvent the curse of dimensionality.

★ In practice, there exist mainly two estimation procedures, backfitting and marginal integration. If the real model is additive, then there are many similarities in terms of what they do to the data. Otherwise their behavior and interpretation are rather different.

★ The backfitting estimator is an iterative procedure of the kind:

$$\widehat{g}_\alpha^{(l)} = \mathbf{S}_\alpha \left\{ \boldsymbol{Y} - \sum_{j \neq \alpha} \widehat{g}_j^{(l-1)} \right\}, \quad l = 1, 2, 3, \ldots$$

until some prespecified tolerance is reached. This is a successive one-dimensional regression of the partial residuals on the corresponding X_α.

★ Usually backfitting fits the regression better than the integration estimator. But it pays for a low MSE (or MASE) for the regression with high MSE (MASE respectively) in the additive function estimation. Furthermore, it is rather sensitive against the bandwidth choice. Also, an increase in the correlation ρ of the design leads to a worse estimate.

★ The marginal integration estimator is based on the idea that

$$E_{\mathbf{X}_{\underline{\alpha}}}\{m(X_\alpha, \mathbf{X}_{\underline{\alpha}}\} = c + g_\alpha(X_\alpha).$$

Replacing m by a pre-estimator and the expectation by averaging defines a consistent estimate.

★ Marginal integration suffers more from sparseness of observations than the backfitting estimator does. So for example the boundary effects are worse for the integration estimator. In the center of the support of $\mathbf{X}$ this estimator mostly has lower MASE for the estimators of the additive component functions. An increasing covariance of the explanatory variables affects the MASE strongly in a negative sense. Regarding the bandwidth this estimator seems to be quite robust.

★ If the real model is not additive, the integration estimator is estimating the marginals by integrating out the directions of no interest. In contrast, the backfitting estimator is looking in the space of additive models for the best fit of the response Y on $\mathbf{X}$.

9

Generalized Additive Models

In Chapter 8 we discussed additive models (AM) of the form

$$E(Y|\boldsymbol{X}) = c + \sum_{\alpha=1}^{d} g_\alpha(x_\alpha)\,. \tag{9.1}$$

Note that we put $EY = c$ and $E\{g_\alpha(X_\alpha)\} = 0$ for identification.

In this chapter we discuss variants of the model (9.1). Recall that the main advantage of an additive model is that (compared to a fully nonparametric model) it avoids the curse of dimensionality and that the component functions are easy to interpret. Stone (1986) proved that this also holds for generalized additive models.

We will focus here on the modification of (9.1) by additional parametric, in particular linear, components. The resulting additive partial linear model (APLM) can be written as

$$E(Y|\boldsymbol{U},\boldsymbol{T}) = c + \boldsymbol{U}^\top\beta + \sum_{\alpha=1}^{q} g_\alpha(T_\alpha)\,, \tag{9.2}$$

where we again use the partitioning $\boldsymbol{X} = (\boldsymbol{U},\boldsymbol{T})$ introduced in Chapter 7. Moreover, we will extend AM and APLM by a possibly nontrivial link function G. This leads to the generalized additive model (GAM)

$$E(Y|\boldsymbol{X}) = G\left\{c + \sum_{\alpha=1}^{d} g_\alpha(X_\alpha)\right\} \tag{9.3}$$

or to the generalized additive partial linear model (GAPLM)

$$E(Y|\boldsymbol{U},\boldsymbol{T}) = G\left\{c + \boldsymbol{U}^\top\beta + \sum_{\alpha=1}^{q} g_\alpha(T_\alpha)\right\}\,. \tag{9.4}$$

It is obvious that model (9.4) is a modification of the GPLM. Consequently we will base the estimation of the GAPLM to a significant extent on the estimation algorithms for the GPLM. In fact, most of the methodology introduced in this chapter comes from the combination of previously introduced techniques for GLM, GPLM, and AM. We will therefore frequently refer to the previous Chapters 5, 7 and 8.

9.1 Additive Partial Linear Models

The APLM can be considered as a modification of the AM by a parametric (linear) part or as a nontrivial extension of the linear model by additive components. The main motivation for this partial linear approach is that explanatory variables are often of mixed discrete-continuous structure. Apart from that, sometimes (economic) theory may guide us how some effects inside the regression have to be modeled. As a positive side effect we will see that parametric terms can be typically estimated more efficiently.

Consider now the APLM from (9.2) in the following way:

$$Y = c + \boldsymbol{U}^\top \beta + \sum_{\alpha=1}^{q} g_\alpha(T_\alpha) + \varepsilon \tag{9.5}$$

with $E(\varepsilon|\boldsymbol{U},\boldsymbol{T}) = 0$ and $\mathrm{Var}(Y|\boldsymbol{U},\boldsymbol{T}) = \mathrm{Var}(\varepsilon|\boldsymbol{U},\boldsymbol{T}) = \sigma^2(\boldsymbol{U},\boldsymbol{T})$. We have already presented estimators for $\boldsymbol{\beta}$ and $m = c + \sum g_\alpha$ in Section 7.2.2:

$$\widehat{\boldsymbol{\beta}} = \left\{\mathbf{U}^\top(\mathbf{I}-\mathbf{S})^2\mathbf{U}\right\}^{-1}\mathbf{U}^\top(\mathbf{I}-\mathbf{S})^2\boldsymbol{Y}, \tag{9.6}$$

$$\widehat{\boldsymbol{m}} = \mathbf{S}(\boldsymbol{Y} - \mathbf{U}\widehat{\boldsymbol{\beta}}). \tag{9.7}$$

As known from Denby (1986) and Speckman (1988), the parameter $\boldsymbol{\beta}$ can be estimated at $\sqrt{n}$-rate. The problematic part is thus the estimation of the additive part, as the estimate of m lacks precision due to the curse of dimensionality. When using backfitting for the additive components, the construction of a smoother matrix is principally possible, however, eluding from any asymptotic theory.

For this reason we consider a procedure based on marginal integration, suggested first by Fan, Härdle & Mammen (1998). Their idea is as follows: Assume $\boldsymbol{U}$ has M realizations being $\boldsymbol{u}^{(1)}, \ldots, \boldsymbol{u}^{(M)}$. We can then calculate the estimates $\widetilde{g}_\alpha^k$ for each of the subsamples $k = 1, \ldots, M$. Now we average over the $\widetilde{g}_\alpha^k$ to obtain the final estimate $\widehat{g}_\alpha$.

Note that this subsampling affects only the pre-estimation, when determining $\widetilde{m}$ at the points over which we have to integrate. To estimate

$\widetilde{m}(t_\alpha, \boldsymbol{T}_{i\underline{\alpha}}|\boldsymbol{u}^{(k)})$, we use a local linear expansion in the direction of interest α (cf. (8.16)), and minimize

$$\sum_{i=1}^{n} \{Y_i - \gamma_0 - \gamma_1(T_{i\alpha} - t_\alpha)\}^2 K_h(T_{i\alpha} - t_\alpha)\mathcal{K}_{\mathbf{H}}(\boldsymbol{T}_{i\underline{\alpha}} - \boldsymbol{T}_{l\underline{\alpha}})\,\mathrm{I}\{\boldsymbol{U}_i = \boldsymbol{u}^{(k)}\} \quad (9.8)$$

for each $\boldsymbol{T}_{l\underline{\alpha}}$. The notation is the same as in Chapter 8. We can estimate $\widetilde{m}(t_\alpha, \boldsymbol{T}_{i\underline{\alpha}}|\boldsymbol{u}^{(k)})$ by the optimized γ_0. Repeating this for all k and using the marginal integration technique, we obtain

$$\widehat{g}_\alpha(\bullet) = \frac{1}{n}\sum_{i=1}^{n} \widetilde{m}(\bullet, \boldsymbol{T}_{i\underline{\alpha}}|\boldsymbol{U}_i) - \frac{1}{n}\sum_{i=1}^{n}\frac{1}{n}\sum_{l=1}^{n} \widetilde{m}(T_{i\alpha}, \boldsymbol{T}_{l\underline{\alpha}}|\boldsymbol{U}_l)\,, \quad (9.9)$$

cf. equation (8.14).

Under the same assumptions as in Sections 8.2.1 and 8.2.2, complemented by regularity conditions for $\boldsymbol{U}$ and adjustments for the pdfs f_α, we obtain the same asymptotic results for $\widehat{g}_\alpha$ as in Chapter 8. The mentioned adjustments are necessary as we have to use conditional densities $f(\bullet|\boldsymbol{u})$, $f_\alpha(\bullet|\boldsymbol{u})$, and $f_{\underline{\alpha}}(\bullet|\boldsymbol{u})$ now; for details we refer to Fan, Härdle & Mammen (1998). For example, when $\boldsymbol{U}$ is discrete, we set $f(\boldsymbol{t}, \boldsymbol{u}) = f(\boldsymbol{t}|\boldsymbol{u})P(\boldsymbol{U} = \boldsymbol{u})$.

Having estimated the nonparametric additive components g_α, we turn to the estimation of the parameters $\boldsymbol{\beta}$ and c. Note that c is now no longer the unconditional expectation of Y, but covers also the parametric linear part:

$$c = E(Y) - E(\boldsymbol{U}^\top \boldsymbol{\beta})\,.$$

Thus, c could be estimated by

$$\widehat{c} = \overline{Y} - \frac{1}{n}\sum_{i=1}^{n} \boldsymbol{U}_i^\top \widehat{\boldsymbol{\beta}}\,. \quad (9.10)$$

This estimator is unbiased with parametric $\sqrt{n}$-rate if this also holds for $\widehat{\boldsymbol{\beta}}$.

How does one to obtain such a $\sqrt{n}$-consistent $\widehat{\boldsymbol{\beta}}$? The solution is an ordinary LS regression of the partial residuals $Y_i - \sum \widehat{g}_\alpha(T_{i\alpha})$ on the $\boldsymbol{U}_i$, i.e., minimizing

$$\sum_{i=1}^{n}\left\{Y_i - \sum_{\alpha}\widehat{g}_\alpha(T_{i\alpha}) - \gamma - \boldsymbol{U}_i^\top \beta\right\}^2 . \quad (9.11)$$

If we define

$$\mathbf{U} = \begin{pmatrix} 1 & \boldsymbol{U}_1^\top \\ 1 & \boldsymbol{U}_2^\top \\ \vdots & \vdots \\ 1 & \boldsymbol{U}_n^\top \end{pmatrix},$$

this writes as

$$\begin{pmatrix} \widehat{\gamma} \\ \widehat{\beta} \end{pmatrix} = (\mathbf{U}^\top \mathbf{U})^{-1} \mathbf{U}^\top \left\{ Y - \sum_\alpha \widehat{g}_\alpha \right\}. \tag{9.12}$$

Note that γ stands for the constant c plus a bias correction caused by the nonparametric estimates $\widehat{g}_\alpha$. Since we have always included an intercept, this bias does not affect the estimation of the slope vector β but it is not recommended that we use $\widehat{\gamma}$ as an estimate for c. Instead, the constant c should be estimated as suggested in equation (9.10). The following theorem gives the asymptotic properties of the estimator $\widehat{\delta} = (\widehat{\gamma}, \widehat{\boldsymbol{\beta}}^\top)^\top$:

Theorem 9.1.
Under the aforementioned assumptions and $h = o(n^{-1/4})$ *it holds*

$$\sqrt{n}(\widehat{\delta} - \delta) \xrightarrow{L} N(0, \mathbf{V}^{-1} \mathbf{\Sigma} \mathbf{V}^{-1})$$

where

$$\mathbf{V} = E\boldsymbol{U}\boldsymbol{U}^\top, \quad \mathbf{\Sigma} = E\left\{ \sigma(\boldsymbol{T}, \boldsymbol{U}) \boldsymbol{Z}\boldsymbol{Z}^\top \right\} + \mathrm{Var}\left(\sum_\alpha \boldsymbol{V}_\alpha \right)$$

and

$$\boldsymbol{Z} = \boldsymbol{U} - \sum_\alpha \frac{f_{\underline{\alpha}}(\boldsymbol{T}_{\underline{\alpha}})}{f(\boldsymbol{T}, \boldsymbol{U})} f_\alpha(T_\alpha)\, E\left(\boldsymbol{U}|T_\alpha\right),$$

$$\boldsymbol{V}_\alpha = E(\boldsymbol{U}) \left[\left\{ \sum_{j \neq \alpha} g_j + \boldsymbol{U}^\top \delta \right\} - E\left\{ \sum_{j \neq \alpha} g_j + \boldsymbol{U}^\top \delta \right\} \right].$$

The condition on the bandwidth means that to obtain $\sqrt{n}$-consistent estimator of $\widehat{\boldsymbol{\beta}}$, we have to undersmooth in the nonparametric part. A careful study of the proof reveals a bias term of order h^2, which with $h = o(n^{-1/4})$ is faster than $\sqrt{n}$. Otherwise $\widehat{\beta}$ would inherit the bias from the $\widehat{g}_\alpha$s as it is based on a regression of the partial residuals $Y - \sum_\alpha \widehat{g}_\alpha$.

Unfortunately, in practice this procedure is hardly feasible if $\boldsymbol{U}$ has too many distinct realizations. For that case, Fan, Härdle & Mammen (1998) suggest a modification that goes back to an idea of Carroll, Fan, Gijbels & Wand (1997). Their approach leads to much more complicated asymptotic expressions, so that we only sketch the algorithm.

The problem for (quasi-)continuous $\boldsymbol{U}$ is that we cannot longer work on subsamples. Instead, we simultaneously estimate the impact of $\boldsymbol{U}$ and $\boldsymbol{T}$. More precisely, $\boldsymbol{\beta}$ is estimated by minimizing

$$\sum_{i=1}^n \{Y_i - \gamma_0 - \gamma_1(T_{i\alpha} - t_\alpha) - \gamma_2^\top \boldsymbol{U}_i\}^2 K_h(T_{i\alpha} - t_\alpha) \mathcal{K}_{\mathbf{H}}(\boldsymbol{T}_{i\underline{\alpha}} - \boldsymbol{T}_{l\underline{\alpha}}) \tag{9.13}$$

when calculating the pre-estimate $\widetilde{m}$ for all $t_\alpha, \boldsymbol{T}_{l\underline{\alpha}}$. Note that $\widehat{\gamma}_0$ is not really an estimate for m but rather for $c + \sum_\alpha g_\alpha$ at the point $(t_\alpha, \boldsymbol{T}_{l\underline{\alpha}})$. Consequently, we will use further $\widehat{\gamma}_0$ at the place off $\widetilde{m}$. The centered average over the $\widehat{\gamma}_0(t_\alpha, \boldsymbol{T}_{l\underline{\alpha}})$ is then the estimate for g_α:

$$\widehat{g}_\alpha(\bullet) = \frac{1}{n}\sum_{i=1}^{n} \widehat{\gamma}_0(\bullet, \boldsymbol{T}_{i\underline{\alpha}}) - \frac{1}{n}\sum_{i=1}^{n}\frac{1}{n}\sum_{l=1}^{n} \widehat{\gamma}_0(T_{i\alpha}, \boldsymbol{T}_{l\underline{\alpha}}).$$

The estimation of the parameter $\boldsymbol{\beta}$ can again be done by ordinary LS as in (9.11) and (9.12).

Example 9.1.
We consider data from 1992 on female labor supply in Eastern Germany. The aim of this example to study the impact of various factors on female labor supply which is measured by weekly hours of work (Y). The underlying data set is a subsample from the German Socio Economic Panel (GSOEP) of $n = 607$ women having a job and living together with a partner. The explanatory variables are:

U_1 female has children of less than 16 years old (1 if yes),
U_2 unemployment rate in the state where she lives,
T_1 age (in years),
T_2 wages (per hour),
T_3 "Treiman prestige index" of the job (Treiman, 1975),
T_4 rent or redemption for the flat,
T_5 years of education,
T_6 monthly net income of husband or partner.

Here, variable T_5 reflects the human capital, T_2, T_3 represent the particular attractiveness of the job, and T_4 is the main part of the household's expenditures. Note that since there are only five East German states, U_2 can not take on more than 5 different values.

The parametric coefficients are estimated as $\widehat{\beta}_1 = -1.457$ (female has children) and $\widehat{\beta}_2 = 0.5119$ (unemployment rate). Figures 9.1 and 9.2 show the estimated curves $\widehat{g}_\alpha$. The approximate confidence intervals are constructed as 1.64 times the (estimated) pointwise standard deviation of the curve estimates. The displayed point clouds are the logarithms of the distance to the partial residuals, i.e. $Y - \boldsymbol{U}^\top\widehat{\boldsymbol{\beta}} - \sum_{j\neq\alpha}\widehat{g}_j(T_j)$.

The plots show clear nonlinearities, a fact that has been confirmed in Härdle, Sperlich & Spokoiny (2001) when testing these additive functions for linearity. If the model is chosen correctly, the results quantify the extent to which each variable affects the female labor supply. For a comparison with

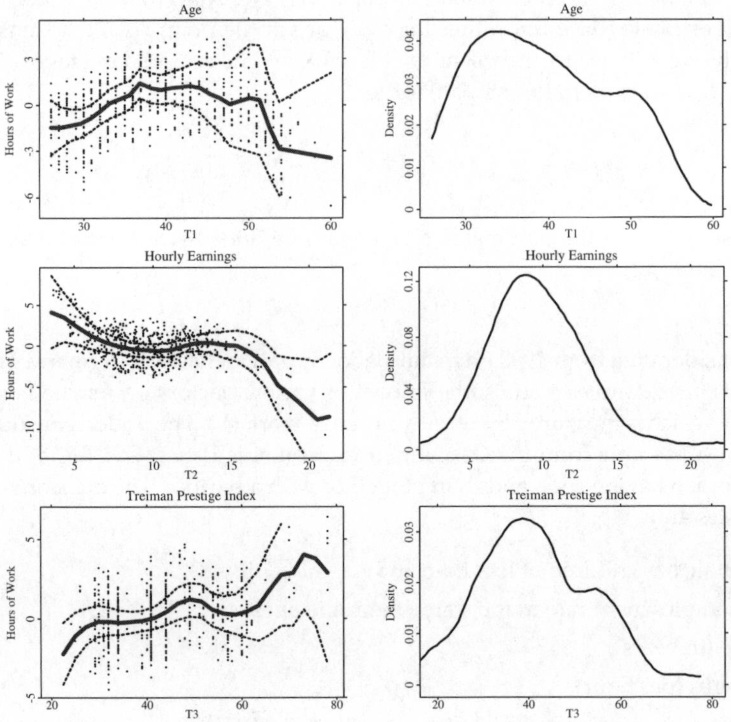

Figure 9.1. Estimates of additive components with approximate 90% confidence intervals (left), density estimates of the regressor (right), female labor supply data, variables T_1 to T_3

a parametric model we show in Table 9.1 the results of a parametric least squares analysis.

As can be seen from the table, T_1^2 (squared AGE) and T_2^2 (squared WAGE) have been added, and indeed their presence is highly significant. Their introduction was motivated by the nonparametric estimates $\widehat{g}_3$, $\widehat{g}_4$. Clearly, the piecewise linear shapes of g_5 and g_7 are harder to model in a parametric model. Here, at least the signs of the estimated parameters agree with the slopes of the nonparametric estimate in the upper part. Both factors are highly significant but the nonparametric analysis suggests that both factors are less influential for young women and the low income group. □

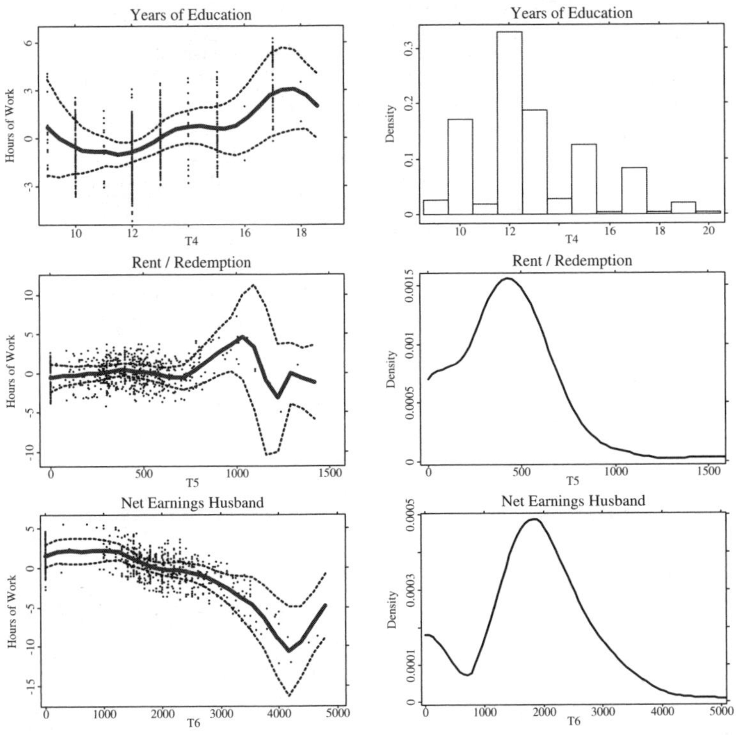

Figure 9.2. Estimates of additive components with approximate 90% confidence intervals (left), density estimates of the regressor (right), female labor supply data, variables T_4 to T_6

9.2 Additive Models with Known Link

It is clear that additive models are just a special case of APLM or GAM (without a parametric linear part and a trivial link function G = identity). The other way around, what we said about the advantages and motivations for additive model holds also for APLM and GAM. In Chapter 5, models with link function have been introduced and motivated, in particular for latent variables or binary choice models. Hence, an additive modeling of the index function is analogous to extending linear regressions models to additive models.

Again it was Stone (1986) who showed that GAM has the favorable property of circumventing the curse of dimensionality. Doing the estimation

Table 9.1. OLS coefficients for female labor supply data

Source	Sum of Squares	df	Mean square	F-ratio
Regression	6526.3	10	652.6	9.24
Residual	42101.1	596	70.6	
R^2 = 13.4%		R^2(adjusted) = 12,0%		

Variable	Coefficients	S.E.	t-values	p-values
constant	1.36	8.95	0.15	0.8797
CHILDREN	-2.63	1.09	-2.41	0.0163
UNEMPLOYMENT	0.48	0.22	2.13	0.0333
AGE	1.63	0.43	3.75	0.0002
AGE^2	-0.021	0.0054	-3.82	0.0001
WAGES	-1.07	0.18	-6.11	$\leq$ 0.0001
WAGES^2	0.017	0.0033	4.96	$\leq$ 0.0001
PRESTIGE	0.13	0.034	3.69	0.0002
EDUCATION	0.66	0.19	3.58	0.0004
HOUSING	0.0018	0.0012	1.56	0.1198
NETINCOME	-0.0016	0.0003	-4.75	$\leq$ 0.0001

properly, the rate of convergence can reach the same degree that is typical for the univariate case.

9.2.1 GAM using Backfitting

As we have discussed for additive models, there are two alternative approaches estimating the component functions, backfitting and marginal integration. However, recall that in models with a nontrivial link G, the response Y is not directly related to the index function. This fact must now be taken into account in the estimation procedure. For example, consider the partial residual

$$\boldsymbol{r}_\alpha = \boldsymbol{Y} - \widehat{c} - \sum_{j \neq \alpha} \widehat{\boldsymbol{g}}_j.$$

This $\boldsymbol{r}_\alpha$ is not appropriate for the generalized model as it ignores the link function G. So both methods, backfitting and marginal integration need to be extended for the link function. Instead of using Y we consider a transformation of Y, which is essentially the inverse of the link function applied to Y.

As before, we denote this adjusted dependent variable with Z. After conducting a complete backfitting with partial residuals based on Z we obtain a set of estimated functions $\widehat{g}_\alpha(\bullet)$ that explain the variable Z. But how good do these functions explain the untransformed dependent variable Y? The fit of the overall model in this sense is assessed by the local scoring algorithm.

The complete estimation procedure for GAM thus consists of two iterative algorithms: backfitting and local scoring. Backfitting is the "inner" iteration, whereas local scoring can bee seen as the "outer" iteration.

We summarized the final algorithm as given by Buja, Hastie & Tibshirani (1989) and Hastie & Tibshirani (1990). For the presentation keep in mind that local scoring corresponds to Fisher scoring in the IRLS algorithm for the GLM. Then, backfitting fits the index by additive instead of linear components. The inner backfitting algorithm is thus completely analogous to that in Chapter 8.

Local Scoring Algorithm for GAM	
initialization	$\widehat{c} = \overline{Y},\ \widehat{g}_{\alpha}^{(0)} \equiv 0$ for $\alpha = 1, \ldots, d,$
loop	over outer iteration counter m
	$\widehat{\eta}_i^{(m)} = \widehat{c}^{(m)} + \sum \widehat{g}_{\alpha}^{(m)}(x_{i\alpha})$
	$\widehat{\mu}_i^{(m)} = G(\widehat{\eta}_i^{(m)})$
	$Z_i = \widehat{\eta}_i^{(m)} + \left(Y_i - \widehat{\mu}_i^{(m)}\right)\left\{G'(\widehat{\eta}_i^{(m)})\right\}^{-1}$
	$w_i = \left\{G'(\widehat{\eta}_i^{(m)})\right\}^2 \left\{V(\widehat{\mu}_i^{(m)})\right\}^{-1}$
	obtain $\widehat{c}^{(m+1)}, \widehat{g}_{\alpha}^{(m+1)}$ by applying backfitting to Z_i with regressors x_i and weights w_i
until	convergence is reached

Recall that $V(\bullet)$ is the variance function of Y in a generalized model, see Subsections 5.2.1 and 5.2.3. For the definition of Z_i and w_i we refer to IRLS algorithm in Subsection 5.2.3 and the GPLM algorithms in Chapter 7,

Next, we present the backfitting routine which is applied inside local scoring to fit the additive component functions. This backfitting differs from that in the AM case in that firstly we use the adjusted dependent variable Z instead of Y and secondly we use a weighted smoothing. For this purpose we introduce the weighted smoother matrix $\mathbf{S}_{\alpha}(\bullet|w)$ defined as $\boldsymbol{D}^{-1}\mathbf{S}_{\alpha}\mathbf{W}$ where $\mathbf{W} = \text{diag}(w_1, \ldots w_n)^{\top}$ (w_i from the local scoring) and $\mathbf{D} = \text{diag}(\mathbf{S}_{\alpha}\mathbf{W})$.

Backfitting Algorithm for GAM	
initialization	$\overline{Z} = \frac{1}{n}\sum_{i=1}^{n} Z_i,\ \widehat{g}_\alpha^{(0)} \equiv 0$ for $\alpha = 1,\ldots,d$,
repeat	for $\alpha = 1,\ldots,d$ the cycles:
	$\boldsymbol{r}_\alpha = \boldsymbol{Z} - \overline{Z} - \sum\limits_{k=1}^{\alpha-1} \widehat{\boldsymbol{g}}_\alpha^{(l+1)} - \sum\limits_{k=\alpha+1}^{d} \widehat{\boldsymbol{g}}_k^{(l)}$,
	$\widehat{g}_\alpha^{(l+1)}(\bullet) = \mathbf{S}_\alpha(\boldsymbol{r}_\alpha \vert w)$
until	convergence is reached

Here, we use again vector notation: $\boldsymbol{r}_\alpha = (r_{1\alpha},\ldots,r_{n\alpha})^\top$, $\boldsymbol{Z} = (Z_1,\ldots,Z_n)^\top$, and $\widehat{\boldsymbol{g}}_\alpha = (\widehat{g}_\alpha(X_{1\alpha}),\ldots,g_\alpha(X_{n\alpha}))^\top$.

The theoretical properties of these iterative procedures are even complicated when the index consists of known (up to some parameters) component functions. For the general case of nonparametric index functions asymptotic results have only been be developed for special cases. The situation is different for the marginal integration approach which we study in the following subsection.

9.2.2 GAM using Marginal Integration

When using the marginal integration approach to estimate a GAM, the local scoring loop is not needed. Here, the extension from the additive model to the to GAM is straight forward. Recall that we consider

$$m(\boldsymbol{X}) = G\left\{c + \sum_{\alpha=1}^{d} g_\alpha(X_\alpha)\right\}$$

with a known link function and existing inverse G^{-1} and only continuous explanatory variables $\boldsymbol{X}$. Hence, we can write

$$G^{-1}\{m(\boldsymbol{X})\} = c + \sum_{\alpha=1}^{d} g_\alpha(X_\alpha).$$

As we have seen for the additive model, the component function $g_\alpha(x_\alpha)$ is up to a constant equal to

$$\int G^{-1}\{m(x_\alpha, \boldsymbol{x}_{\underline{\alpha}})\}\, f_{\underline{\alpha}}(\boldsymbol{x}_{\underline{\alpha}}) d\boldsymbol{x}_{\underline{\alpha}}$$

when we use the identifying condition $E\{g_\alpha(X_\alpha)\} = 0$ for all α. Thus we obtain an explicit expression for its estimator by

$$\widehat{g}_\alpha(x_\alpha) = \frac{1}{n}\sum_{l=1}^{n} G^{-1}\left\{\widetilde{m}(x_\alpha, \mathbf{X}_{l\underline{\alpha}})\right\} - \frac{1}{n}\sum_{i=1}^{n}\frac{1}{n}\sum_{l=1}^{n} G^{-1}\left\{\widetilde{m}(X_{i\alpha}, \mathbf{X}_{l\underline{\alpha}})\right\}. \quad (9.14)$$

Using a kernel smoother for the pre-estimator $\widetilde{m}$, this estimator has similar asymptotical properties as we found for the additive model, cf. Subsection 8.2.2. We remark that using a local polynomial smoother for $\widetilde{m}$ can yield simultaneously estimates for the derivatives of the component functions g_α. However, due to the existence of a nontrivial link G, all expressions become more complicated. Let us note that for models of the form

$$G^{-1}\{m(\mathbf{x})\} = c + \sum_{\alpha=1}^{d} g_\alpha(x_\alpha)$$

using the definitions

$$J_v = \{(j_1, j_2 \ldots, j_v) | 0 \le j_1, j_2, \ldots, j_v \le v, \text{ and } j_1 + 2j_2 + \cdots + vj_v = v\}$$

and

$$\partial_\alpha^{(\lambda)} m(\boldsymbol{t}) = \partial^\lambda m(x) / \partial x_\alpha^\lambda,$$

it holds

$$g_\alpha^{(v)}(x_\alpha) = v! \sum_{(j_1, j_2, \ldots, j_v) \in J_v} G^{-1^{(j_1+j_2+\ldots+j_v)}}\{m(x_\alpha, \boldsymbol{x}_{\underline{\alpha}})\} \prod_{\lambda=1}^{v} \frac{\{\partial_\alpha^{(\lambda)} m(x_\alpha, \boldsymbol{x}_{\underline{\alpha}})\}^{j_\lambda}}{(\lambda!)^{j_\lambda} j_\lambda!} \quad (9.15)$$

with $G^{-1^{(\kappa)}}$ being the κth derivative of G^{-1}. For example, if we are interested in the first derivative of g_α, equation (9.14) with (9.15) gives

$$\widehat{g}_\alpha^{(1)}(x_\alpha) = \frac{1}{n}\sum_{l=1}^{n} G^{-1^{(1)}}\{\widetilde{m}(x_\alpha, \mathbf{X}_{l\underline{\alpha}})\} \widetilde{\partial}_\alpha^{(1)} m(x_\alpha, \mathbf{X}_{l\underline{\alpha}}),$$

where we used both $\widetilde{m}$ as well as $\widetilde{\partial}_\alpha^{(1)} m$, from the local linear or higher order polynomial regression.

The expression for the second derivative is additionally complicated. Yang, Sperlich & Härdle (2003) provide asymptotic theory and simulations for this procedure. For the sake of simplicity we restrict the following theorem to the local constant estimation of the pre-estimate $\widetilde{m}$, see also Linton & Härdle (1996). As introduced previously, the integration estimator requires to choose two bandwidths, h for the direction of interest and $\widetilde{h}$ for the nuisance direction.

Theorem 9.2.
Assume the bandwidths fulfill $h = O(n^{-1/5})$ and that $n^{2/5}\widetilde{h}^d \to 0$ and $n^{2/5}\widetilde{h}^{d-1} \to \infty$. Then, under smoothness and regularity conditions we have

$$n^{2/5}\{\widehat{g}_\alpha(x_\alpha) - g_\alpha(x_\alpha)\} \xrightarrow{L} N\{b_\alpha(t_\alpha), v_\alpha(t_\alpha)\}.$$

It is obvious that marginal integration leads to estimates of the same rate as for univariate Nadaraya-Watson regression. This is the same rate that we obtained for the additive model without link function. Let us mention that in general the properties of backfitting and marginal integration found in Chapter 8 carry over to the GAM. However, detailed simulation studies do not yet exist for this case and a theoretical comparison is not possible due to the lack of asymptotical results for backfitting in GAM.

9.3 Generalized Additive Partial Linear Models

The most complex form of generalized models that we consider here is the GAPLM

$$E(Y|\boldsymbol{U},\boldsymbol{T}) = G\left\{\boldsymbol{U}^\top\beta + c + \sum_{\alpha=1}^{q} g_\alpha(T_\alpha)\right\}. \tag{9.16}$$

The estimation of this model involves all techniques that were previously introduced. We are particularly interested in estimating the parameter β at $\sqrt{n}$-rate and the component functions g_α with the rate that is typical for the dimension of T_α.

9.3.1 GAPLM using Backfitting

For model (9.16), the backfitting and local scoring procedures from Subsection 9.2.1 can be directly used. Since we have a link function G, the local scoring algorithm is used without any changes. For the "inner" backfitting iteration, the algorithm is adapted to a combination of parametric (linear) regression and additive modeling. Essentially, the weighted smoother matrix $\mathbf{S}_\alpha(\bullet|w)$ is replaced by a weighted linear projection matrix

$$\mathbf{P}_w = \mathbf{U}(\mathbf{U}^\top\mathbf{W}\mathbf{U})^{-1}\mathbf{U}^\top\mathbf{W}$$

for the linear component. Here $\mathbf{W}$ is the weight matrix from Subsection 9.2.1 and $\mathbf{U}$ is the design matrix obtained from the observations of $\boldsymbol{U}$. We refer for further details to Hastie & Tibshirani (1990).

9.3.2 GAPLM using Marginal Integration

The marginal integration approach for (9.16) is a subsequent application of the semiparametric ML procedure for the GPLM (see Chapter 7), followed by marginal integration (as introduced in Chapter 8) applied on the nonparametric component of the GPLM. For this reason we only sketch the complete procedure and refer for the details to Härdle, Huet, Mammen & Sperlich (2004).

The key idea for estimating the GAPLM is the following: We use the profile likelihood estimator for the GPLM with a modification of the local likelihood function (7.8):

$$\ell_{h,\mathbf{H}}(\mathbf{Y}, \boldsymbol{\mu}_{m(\mathbf{T})}, \phi) \tag{9.17}$$
$$= \sum_{i=1}^{n} K_h(t_\alpha - T_\alpha) \mathcal{K}_{\mathbf{H}}(\boldsymbol{t}_{\underline{\alpha}} - \boldsymbol{T}_{\underline{\alpha}})\, \ell\left(Y_i, G\{\boldsymbol{U}_i^\top \boldsymbol{\beta} + m(t_\alpha, \boldsymbol{t}_{\underline{\alpha}})\}, \phi\right).$$

As for the GPLM this local likelihood is maximized with respect to the nonparametric component $m_{\boldsymbol{\beta}}(t_\alpha, \boldsymbol{t}_{\underline{\alpha}})$, this gives an estimate that does not (yet) make use of th additive structure.

We apply the now marginal integration method to this pre-estimate. The final estimator is

$$\widehat{g}_\alpha(t_\alpha) = \frac{1}{n}\sum_{l=1}^{n} \widehat{m}(t_\alpha, \boldsymbol{T}_{l\underline{\alpha}}) - \frac{1}{n}\sum_{i=1}^{n} \frac{1}{n}\sum_{l=1}^{n} \widehat{m}(T_{i\alpha}, \boldsymbol{T}_{l\underline{\alpha}}). \tag{9.18}$$

To avoid numerical problems, in particular at boundary regions or in regions of sparse data, a weight function should be applied inside the averaging. More precisely, the final estimate should calculated by:

$$\widehat{g}_\alpha(t_\alpha) = \frac{\frac{1}{n}\sum_{l=1}^{n} w_{\underline{\alpha}}(\boldsymbol{T}_{l\underline{\alpha}})\widehat{m}(t_\alpha, \boldsymbol{T}_{l\underline{\alpha}})}{\frac{1}{n}\sum_{l=1}^{n} w_{\underline{\alpha}}(\boldsymbol{T}_{l\underline{\alpha}})} - \frac{\frac{1}{n}\sum_{i=1}^{n} w_\alpha(T_{i\alpha})\widetilde{g}_\alpha(T_{i\alpha})}{\frac{1}{n}\sum_{i=1}^{n} w_\alpha(T_{i\alpha})}.$$

Finally, the constant c is estimated by

$$\widehat{c} = \frac{1}{n}\sum_{i=1}^{n} \widehat{m}(\boldsymbol{T}_i) \frac{1}{n}\sum_{i=1}^{n} \widehat{m}(T_{i\alpha}, \boldsymbol{T}_{i\underline{\alpha}}). \tag{9.19}$$

For these estimators we have asymptotic properties according to the following theorem.

Theorem 9.3.
Suppose the bandwidths tend to zero and fulfill $\mathbf{H} = \widetilde{h}\mathbf{I}$, $nh\widetilde{h}^{2(q-1)} / \log^2(n) \to \infty$, *then under some smoothness and regularity conditions*

$$\sqrt{nh}\{\widehat{g}_\alpha(t_\alpha) - g_\alpha(t_\alpha)\} \xrightarrow{L} N(b_\alpha(t_\alpha), v_\alpha(t_\alpha)).$$

The expressions for bias and variance are quite complex such that we omit them here. We remark that the correlation between the estimates of the components are of higher order rate. Consequently, summing up the estimates would give us a consistent estimate of the index function m with the one-dimensional nonparametric rate.

Härdle, Huet, Mammen & Sperlich (2004) also state that the bias for the estimates $\widehat{g}_\alpha$ is not negligible. Therefore they propose a bias correction procedure using (wild) bootstrap.

Example 9.2.
To illustrate the GAPLM estimation we use the data set as in Example 5.1 selecting the most southern state (Sachsen) of East Germany. Recall that the data comprise the following explanatory variables:

U_1 family/friend in West,
U_2 unemployed/job loss certain,
U_3 middle sized city (10,000–100,000 habitants),
U_4 female (1 if yes),
T_1 age of person (in years),
T_2 household income (in DM).

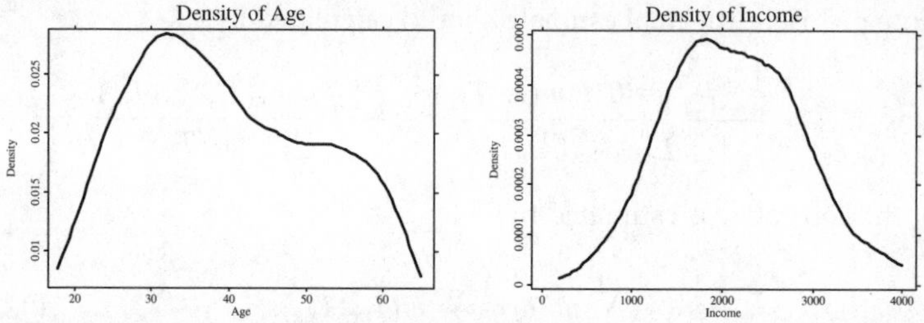

Figure 9.3. Density plots for migration data (subsample from Sachsen), AGE on the left, HOUSEHOLD INCOME on the right

We first show the density plots for the two continuous variables in Figure 9.3. Table 9.2 gives descriptive statistics for the data. In the following, AGE and INCOME have been standardized which corresponds to multiplying the bandwidths with the empirical standard deviations.

Table 9.3 presents on the left the results of a parametric logit estimation. Obviously, AGE has a significant linear impact on the migration intention whereas this does not hold for household income. On the right hand side of Table 9.3 we have listed the results for the linear part of the GAPLM. Since

Table 9.2. Descriptive statistic for migration data (subsample from Sachsen, $n = 955$)

		Yes	No	(in %)	
Y	MIGRATION INTENTION	39.6	60.4		
U_1	FAMILY/FRIENDS	82.4	27.6		
U_2	UNEMPLOYED/JOB LOSS	18.3	81.7		
U_3	CITY SIZE	26.0	74.0		
U_4	FEMALE	51.6	48.4		
		Min	Max	Mean	S.D.
T_1	AGE	18	65	40.37	12.69
T_2	INCOME	200	4000	2136.31	738.72

Table 9.3. Logit and GAPLM coefficients for migration data

	GLM			GAPLM	
	Coefficients	S.E.	p-values	Coefficients	
				$h = 0.75$	$h = 1.00$
FAMILY/FRIENDS	0.7604	0.1972	<0.001	0.7137	0.7289
UNEMPLOYED/JOB LOSS	0.1354	0.1783	0.447	0.1469	0.1308
CITY SIZE	0.2596	0.1556	0.085	0.3134	0.2774
FEMALE	-0.1868	0.1382	0.178	-0.1898	-0.1871
AGE (stand.)	-0.5051	0.0728	<0.001	—	—
INCOME (stand.)	0.0936	0.0707	0.187	—	—
constant	-1.0924	0.2003	<0.001	-1.1045	-1.1007

the choice of bandwidths can be crucial, we used two different bandwidths for the estimation. We see that the coefficients for the GAPLM show remarkable differences with respect to the logit coefficients. We can conclude that the impact of family/friends in the West seems to be overestimated by the parametric logit whereas the city size effect is larger for the semiparametric model. The nonparametric function estimates for AGE and INCOME are displayed in Figure 9.4.

In contrast to Example 7.1 the GAPLM allows us to include both, AGE and INCOME, as univariate nonparametric functions. The interpretation of these functions is much easier. We can easily see that the component function for AGE is clearly monotone decreasing. The nonparametric impact of INCOME, however, does not vanish when the bandwidth is increased. We will come back to this point when testing functional forms in such models in the following section. □

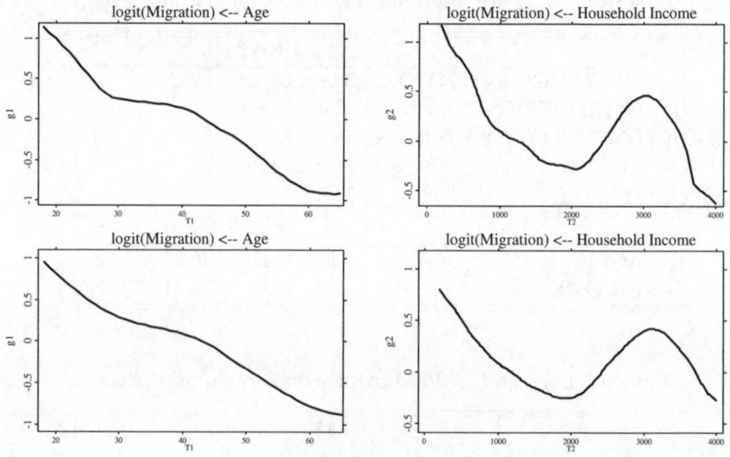

Figure 9.4. Additive curve estimates for AGE (left) and INCOME (right) in Sachsen (upper plots with $h = 0.75$, lower with $h = 1.0$)

9.4 Testing in Additive Models, GAM, and GAPLM

For testing the GAM or GAPLM specification we concentrate on the presentation of a general approach and discuss a typical testing approach which is similar to that in Chapter 7:

- We are interested in checking for a specific parametric specification (including the problem of checking for significance at all).
- We estimate the model under the null hypothesis and compare with a nonparametric estimate for the alternative. The distance between both functions of interest can be measured by the integrated squared difference, eventually using a weight function w.
- The asymptotic distribution of the test statistic is too complex and thus not useful in practice. Therefore we apply bootstrap to generate samples $(\boldsymbol{U}_i, \boldsymbol{T}_i, Y_i^*)$ under H_0. These samples can be used to approximate the distribution of the test statistic under H_0.

In order to cover the most complex case, we now focus on the GAPLM. The modification of the test procedure to simpler models is straightforward.

Certainly, the most interesting testing problem is that of testing the specification of a single component function. Let us consider the null hypotheses which assumes a polynomial structure for g_α (α fixed). For example, the null function is the most simple polynomial. Testing $H_0 : g_\alpha \equiv 0$ means to test for significant impact of T_α on the response. The alternative is an arbitrary functional form.

We explain the test procedure using the example of a linear hypothesis for the first component function g_1. This means

$$H_0: \; g_1(t_1) = \gamma \cdot t_1, \qquad \text{for all } t_1.$$

As the procedure and motivation for each step are the same as in Chapter 7, we condense our presentation to the most essential steps. Recall the GAPLM:

$$E(Y|\boldsymbol{U},\boldsymbol{T}) = G\{\boldsymbol{U}^\top \boldsymbol{\beta} + c + g_1(T_1) + g_2(T_2) + \cdots + g_q(T_q)\}.$$

We know already that a direct comparison of the parametric estimate

$$\widetilde{g}_1(t_1) = \widehat{\gamma} u$$

and the nonparametric estimate $\widehat{g}_1(u)$ causes the problem of comparing two functional estimates with biases of different magnitude. To avoid this discrepancy we replace $\widetilde{g}_1$ with a bootstrap estimate $E^*\widehat{g}_1^*$ that takes the bias into account. $E^*\widehat{g}_1^*$ is the bootstrap expectation given the data $(Y_i, \boldsymbol{U}_i, \boldsymbol{T}_i)$ $(i = 1, \ldots, n)$ under H_0 over nonparametric estimates of g_1 from bootstrap samples $(Y_i^*, \boldsymbol{U}_i, \boldsymbol{T}_i)$ $(i = 1, \ldots, n)$. The Y_i^* are generated under the H_0 model as in Chapter 7.

Estimation under H_0 means that we consider the model

$$E(Y|\boldsymbol{U},\boldsymbol{T}) = G\{\boldsymbol{U}^\top \boldsymbol{\beta} + b + \gamma T_1 + g_2(T_2) + \cdots + g_q(T_q)\}.$$

The constant b in this equation can be different from c, because the function γT_1 is not necessarily centered as we assumed for $g_1(T_1)$. The estimation of the parametric components $\boldsymbol{\beta}$, b and γ as well as of the components $g_2, \ldots, g_q$ is performed as presented in Section 9.3.2.

We define the test statistic in analogy to (7.34):

$$\widetilde{LR} = \sum_{i=1}^{n} w(\boldsymbol{T}_i) \frac{[G'\{\boldsymbol{U}_i^\top \boldsymbol{\beta} + \widehat{m}(\boldsymbol{T}_i)\}]^2}{V\{G(\widehat{\mu}_i)\}} \{\widehat{g}_1(T_{i1}) - E^*\widehat{g}_1^*(T_{i1})\}^2 .$$

where $\widehat{m}(\boldsymbol{T}_i) = \widehat{c} + \widehat{g}_1(T_{i1}) + \ldots + \widehat{g}_q(T_{iq})$ and $\widehat{\mu}_i = G\{\boldsymbol{U}_i^\top \widehat{\boldsymbol{\beta}} + \widehat{m}(\boldsymbol{T}_i)\}$. The function w defines trimming weights to obtain numerically stable estimators on the range of $\boldsymbol{T}$ that is of interest.

Härdle, Huet, Mammen & Sperlich (2004) prove that (under some regularity assumptions) the test statistic $\widetilde{LR}$ has an asymptotic normal distribution under H_0. As in the GPLM case, the convergence of the test statistic is very slow. Therefore we prefer the bootstrap approximation of the quantiles of $\widetilde{LR}$. The approach here is analog to that in Subsection 7.3.2. We will now study what the test implies for our example on migration intention.

Example 9.3.
We continue Example 9.2 but concentrate now on testing whether the non-linearities found for the impact of AGE and INCOME are significant.

As a test statistic we compute $\widetilde{LR}$ and derive its critical values from the bootstrap test statistics $\widetilde{LR}^*$. The bootstrap sample size is set to $n_{boot} = 499$ replications, all other parameters were set to the values of Example 9.2. We find the following results: For AGE linearity has always been rejected at the 1% level, in particular for all bandwidths that we used. This result may be surprising but a closer inspection of the numerical results shows that the hypothesis based bootstrap estimates have almost no deviation. In consequence a slight difference from linearity already leads to the rejection of H_0.

This is different for the variable INCOME. The bootstrap estimates vary in this case. Here, linearity is rejected at the 2% level for $h = 0.75$ and at 1% level for $h = 1.0$. Note that this is not in contradiction to the results in Chapter 7 as the results here are based on different samples and models. □

Let us consider a second example. This example is interesting since some of the results seem to be contradictory at a first glance. However, we have to take into account that nonparametric methods may not reveal their power as the sample size is too small.

Example 9.4.
We use again the data of Proença & Werwatz (1995) which were already introduced in Example 6.1. The data are a subsample of 462 individuals from the first nine waves of the GSOEP, including all individuals who have completed an apprenticeship in the years between 1985 and 1992.

Table 9.4. Logit and GAPLM coefficients for unemployment data

	GLM (logit)		GAPLM
	Coefficients	S.E.	Coefficients
FEMALE	-0.3651	0.3894	-0.3962
AGE	0.0311	0.1144	—
SCHOOL	0.0063	0.1744	0.0452
EARNINGS	-0.0009	0.0010	—
CITY SIZE	-5.e-07	4.e-07	—
FIRM SIZE	-0.0120	0.4686	-0.1683
DEGREE	-0.0017	0.0021	—
URATE	0.2383	0.0656	—
constant	-3.9849	2.2517	-2.8949

We are interested in the question which factors cause unemployment after the apprenticeship. In contrast to Example 6.1 we use a larger set of explanatory variables:

U_1 female (1 if yes),

U_2 years of school education,

U_3 firm size (1 if large firm),

T_1 age of the person,

T_2 earnings as an apprentice (in DM),

T_3 city size (in $100,000$ habitants),

T_4 degree of apprenticeship, i.e., the percentage of people apprenticed in a certain occupation, divided by the percentage of people employed in this occupation in the entire economy, and

T_5 unemployment rate in the particular country the apprentice is living in.

Here, SCHOOL and AGE represent the human capital, EARNINGS represents the value of an apprenticeship, and CITY SIZE could be interesting since larger cities often offer more employment opportunities. To this the variable URATE also fits. FIRM SIZE tells us whether e.g. in small firms the number of apprenticeship positions provided exceeds the number of workers retained after the apprenticeship is completed. Density plots of the continuous variables are given in Figure 9.5.

We estimate a parametric logit model and the corresponding GAPLM to compare the results. Table 9.4 reports the coefficient estimates for both models, for the parametric model standard deviations are also given. The nonparametric function estimates are plotted in Figure 9.6. We used bandwidths which are inflated from the standard deviations of the variables by certain factors.

Looking at the function estimates we get the impression of strong nonlinearity in all variables except for URATE. But in contrast, the test results show that the linearity hypothesis cannot be rejected for all variables and significant levels from 1% to 20%.

What does this mean? It turns out that the parametric logit coefficients for all variables (except the constant and URATE) are already insignificant, see Table 9.4. As we see now this is not because of the misspecification of the parametric logit model. It seems that the data can be explained by neither the parametric logit model nor the semiparametric GAPLM. Possible reasons may be the insignificant sample size or a lack of the appropriate explanatory variables. Let us also remark that from the density plots we find that applying nonparametric function estimates could be problematic except for DEGREE and URATE. □

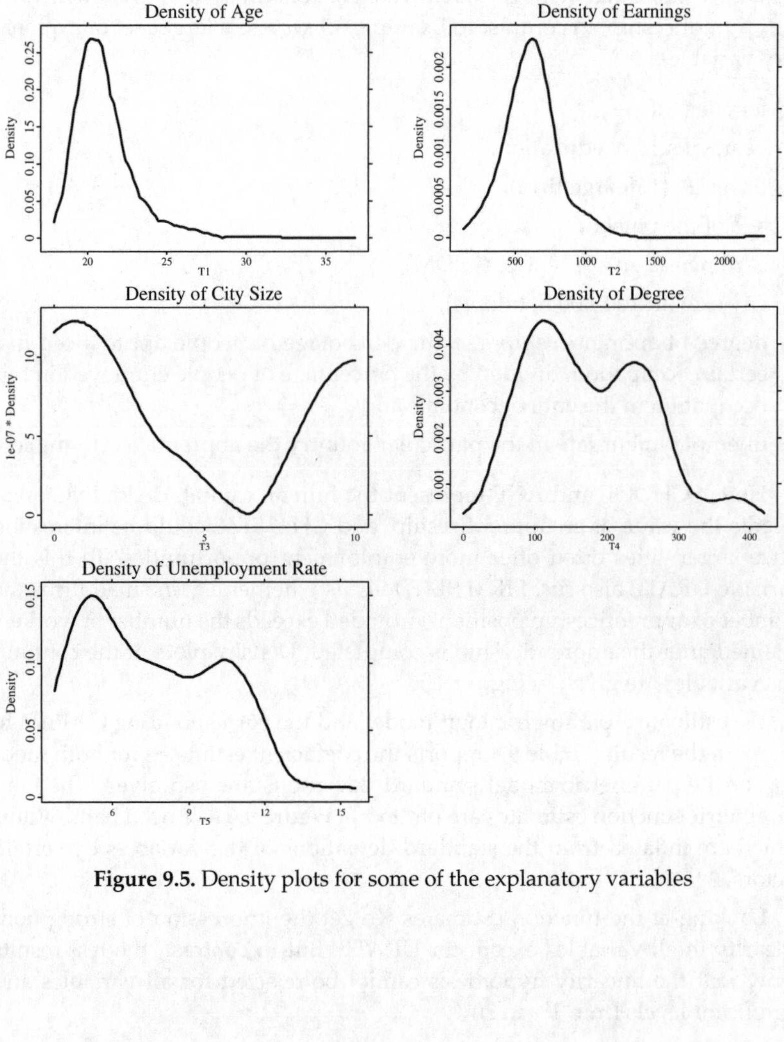

Figure 9.5. Density plots for some of the explanatory variables

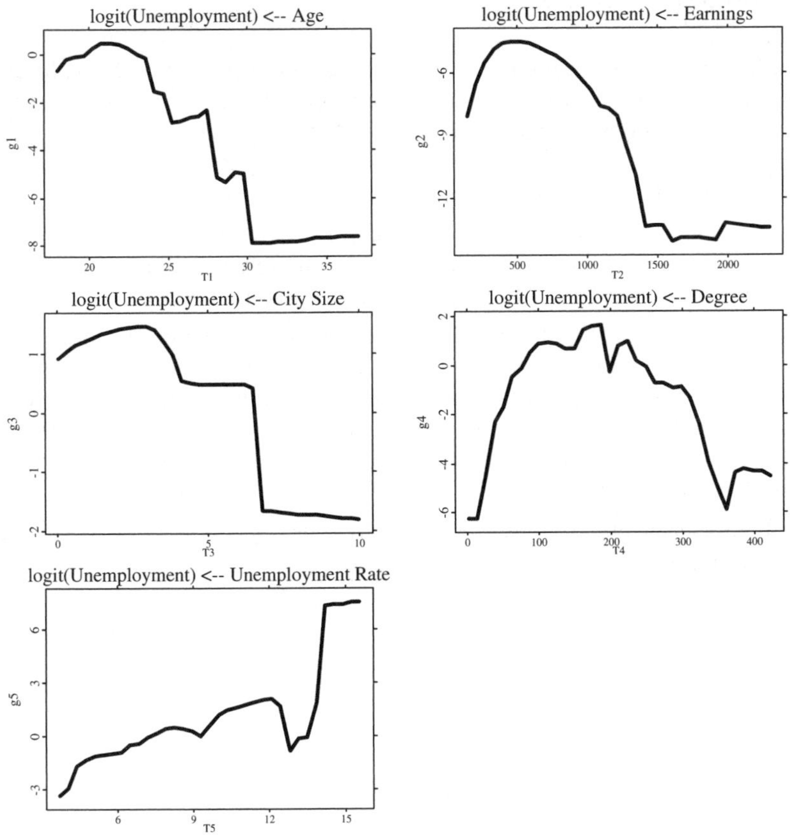

Figure 9.6. Estimates of additive components for unemployment data

Bibliographic Notes

The semiparametric approach to partial linear models can already be found in Green & Yandell (1985). The way we present it here was developed by Speckman (1988) and Robinson (1988b). A variety of generalized models can be found in the monograph of Hastie & Tibshirani (1990). This concerns in particular backfitting algorithms.

The literature on marginal integration for generalized models is very recent. Particularly interesting is the combination of marginal integration and backfitting to yield efficient estimators as discussed in Linton (2000).

Variants of GAM and GAPLM are the extension to parametric nonlinear components (Carroll, Härdle & Mammen, 2002), nonparametric components of single index form (Carroll, Fan, Gijbels & Wand, 1997), models with nonparametric link and nonparametric component functions (Horowitz, 1998a), and weak separable models (Rodríguez-Póo, Sperlich & Vieu, 2003; Mammen & Nielsen, 2003).

For the issue of hypothesis testing we refer for further reading to Gozalo & Linton (2001) and Yang, Sperlich & Härdle (2003).

Exercises

Exercise 9.1. What would be a proper algorithm using backfitting in the semiparametric APLM when the dimension q (of the nonparametric part) is bigger than one?

Exercise 9.2. Discuss for the semiparametric APLM the following question: Why don't we use the approach of Speckman (1988) (Chapter 7) to get an estimate for β? Afterwards we could apply backfitting and/or marginal integration on the nonparametric part. a) Does this work? b) What follows for the properties of β?

Exercise 9.3. Recall equation (9.13). Why is $\widehat{\gamma}_2$ from (9.13) inefficient for estimating β?

Exercise 9.4. Recall Subsection 8.2.3 (estimation of interaction terms) and Section 9.4 (testing). Construct a test for additivity in additive models with only using procedures from these two sections.

Exercise 9.5. Again recall the estimation of interaction terms from Subsection 8.2.3. In Subsection 9.2.2 we introduced a direct extension for marginal integration to marginal integration in GAM. In Section 9.3.2 we extended it, less straight forward, to GAPLM. How can the interaction be incorporated into these models? What would a test for interaction look like?

Exercise 9.6. In Chapter 8 we discussed intensively the difference between backfitting and marginal integration and indicated that their interpretations carry over to the GAM and GAPLM. Based on these differences, construct a test on separability of the impact of T_1 and T_2. What would a general test on additivity look like?

Exercise 9.7. Think of a general test to compare multidimensional regression against an additive model structure. Discuss the construction of such a test, its advantages and disadvantages.

Summary

- ⋆ The nonparametric additive components in all these extensions of simple additive models can be estimated with the rate that is typical for one dimensional smoothing.

- ⋆ An additive partial linear model (APLM) is of the form

$$E(Y|\boldsymbol{U},\boldsymbol{T}) = \boldsymbol{U}^\top \boldsymbol{\beta} + c + \sum_{\alpha=1}^{q} g_\alpha(T_\alpha)\,.$$

Here, $\boldsymbol{\beta}$ and c can be estimated with the parametric rate $\sqrt{n}$. While for the marginal integration estimator in the suggested procedure it is necessary to undersmooth, it is still not clear for the backfitting what to do when $q > 1$.

- ⋆ A generalized additive model (GAM) has the form

$$E(Y|\boldsymbol{X}) = G\{c + \sum_{\alpha=1}^{d} g_\alpha(X_\alpha)\}$$

with a (known) link function G. To estimate this using the backfitting we combine the local scoring and the Gauss-Seidel algorithm. Theory is lacking here. Using the marginal integration we get a closed formula for the estimator for which asymptotic theory can be also derived.

- ⋆ The generalized additive partial linear model (GAPLM) is of the form

$$E(Y|\boldsymbol{U},\boldsymbol{T}) = G\left\{\boldsymbol{U}^\top \boldsymbol{\beta} + c + \sum_{\alpha=1}^{q} g_\alpha(T_\alpha)\right\}$$

with a (known) link function G. In the parametric part $\boldsymbol{\beta}$ and c can be estimated again with the $\sqrt{n}$-rate. For the backfitting we combine estimation in APLM with the local scoring algorithm. But again the the case $q > 1$ is not clear and no theory has been provided. For the marginal integration approach we combine the quasi-likelihood procedure with the marginal integration afterwards.

---

* ⋆ In all considered models, we have only developed theory for the marginal integration so far. Interpretation, advantages and drawbacks stay the same as mentioned in the context of additive models (AM).

---

* ⋆ We can perform test procedures on the additive components separately. Due to the complex structure of the estimation procedures we have to apply (wild) bootstrap methods.

References

Achmus, S. (2000). Nichtparametrische additive Modelle, *Doctoral Thesis*, Technical University of Braunschweig, Germany.

Ahn, H. & Powell, J. L. (1993). Semiparametric selection of censored selection models with a nonparametric selection mechanism, *Econometrica* **58**: 3–29.

Amemiya, T. (1984). Tobit models: A survey, *Journal of Econometrics* **24**: 3–61.

Andrews, D. W. K. & Whang, Y.-J. (1990). Additive interactive regression models: circumvention of the curse of dimensionality, *Econometric Theory* **6**: 466–479.

Begun, J., Hall, W., Huang, W. & Wellner, J. (1983). Information and asymptotic efficiency in parametric–nonparametric models, *Annals of Statistics* **11**: 432–452.

Berndt, E. (1991). *The Practice of Econometrics*, Addison–Wesley.

Bickel, P. & Doksum, K. (1981). An analysis of transformations revisited, *Journal of the American Statistical Association* **76**: 296–311.

Bickel, P., Klaassen, C., Ritov, Y. & Wellner, J. (1993). *Efficient and Adaptive Estimation for Semiparametric Models*, The Johns Hopkins University Press.

Bickel, P. & Rosenblatt, M. (1973). On some global measures of the deviations of density function estimators, *Annals of Statistics* **1**: 1071–1095.

Bierens, H. (1990). A consistent conditional moment test of functional form, *Econometrica* **58**: 1443–1458.

Bierens, H. & Ploberger, W. (1997). Asymptotic theory of integrated conditional moment tests, *Econometrica* **65**: 1129–1151.

Bonneu, M. & Delecroix, M. (1992). Estimation semiparamétrique dans les modèles explicatifs conditionnels à indice simple, *Cahier de gremaq, 92.09.256*, GREMAQ, Université Toulouse I.

Bonneu, M., Delecroix, M. & Malin, E. (1993). Semiparametric versus nonparametric estimation in single index regression model: A computational approach, *Computational Statistics* **8**: 207–222.

Bossaerts, P., Hafner, C. & Härdle, W. (1996). Foreign exchange rates have surprising volatility, *in* P. M. Robinson & M. Rosenblatt (eds), *Athens Conference on Applied Probability and Time Series Analysis. Volume II: Time Series Analysis. In Memory of E.J. Hannan*, Lecture Notes in Statistics, Springer, pp. 55–72.

Boularan, J., Ferré, L. & Vieu, P. (1994). Growth curves: a two-stage nonparametric approach, *Journal of Statistical Planning and Inference* **38**: 327–350.

Bowman, A. & Azzalini, A. (1997). *Applied Smoothing Techniques for Data Analysis*, Oxford University Press, Oxford, UK.

Box, G. & Cox, D. (1964). An analysis of transformations, *Journal of the Royal Statistical Society, Series B* **26**: 211–243.

Breiman, L. & Friedman, J. H. (1985). Estimating optimal transformations for multiple regression and correlations (with discussion), *Journal of the American Statistical Association* **80**(391): 580–619.

Buja, A., Hastie, T. J. & Tibshirani, R. J. (1989). Linear smoothers and additive models (with discussion), *Annals of Statistics* **17**: 453–555.

Burda, M. (1993). The determinants of East–West German migration, *European Economic Review* **37**: 452–461.

Cao, R., Cuevas, A. & González Manteiga, W. (1994). A comparative study of several smoothing methods in density estimation, *Computational Statistics & Data Analysis* **17**(2): 153–176.

Carroll, R. J., Fan, J., Gijbels, I. & Wand, M. P. (1997). Generalized partially linear single–index models, *Journal of the American Statistical Association* **92**: 477–489.

Carroll, R. J., Härdle, W. & Mammen, E. (2002). Estimation in an additive model when the components are linked parametrically, *Econometric Theory* **18**(4): 886–912.

Chaudhuri, P. & Marron, J. S. (1999). SiZer for exploration of structures in curves, *Journal of the American Statistical Association* **94**: 807–823.

Chen, R., Liu, J. S. & Tsay, R. S. (1995). Additivity tests for nonlinear autoregression, *Biometrika* **82**: 369–383.

Cleveland, W. S. (1979). Robust locally-weighted regression and smoothing scatterplots, *Journal of the American Statistical Association* **74**: 829–836.

Collomb, G. (1985). Nonparametric regression – an up-to-date bibliography, *Statistics* **2**: 309–324.

Cosslett, S. (1983). Distribution–free maximum likelihood estimation of the binary choice model, *Econometrica* **51**: 765–782.

Cosslett, S. (1987). Efficiency bounds for distribution–free estimators of the binary choice model, *Econometrica* **55**: 559–586.

Dalelane, C. (1999). Bootstrap confidence bands for the integration estimator in additive models, *Diploma thesis*, Department of Mathematics, Humboldt-Universität zu Berlin.

Daubechies, I. (1992). *Ten Lectures on Wavelets*, SIAM, Philadelphia, Pennsylvania.

Deaton, A. & Muellbauer, J. (1980). *Economics and Consumer Behavior*, Cambridge University Press, Cambridge.

Delecroix, M., Härdle, W. & Hristache, M. (2003). Efficient estimation in conditional single-index regression, *Journal of Multivariate Analysis* **86**(2): 213–226.

Delgado, M. A. & Mora, J. (1995). Nonparametric and semiparametric estimation with discrete regressors, *Econometrica* **63**(6): 1477–1484.

Denby, L. (1986). Smooth regression functions, *Statistical report 26*, AT&T Bell Laboratories.

Dette, H. (1999). A consistent test for the functional form of a regression based on a difference of variance estimators, *Annals of Statistics* **27**: 1012–1040.

Dette, H., von Lieres und Wilkau, C. & Sperlich, S. (2004). A comparison of different nonparametric methods for inference on additive models, *Nonparametric Statistics* **16**. forthcoming.

Dobson, A. J. (2001). *An Introduction to Generalized Linear Models*, second edn, Chapman and Hall, London.

Donoho, D. L. & Johnstone, I. M. (1994). Ideal spatial adaptation by wavelet shrinkage, *Biometrika* **81**: 425–455.

Donoho, D. L. & Johnstone, I. M. (1995). Adapting to unknown smoothness via wavelet shrinkage, *Journal of the American Statistical Association* **90**: 1200–1224.

Donoho, D. L., Johnstone, I. M., Kerkyacharian, G. & Picard, D. (1995). Wavelet shrinkage: Asymptopia? (with discussion), *Journal of the Royal Statistical Society, Series B* **57**: 301–369.

Duan, N. & Li, K.-C. (1991). Slicing regression: A link-free regression method, *Annals of Statistics* **19**(2): 505–530.

Duin, R. P. W. (1976). On the choice of smoothing parameters of Parzen estimators of probability density functions, *IEEE Transactions on Computers* **25**: 1175–1179.

Eilers, P. H. C. & Marx, B. D. (1996). Flexible smoothing with b-splines and penalties (with discussion, *Statistical Science* **11**: 89–121.

Epanechnikov, V. (1969). Nonparametric estimation of a multidimensional probability density, *Teoriya Veroyatnostej i Ee Primeneniya* **14**: 156–162.

Eubank, R. L. (1999). *Nonparametric Regression and Spline Smoothing*, Marcel Dekker, New York.

Eubank, R. L., Hart, J. D., Simpson, D. G. & Stefanski, L. A. (1995). Testing for additivity in nonparametric regression, *Annals of Statistics* **23**: 1896–1920.

Eubank, R. L., Kambour, E. L., Kim, J. T., Klipple, K., Reese, C. S. & Schimek, M. G. (1998). Estimation in partially linear models, *Computational Statistics & Data Analysis* **29**: 27–34.

Fahrmeir, L. & Tutz, G. (1994). *Multivariate Statistical Modelling Based on Generalized Linear Models*, Springer.

Fan, J. & Gijbels, I. (1996). *Local Polynomial Modelling and Its Applications*, Vol. 66 of *Monographs on Statistics and Applied Probability*, Chapman and Hall, New York.

Fan, J., Härdle, W. & Mammen, E. (1998). Direct estimation of low-dimensional components in additive models, *Annals of Statistics* **26**: 943–971.

Fan, J. & Li, Q. (1996). Consistent model specification test: Omitted variables and semiparametric forms, *Econometrica* **64**: 865–890.

Fan, J. & Marron, J. S. (1992). Best possible constant for bandwidth selection, *Annals of Statistics* **20**: 2057–2070.

Fan, J. & Marron, J. S. (1994). Fast implementations of nonparametric curve estimators, *Journal of Computational and Graphical Statistics* **3**(1): 35–56.

Fan, J. & Müller, M. (1995). Density and regression smoothing, *in* W. Härdle, S. Klinke & B. A. Turlach (eds), *XploRe – an interactive statistical computing environment*, Springer, pp. 77–99.

Friedman, J. H. (1987). Exploratory projection pursuit, *Journal of the American Statistical Association* **82**: 249–266.

Friedman, J. H. & Stuetzle, W. (1981). Projection pursuit regression, *Journal of the American Statistical Association* **76**(376): 817–823.

Friedman, J. H. & Stuetzle, W. (1982). Smoothing of scatterplots, *Technical report*, Department of Statistics, Stanford.

Fuss, M., McFadden, D. & Mundlak, Y. (1978). A survey of functional forms in the economic analysis of production, *in* M. Fuss & D. McFadden (eds), *Production Economics: A Dual Approach to Theory and Applications*, North-Holland, Amsterdam, pp. 219–268.

Gallant, A. & Nychka, D. (1987). Semi-nonparametric maximum likelihood estimation, *Econometrica* **55**(2): 363–390.

Gasser, T. & Müller, H. G. (1984). Estimating regression functions and their derivatives by the kernel method, *Scandinavian Journal of Statistics* **11**: 171–185.

Gill, J. (2000). *Generalized Linear Models: A Unified Approach*, Sage University Paper Series on Quantitative Applications in the Social Sciences, 07-134, Thousand Oaks, CA.

Gill, R. D. (1989). Non- and semi-parametric maximum likelihood estimators and the von Mises method (Part I), *Scandinavian Journal of Statistics* **16**: 97–128.

Gill, R. D. & van der Vaart, A. W. (1993). Non- and semi-parametric maximum likelihood estimators and the von Mises method (Part II), *Scandinavian Journal of Statistics* **20**: 271–288.

González Manteiga, W. & Cao, R. (1993). Testing hypothesis of general linear model using nonparametric regression estimation, *Test* **2**: 161–189.

Gozalo, P. L. & Linton, O. (2001). A nonparametric test of additivity in generalized nonparametric regression with estimated parameters, *Journal of Econometrics* **104**: 1–48.

Grasshoff, U., Schwalbach, J. & Sperlich, S. (1999). Executive pay and corporate financial performance. an explorative data analysis, *Working paper 99-84 (33)*, Universidad Carlos III de Madrid.

Green, P. J. & Silverman, B. W. (1994). *Nonparametric Regression and Generalized Linear Models*, Vol. 58 of *Monographs on Statistics and Applied Probability*, Chapman and Hall, London.

Green, P. J. & Yandell, B. S. (1985). Semi-parametric generalized linear models, *Proceedings 2nd International GLIM Conference*, Vol. 32 of *Lecture Notes in Statistics 32*, Springer, New York, pp. 44–55.

GSOEP (1991). *Das Sozio-ökonomische Panel (SOEP) im Jahre 1990/91*, Projektgruppe "Das Sozio-ökonomische Panel", Deutsches Institut für Wirtschaftsforschung. Vierteljahreshefte zur Wirtschaftsforschung, pp. 146–155.

Habbema, J. D. F., Hermans, J. & van den Broek, K. (1974). A stepwise discrimination analysis program using density estimation, *COMPSTAT '74. Proceedings in Computational Statistics*, Physica, Vienna.

Hall, P. & Marron, J. S. (1991). Local minima in cross–validation functions, *Journal of the Royal Statistical Society, Series B* **53**: 245–252.

Hall, P., Marron, J. S. & Park, B. U. (1992). Smoothed cross-validation, *Probability Theory and Related Fields* **92**: 1–20.

Hall, P., Sheather, S. J., Jones, M. C. & Marron, J. S. (1991). On optimal data–based bandwidth selection in kernel density estimation, *Biometrika* **78**: 263–269.

Han, A. (1987). Non–parametric analysis of a generalized regression model, *Journal of Econometrics* **35**: 303–316.

Hardin, J. & Hilbe, J. (2001). *Generalized Linear Models and Extensions*, Stata Press.

Härdle, W. (1990). *Applied Nonparametric Regression*, Econometric Society Monographs No. 19, Cambridge University Press.

Härdle, W. (1991). *Smoothing Techniques, With Implementations in S*, Springer, New York.

Härdle, W., Huet, S., Mammen, E. & Sperlich, S. (2004). Bootstrap inference in semiparametric generalized additive models, *Econometric Theory* **20**: to appear.

Härdle, W., Kerkyacharian, G., Picard, D. & Tsybakov, A. B. (1998). *Wavelets, Approximation, and Statistical Applications*, Springer, New York.

Härdle, W. & Mammen, E. (1993). Testing parametric versus nonparametric regression, *Annals of Statistics* **21**: 1926–1947.

Härdle, W., Mammen, E. & Müller, M. (1998). Testing parametric versus semiparametric modelling in generalized linear models, *Journal of the American Statistical Association* **93**: 1461–1474.

Härdle, W. & Müller, M. (2000). Multivariate and semiparametric kernel regression, *in* M. Schimek (ed.), *Smoothing and Regression*, Wiley, New York, pp. 357–391.

Härdle, W. & Scott, D. W. (1992). Smoothing in by weighted averaging using rounded points, *Computational Statistics* **7**: 97–128.

Härdle, W., Sperlich, S. & Spokoiny, V. (2001). Structural tests in additive regression, *Journal of the American Statistical Association* **96**(456): 1333–1347.

Härdle, W. & Stoker, T. M. (1989). Investigating smooth multiple regression by the method of average derivatives, *Journal of the American Statistical Association* **84**: 986–995.

Härdle, W. & Tsybakov, A. B. (1997). Local polynomial estimators of the volatility function in nonparametric autoregression, *Journal of Econometrics* **81**(1): 223–242.

Harrison, D. & Rubinfeld, D. L. (1978). Hedonic prices and the demand for clean air, *J. Environ. Economics and Management* **5**: 81–102.

Hart, J. D. (1997). *onparametric Smoothing and Lack-of-Fit Tests*, Springer, New York.

Hastie, T. J. & Tibshirani, R. J. (1986). Generalized additive models (with discussion), *Statistical Science* **1**(2): 297–318.

Hastie, T. J. & Tibshirani, R. J. (1990). *Generalized Additive Models*, Vol. 43 of *Monographs on Statistics and Applied Probability*, Chapman and Hall, London.

Heckman, J. (1976). The common structure of statistical models of truncation, sample selection and limited dependent variables and a simple estimator for such model, *Annals of Economic and Social Measurement* **5**: 475–492.

Heckman, J. (1979). Sample selection bias as a specification error, *Econometrica* **47**: 153–161.

Hengartner, N., Kim, W. & Linton, O. (1999). A computationally efficient oracle estimator for additive nonparametric regression with bootstrap confidence intervals, *Journal of Computational and Graphical Statistics* **8**: 1–20.

Honoré, B. E. & Powell, J. L. (1994). Pairwise difference estimators of censored and truncated regression models, *Journal of Econometrics* **64**: 241–278.

Horowitz, J. L. (1993). Semiparametric and nonparametric estimation of quantal response models, *in* G. S. Maddala, C. R. Rao & H. D. Vinod (eds), *Handbook of Statistics*, Elsevier Science Publishers, pp. 45–72.

Horowitz, J. L. (1996). Semiparametric estimation of a regression model with an unknown transformation of the dependent variable, *Econometrica* **64**: 103–137.

Horowitz, J. L. (1998a). Nonparametric estimation of a generalized additive model with an unknown link function, *Technical report*, University of Iowa.

Horowitz, J. L. (1998b). *Semiparametric Methods in Econometrics*, Springer.

Horowitz, J. L. & Härdle, W. (1994). Testing a parametric model against a semiparametric alternative, *Econometric Theory* **10**: 821–848.

Horowitz, J. L. & Härdle, W. (1996). Direct semiparametric estimation of single-index models with discrete covariates, *Journal of the American Statistical Association* **91**(436): 1632–1640.

Hsing, T. & Carroll, R. J. (1992). An asymptotic theory for sliced inverse regression, *Annals of Statistics* **20**(2): 1040–1061.

Ichimura, H. (1993). Semiparametric least squares (SLS) and weighted SLS estimation of single–index models, *Journal of Econometrics* **58**: 71–120.

Ingster, Y. I. (1993). Asymptotically minimax hypothesis testing for nonparametric alternatives. I - III, *Math. Methods of Statist.* **2**: 85 – 114, 171 – 189, 249 – 268.

Jones, M. C., Marron, J. S. & Sheather, S. J. (1996). Progress in data-based bandwidth selection for kernel density estimation, *Computational Statistics* **11**(3): 337–381.

Kallenberg, W. C. M. & Ledwina, T. (1995). Consistency and Monte-Carlo simulations of a data driven version of smooth goodness-of-fit tests, *Annals of Statistics* **23**: 1594–1608.

Klein, R. & Spady, R. (1993). An efficient semiparametric estimator for binary response models, *Econometrica* **61**: 387–421.

Korostelev, A. & Müller, M. (1995). Single index models with mixed discrete-continuous explanatory variables, *Discussion Paper 26*, Sonderforschungsbereich 373, Humboldt-Universität zu Berlin.

Ledwina, T. (1994). Data-driven version of Neyman's smooth test of fit, *Journal of the American Statistical Association* **89**: 1000–1005.

Lejeune, M. (1985). Estimation non-paramétrique par noyaux: régression polynomiale mobile, *Revue de Statistique Appliqueés* **33**: 43–67.

Leontief, W. (1947a). Introduction to a theory of the internal structure of functional relationships, *Econometrica* **15**: 361–373.

Leontief, W. (1947b). A note on the interrelation of subsets of independent variables of a continuous function with continuous first derivatives., *Bulletin of the American Mathematical Society* **53**: 343–350.

Lewbel, A. & Linton, O. (2002). Nonparametric censored and truncated regression, *Econometrica* **70**: 765–780.

Li, K.-C. (1991). Sliced inverse regression for dimension reduction (with discussion), *Journal of the American Statistical Association* **86**(414): 316–342.

Linton, O. (1997). Efficient estimation of additive nonparametric regression models, *Biometrika* **84**: 469–473.

Linton, O. (2000). Efficient estimation of generalized additive nonparametric regression models, *Econometric Theory* **16**(4): 502–523.

Linton, O. & Härdle, W. (1996). Estimation of additive regression models with known links, *Biometrika* **83**(3): 529–540.

Linton, O. & Nielsen, J. P. (1995). A kernel method of estimating structured nonparametric regression based on marginal integration, *Biometrika* **82**: 93–101.

Loader, C. (1999). *Local Regression and Likelihood*, Springer, New York.

Mack, Y. P. (1981). Local properties of k-nn regression estimates, *SIAM J. Alg. Disc. Math.* **2**: 311–323.

Maddala, G. S. (1983). *Limited-dependent and qualitative variables in econometrics*, Econometric Society Monographs No. 4, Cambridge University Press.

Mammen, E., Linton, O. & Nielsen, J. P. (1999). The existence and asymptotic properties of a backfitting projection algorithm under weak conditions, *Annals of Statistics* **27**: 1443–1490.

Mammen, E. & Nielsen, J. P. (2003). Generalised structured models, *Biometrika* **90**: 551–566.

Manski, C. (1985). Semiparametric analysis of discrete response: Asymptotic properties of the maximum score estimator, *Journal of Econometrics* **3**: 205–228.

Marron, J. S. (1989). Comments on a data based bandwidth selector, *Computational Statistics & Data Analysis* **8**: 155–170.

Marron, J. S. & Härdle, W. (1986). Random approximations to some measures of accuracy in nonparametric curve estimation, *J. Multivariate Anal.* **20**: 91–113.

Marron, J. S. & Nolan, D. (1988). Canonical kernels for density estimation, *Statistics & Probability Letters* **7**(3): 195–199.

Masry, E. & Tjøstheim, D. (1995). Non-parametric estimation and identification of nonlinear arch time series: strong convergence properties and asymptotic normality, *Econometric Theory* **11**: 258–289.

Masry, E. & Tjøstheim, D. (1997). Additive nonlinear ARX time series and projection estimates, *Econometric Theory* **13**: 214–252.

McCullagh, P. & Nelder, J. A. (1989). *Generalized Linear Models*, Vol. 37 of *Monographs on Statistics and Applied Probability*, 2 edn, Chapman and Hall, London.

Müller, M. (2001). Estimation and testing in generalized partial linear models — a comparative study, *Statistics and Computing* **11**: 299–309.

Müller, M. (2004). Generalized linear models, *in* J. Gentle, W. Härdle & Y. Mori (eds), *Handbook of Computational Statistics (Volume I). Concepts and Fundamentals*, Springer, Heidelberg.

Nadaraya, E. A. (1964). On estimating regression, *Theory of Probability and its Applications* **10**: 186–190.

Nelder, J. A. & Wedderburn, R. W. M. (1972). Generalized linear models, *Journal of the Royal Statistical Society, Series A* **135**(3): 370–384.

Newey, W. K. (1990). Semiparametric efficiency bounds, *Journal of Applied Econometrics* **5**: 99–135.

Newey, W. K. (1994). The asymptotic variance of semiparametric estimation, *Econometrica* **62**: 1349–1382.

Newey, W. K. (1995). Convergence rates for series estimators, *in* G. Maddala, P. Phillips & T. Srinavsan (eds), *Statistical Methods of Economics and Quantitative Economics: Essays in Honor of C.R. Rao*, Blackwell, Cambridge, pp. 254–275.

Newey, W. K., Powell, J. L. & Vella, F. (1999). Nonparametric estimation of triangular simultaneous equation models, *Econometrica* **67**: 565–603.

Nielsen, J. P. & Linton, O. (1998). An optimization interpretation of integration and backfitting estimators for separable nonparametric models, *Journal of the Royal Statistical Society, Series B* **60**: 217–222.

Nielsen, J. P. & Sperlich, S. (2002). Smooth backfitting in practice, *Working paper 02-59*, Universidad Carlos III de Madrid.

Opsomer, J. & Ruppert, D. (1997). Fitting a bivariate additive model by local polynomial regression, *Annals of Statistics* **25**: 186–211.

Pagan, A. & Schwert, W. (1990). Alternative models for conditional stock volatility, *Journal of Econometrics* **45**: 267–290.

Pagan, A. & Ullah, A. (1999). *Nonparametric Econometrics*, Cambridge University Press.

Park, B. U. & Marron, J. S. (1990). Comparison of data–driven bandwidth selectors, *Journal of the American Statistical Association* **85**: 66–72.

Park, B. U. & Turlach, B. A. (1992). Practical performance of several data driven bandwidth selectors, *Computational Statistics* **7**: 251–270.

Powell, J. L., Stock, J. H. & Stoker, T. M. (1989). Semiparametric estimation of index coefficients, *Econometrica* **57**(6): 1403–1430.

Proença, I. & Werwatz, A. (1995). Comparing parametric and semiparametric binary response models, *in* W. Härdle, S. Klinke & B. Turlach (eds), *XploRe: An Interactive Statistical Computing Environment*, Springer, pp. 251–274.

Robinson, P. M. (1988a). Root n–consistent semiparametric regression, *Econometrica* **56**: 931–954.

Robinson, P. M. (1988b). Semiparametric econometrics: A survey, *Journal of Applied Econometrics* **3**: 35–51.

Rodríguez-Póo, J. M., Sperlich, S. & Fernández, A. I. (1999). Semiparametric three step estimation methods for simultaneous equation systems, *Working paper 99-83 (32)*, Universidad Carlos III de Madrid.

Rodríguez-Póo, J. M., Sperlich, S. & Vieu, P. (2003). Semiparametric estimation of weak and strong separable models, *Econometric Theory* **19**: 1008–1039.

Ruppert, D. & Wand, M. P. (1994). Multivariate locally weighted least squares regression, *Annals of Statistics* **22**(3): 1346–1370.

Ruppert, D., Wand, M. P. & Carroll, R. J. (1990). *Semiparametric Regression*, Cambridge University Press.

Schimek, M. G. (2000a). Estimation and inference in partially linear models with smoothing splines, *Journal of Statistical Planning and Inference* **91**: 525–540.

Schimek, M. G. (ed.) (2000b). *Smoothing and Regression*, Wiley, New York.

Scott, D. W. (1992). *Multivariate Density Estimation: Theory, Practice, and Visualization*, John Wiley & Sons, New York, Chichester.

Scott, D. W. & Terrell, G. R. (1987). Biased and unbiased cross-validation in density estimation, *Journal of the American Statistical Association* **82**(400): 1131–1146.

Scott, D. W. & Wand, M. P. (1991). Feasibility of multivariate density estimates, *Biometrika* **78**: 197–205.

Severance-Lossin, E. & Sperlich, S. (1999). Estimation of derivatives for additive separable models, *Statistics* **33**: 241–265.

Severini, T. A. & Staniswalis, J. G. (1994). Quasi-likelihood estimation in semiparametric models, *Journal of the American Statistical Association* **89**: 501–511.

Severini, T. A. & Wong, W. H. (1992). Generalized profile likelihood and conditionally parametric models, *Annals of Statistics* **20**: 1768–1802.

Sheather, S. J. & Jones, M. C. (1991). A reliable data–based bandwidth selection method for kernel density estimation, *Journal of the Royal Statistical Society, Series B* **53**: 683–690.

Silverman, B. W. (1984). Spline smoothing: the equivalent variable kernel method, *Annals of Statistics* **12**: 898–916.

Silverman, B. W. (1986). *Density Estimation for Statistics and Data Analysis*, Vol. 26 of *Monographs on Statistics and Applied Probability*, Chapman and Hall, London.

Simonoff, J. (1996). *Smoothing Methods in Statistics*, Springer, New York.

Speckman, P. E. (1988). Regression analysis for partially linear models, *Journal of the Royal Statistical Society, Series B* **50**: 413–436.

Sperlich, S. (1998). *Additive Modelling and Testing Model Specification*, Shaker Verlag.

Sperlich, S., Linton, O. & Härdle, W. (1999). Integration and backfitting methods in additive models: Finite sample properties and comparison, *Test* **8**: 419–458.

Sperlich, S., Tjøstheim, D. & Yang, L. (2002). Nonparametric estimation and testing of interaction in additive models, *Econometric Theory* **18**(2): 197–251.

Spokoiny, V. (1996). Adaptive hypothesis testing using wavelets, *Annals of Statistics* **24**: 2477–2498.

Spokoiny, V. (1998). Adaptive and spatially adaptive testing of a nonparametric hypothesis, *Math. Methods of Statist.* **7**: 245–273.

Staniswalis, J. G. & Thall, P. F. (2001). An explanation of generalized profile likelihoods, *Statistics and Computing* **11**: 293–298.

Stone, C. J. (1977). Consistent nonparametric regression, *Applied Statistics* **5**: 595–635.

Stone, C. J. (1984). An asymptotically optimal window selection rule for kernel density estimates, *Annals of Statistics* **12**(4): 1285–1297.

Stone, C. J. (1985). Additive regression and other nonparametric models, *Annals of Statistics* **13**(2): 689–705.

Stone, C. J. (1986). The dimensionality reduction principle for generalized additive models, *Annals of Statistics* **14**(2): 590–606.

Stone, C. J., Hansen, M. H., Kooperberg, C. & Truong, Y. (1997). Polynomial splines and their tensor products in extended linear modeling (with discussion), *Annals of Statistics* **25**: 1371–1470.

Stute, W. (1997). Nonparametric model checks for regression, *Annals of Statistics* **25**: 613–641.

Stute, W., González Manteiga, W. & Presedo-Quindimi, M. (1998). Bootstrap approximations in model checks for regression, *Journal of the American Statistical Association* **93**: 141–149.

Tjøstheim, D. & Auestad, B. (1994a). Nonparametric identification of nonlinear time series: Projections, *Journal of the American Statistical Association* **89**: 1398–1409.

Tjøstheim, D. & Auestad, B. (1994b). Nonparametric identification of nonlinear time series: Selecting significant lags, *Journal of the American Statistical Association* **89**: 1410–1430.

Treiman, D. J. (1975). Problems of concept and measurement in the comparative study of occupational mobility, *Social Science Research* **4**: 183–230.

Vella, F. (1998). Estimating models with sample selection bias: A survey, *The Journal of Human Resources* **33**: 127–169.

Venables, W. N. & Ripley, B. (2002). *Modern Applied Statistics with S*, fourth edn, Springer, New York.

Vieu, P. (1994). Choice of regressors in nonparametric estimation, *Computational Statistics & Data Analysis* **17**: 575–594.

Wahba, G. (1990). *Spline models for observational data*, 2 edn, SIAM, Philadelphia, Pennsylvania.

Wand, M. P. & Jones, M. C. (1995). *Kernel Smoothing*, Vol. 60 of *Monographs on Statistics and Applied Probability*, Chapman and Hall, London.

Watson, G. S. (1964). Smooth regression analysis, *Sankhyā, Series A* **26**: 359–372.

Wecker, W. & Ansley, C. (1983). The signal extraction approach to nonlinear regression and spline smoothing, *Journal of the American Statistical Association* **78**: 351–365.

Weisberg, S. & Welsh, A. H. (1994). Adapting for the missing link, *Annals of Statistics* **22**: 1674–1700.

Wu, C. (1986). Jackknife, bootstrap and other resampling methods in regression analysis (with discussion), *Annals of Statistics* **14**: 1261–1350.

Yang, L., Sperlich, S. & Härdle, W. (2003). Derivative estimation and testing in generalized additive models, *Journal of Statistical Planning and Inference* **115**(2): 521–542.

Yatchew, A. (2003). *Semiparametric Regression for the Applied Econometrician*, Cambridge University Press.

Zheng, J. (1996). A consistent test of a functional form via nonparametric estimation techniques, *Journal of Econometrics* **75**: 263–289.

Author Index

Subject Index

Springer Series in Statistics

(continued from p. ii)

Huet/Bouvier/Poursat/Jolivet: Statistical Tools for Nonlinear Regression: A Practical Guide with S-PLUS and R Examples, 2nd edition.
Ibrahim/Chen/Sinha: Bayesian Survival Analysis.
Jolliffe: Principal Component Analysis, 2nd edition.
Knottnerus: Sample Survey Theory: Some Pythagorean Perspectives.
Kolen/Brennan: Test Equating: Methods and Practices.
Kotz/Johnson (Eds.): Breakthroughs in Statistics Volume I.
Kotz/Johnson (Eds.): Breakthroughs in Statistics Volume II.
Kotz/Johnson (Eds.): Breakthroughs in Statistics Volume III.
Küchler/Sørensen: Exponential Families of Stochastic Processes.
Kutoyants: Statistical Influence for Ergodic Diffusion Processes.
Lahiri: Resampling Methods for Dependent Data.
Le Cam: Asymptotic Methods in Statistical Decision Theory.
Le Cam/Yang: Asymptotics in Statistics: Some Basic Concepts, 2nd edition.
Liu: Monte Carlo Strategies in Scientific Computing.
Longford: Models for Uncertainty in Educational Testing.
Manski: Partial Identification of Probability Distributions.
Mielke/Berry: Permutation Methods: A Distance Function Approach.
Pan/Fang: Growth Curve Models and Statistical Diagnostics.
Parzen/Tanabe/Kitagawa: Selected Papers of Hirotugu Akaike.
Politis/Romano/Wolf: Subsampling.
Ramsay/Silverman: Applied Functional Data Analysis: Methods and Case Studies.
Ramsay/Silverman: Functional Data Analysis.
Rao/Toutenburg: Linear Models: Least Squares and Alternatives.
Reinsel: Elements of Multivariate Time Series Analysis, 2nd edition.
Rosenbaum: Observational Studies, 2nd edition.
Rosenblatt: Gaussian and Non-Gaussian Linear Time Series and Random Fields.
Särndal/Swensson/Wretman: Model Assisted Survey Sampling.
Santner/Williams/Notz: The Design and Analysis of Computer Experiments.
Schervish: Theory of Statistics.
Shao/Tu: The Jackknife and Bootstrap.
Simonoff: Smoothing Methods in Statistics.
Singpurwalla and Wilson: Statistical Methods in Software Engineering: Reliability and Risk.
Small: The Statistical Theory of Shape.
Sprott: Statistical Inference in Science.
Stein: Interpolation of Spatial Data: Some Theory for Kriging.
Taniguchi/Kakizawa: Asymptotic Theory of Statistical Inference for Time Series.
Tanner: Tools for Statistical Inference: Methods for the Exploration of Posterior Distributions and Likelihood Functions, 3rd edition.
van der Laan: Unified Methods for Censored Longitudinal Data and Causality.
van der Vaart/Wellner: Weak Convergence and Empirical Processes: With Applications to Statistics.
Verbeke/Molenberghs: Linear Mixed Models for Longitudinal Data.
Weerahandi: Exact Statistical Methods for Data Analysis.
West/Harrison: Bayesian Forecasting and Dynamic Models, 2nd edition.